MEDICINE AND BIOLOGY RESEARCH DEVELOPMENTS

BIOLOGICAL ASPECTS OF HUMAN HEALTH AND WELL-BEING

MEDICINE AND BIOLOGY RESEARCH DEVELOPMENTS

TSISANA SHARTAVA, M.D. – *SERIES EDITOR*

TBILISI, GEORGIA

General Anesthesia Research Developments
Milo Hertzog and Zelig Kuhn (Editors)
2010. ISBN: 978-1-60876-395-5
(Hardcover)
978-1-61761-577-1 (E-book)

Venoms: Sources, Toxicity and Therapeutic Uses
Jonas Gjersoe and Simen Hundstad (Editors)
2010. ISBN: 978-1-60876-448-8

Parasitology Research Trends
Olivier De Bruyn and Stephane Peeters (Editors)
2010. ISBN: 978-1-60741-436-0
(Hardcover)
978-1-61668-716-8 (E-book)

Biomaterials Developments and Applications
Henri Bourg and Amaury Lisle (Editors)
2010. ISBN: 978-1-60876-476-1
(Hardcover)
978-1-61209-862-3 (E-book)

A Guide to Hemorrhoidal Disease
Pravin Jaiprakash Gupta
2010. ISBN: 978-1-60876-431-0
(Hardcover)
978-1-61761-480-4 (E-book)

Type III Secretion Chaperones: A Molecular Toolkit for all Occasions
Matthew S. Francis
2010. ISBN: 978-1-60876-667-3

TRP Channels in Health and Disease: Implications for Diagnosis and Therapy
Arpad Szallasi (Editor)
2010. ISBN: 978-1-61668-337-5

Recent Advances in BIS Guided TCI Anesthesia
David A. Ferreira, Luís Antunes, Pedro Amorim and Catarina Nunes
2010. ISBN: 978-1-61668-627-7
(Softcover)
978-1-61668-888-2 (E-book)

Tracing the Drainage Divide: The Future Challenge for Predictive Medicine
Enzo Grossi
ISBN: 978-1-61209-232-4 (E-book)

Nanotechnology and Advances in Medicine
Maysaa El Sayed Zaki
2011. ISBN: 978-1-61209-640-7

Biological Aspects of Human Health and Well-Being
Tsisana Shartava- (Editor)
2011. ISBN: 978-1-61209-134-1

MEDICINE AND BIOLOGY RESEARCH DEVELOPMENTS

BIOLOGICAL ASPECTS OF HUMAN HEALTH AND WELL-BEING

TSISANA SHARTAVA

EDITOR

Nova Science Publishers, Inc.
New York

NOTICE TO THE READER

The Publisher has taken reasonable care in the preparation of this book, but makes no expressed or implied warranty of any kind and assumes no responsibility for any errors or omissions. No liability is assumed for incidental or consequential damages in connection with or arising out of information contained in this book. The Publisher shall not be liable for any special, consequential, or exemplary damages resulting, in whole or in part, from the readers' use of, or reliance upon, this material. Any parts of this book based on government reports are so indicated and copyright is claimed for those parts to the extent applicable to compilations of such works.

Independent verification should be sought for any data, advice or recommendations contained in this book. In addition, no responsibility is assumed by the publisher for any injury and/or damage to persons or property arising from any methods, products, instructions, ideas or otherwise contained in this publication.

This publication is designed to provide accurate and authoritative information with regard to the subject matter covered herein. It is sold with the clear understanding that the Publisher is not engaged in rendering legal or any other professional services. If legal or any other expert assistance is required, the services of a competent person should be sought. FROM A DECLARATION OF PARTICIPANTS JOINTLY ADOPTED BY A COMMITTEE OF THE AMERICAN BAR ASSOCIATION AND A COMMITTEE OF PUBLISHERS.

Additional color graphics may be available in the e-book version of this book.

LIBRARY OF CONGRESS CATALOGING-IN-PUBLICATION DATA

Biological aspects of human health and well-being / [edited by] Tsisana
Shartava.
        p. cm.
    "INTERNATIONAL JOURNAL OF MEDICAL AND BIOLOGICAL FRONTIERS Volume 16,
Issue 1/2."
    Includes bibliographical references and index.
    ISBN 978-1-61209-134-1 (hardcover : alk. paper)
    1. Biochemistry. 2. Clinical biochemistry. I. Shartava, Tsisana.
    QP514.2.B574 2011
    612'.015--dc22
                    2010047540

Published by Nova Science Publishers, Inc. ✝ New York

CONTENTS

PREFACE

This book presents and discusses current research in the field of biology, with a particular emphasis on biological factors and their role in health and well-being. Topics discussed include the biotechnology of cyanobacteria; the reasons why glucose is the principal source of energy for living beings; post-transcriptional effects of estrogen on gene expression; sialylation mechanism in bacteria and the evolution biology of health and disease clinical medicine from a Darwinian perspective.

Chapter I - Cross infection is the transmission of an infectious agent from one person to another because of a poor barrier protection as in patients and other immunocompromised hosts. This can be a direct transmission or an indirect transmission through instruments, appliances, and surfaces. The most common are nosocomial cross infections, which are acquired at hospitals or other healthcare facilities such as outpatient clinics. Community-acquired cross infections have also been described.

Chapter II - Cyanobacteria, structurally Gram-negative prokaryotes and ancient relatives of chloroplasts, can assist analysis of photosynthesis and its regulation more easily than can studies with higher plants. Many genetic tools have been developed for unicellular and filamentous strains of cyanobacteria during the past three decades. These tools provide abundant opportunity for identifying novel genes; for investigating the structure, regulation and evolution of genes; for understanding the ecological roles of cyanobacteria; and for possible practical applications, such as molecular hydrogen photo production; production of phycobiliproteins to form fluorescent antibody reagents; cyanophycin production; polyhydroxybutyrate biosynthesis; osmolytes production; nanoparticles formation; mosquito control; heavy metal removal; biodegradative ability of cyanobacteria; toxins formation by bloom-forming cyanobacteria; use of natural products of cyanobacteria for medicine and others aspects of cyanobacteria applications have been discussed in this chapter.

Chapter III - Man has attempted to explain the appearance of life on Earth in a very complex manner, therefore the understanding of the diseases that affect human beings has been equally complicated, and thus the treatment of many diseases has had to be very aggressive; it being sufficient to mention the current treatments for cancer, autoimmune diseases, and mitochondrial diseases.

Chapter IV - In its population genetic sense, evolution is defined as the ongoing change of gene frequencies in populations due to one or several of the driving forces of evolution: selection, drift, mutation and migration. Evolution's role is central in the sub-discipline of biology that addresses health and disease in humans and training in evolutionary thinking can both help biomedical researchers and clinicians ask useful questions they might not otherwise pose.

The co-evolution of man and his environment of pathogenic micro-organisms, the rapidly shifting antibiotic resistance of these pathogens and our persistent vulnerability to chronic diseases should all be seen from an evolutionary perspective. These subjects form the core of "evolutionary medicine", which will be illustrated by a number of thought inspiring examples.

The hypothesis that allergy can be viewed like cough and pain as a defence mechanism evolved by natural selection, is gaining support from toxicological studies measuring lower levels of carcinogens in allergic individuals.

Recent research, combining the effects of genes and environment, has provided surprising clues to the cause of atherosclerosis, a major public health problem.

In medical microbiology, the combination of the short generation time of bacteria, the exchange of resistance genes between species and the swift transfer of bacteria from animals to humans and between humans, forms a life threatening cocktail with a critical role for evolutionary mechanisms.

The HLA system which encodes proteins of the immune response, shows the most extensive polymorphism of the whole human genome. The global distribution of HLA alleles illustrates evolution by migration, while the polymorphism itself is promoted by natural selection, operating through pre- and post-conceptual mechanisms.

An example of "recent" evolution in *Homo sapiens* by natural selection and a genetic bottleneck, comes from the relation between *Yersinia pestis* and hemochromatosis. The geographical distribution of the hemochromatosis gene correlates strictly with the area of the 14th century bubonic plague that raged through Europe, which can be explained by a protective mechanism of the hemochromatosis gene against bacterial infection.

The examples above make a strong case for recognizing evolution biology as a basic science for medicine.

Chapter V - Flavonoids are widely distributed in fruit and vegetables and form part of the human diet. These compounds are thought to be a contributing factor to the health benefits of fruit and vegetables in part because of their antioxidant activities. Despite the extensive use of chemical antioxidant assays to assess the activity of flavonoids and other natural products that are safe to consume, their ability to predict an *in vivo* health benefit is debateable. Some are carried out at non-physiological pH and temperature, most take no account of partitioning between hydrophilic and lipophilic environments, and none of them takes into account bioavailability, uptake and metabolism of antioxidant compounds and the biological component that is targeted for protection. However, biological systems are far more complex and dietary antioxidants may function via multiple mechanisms. It is critical to consider moving from using 'the test tube' to employing cell-based assays for screening foods, phytochemicals and other consumed natural products for their potential biological activity. The question then remains as to which cell models to use. Human immortalized cell lines derived from many different cell types from a wide range of anatomical sites are available and are established well-characterized models.

The cytoprotection assay was developed to be a more biologically relevant measurement than the chemically defined antioxidant activity assay because it uses human cells as a substrate and therefore accounts for some aspects of uptake, metabolism and location of flavonoids within cells. Knowledge of structure-activity relationships in the cytoprotection assay may be helpful in assessing potential *in vivo* cellular protective effects of flavonoids.

This study will discuss the cytoprotective properties of flavonoids and focuses on the relationship between their cytoprotective activity, physicochemical properties such as lipophilicity (log P) and bond dissociation enthalpies (BDE), and their chemical structures. The factors underlying the influence the different classes of flavonoids have in modulating their ability to protect human gut cells are discussed and support the contention that the partition coefficients of flavonoids as well as their rate of reaction with the relevant radicals define the protective abilities in cellular environments. By comparing the geometries of several flavonoids, the author were able to explain the structural dependency of the antioxidant action of these flavonoids.

Chapter VI - Sialic acids are important components of carbohydrate chains and are linked to terminal positions of the carbohydrate moiety of glycoconjugates, including glycoproteins and glycolipids. Various studies have focused on clarifying the structure–function relationship of sialic acids and have revealed that N-acetylneuraminic acid (Neu5Ac) is the major sialic acids component of glycoconjugates, and that the sialylated carbohydrate chains of glycoconjugates play significant roles in many biological processes, including immunological responses, viral infections, cell–cell recognition,and inflammation.

Sialylated glycoconjugates are formed by specific sialyltransferases in the cell. All sialyltransferases use cytidinemonophosphate N-acetylneuraminic acid(CMP-Neu5Ac) as the common donor substrate. Up to the present, sialyltransferases have been cloned from various sources, including mammalian organs, bacteria and virus. As to the sialyltransferases, all of the sialyltransferases have been classified into five families in the CAZy (carbohydrate-active enzymes) database (family29, 38, 42, 52 and 80), and all of the marine bacterial sialyltransferases are classified into the family 80.

Generally, the enzymes with a bacterial origin are more stable and productive in *Escherichia coli*protein expression systems than the mammalian-derived enzymes. In addition, the bacterial-derived sialyltransferases show broader acceptor substrate specificity than the mammalian enzymes. These advantages highlight the capacity of bacterial enzymes as efficient tools for the *in vitro* enzymatic synthesis of sialosides.

The recent increase in research focusing on sialyltransferases from a diverse range of bacteria has led to the identification of many bacterial sialyltransferases. Several bacterial CMP-Neu5Ac synthetases have also recently been identified. This article reviews the bacterial CMP-Neu5Ac synthetases and sialyltransferases that show promise as tools for the production of sialosides.

Chapter VII - *Aim:* to understand the 2'-nitroimidazole cytotoxicity and liver cell interaction, the author proposed a "Hapatocellular Dysfunction Criteria". Based on it, forty eight patients with amoebic liver abscess on 2'-nitroimidazole therapy were studied for their carbohydrate metabolizing enzymes in serum and hepatocellular enzymes in liver biopsy tissues. *Materials and Methods:* Proven ten cases were studied for hepatocellular cytomorphology by electron microscopy. The clinical status of amoebiasis was assessed by enzyme linked immunosorbent assay (ELISA) antibody titers and stool examination. *Results and Discussion:* Out of forty eight, forty five patients showed elevated carbohydrate metabolizing enzyme levels in serum. The enzymes hexokinase (in 80% samples), aldolase (in 50% samples), phosphofructokinase (in 60% samples), malate dehydrogenase (in 75% samples), isocitrate dehydrogenase (ICDH) (in 80% patients) were elevated while succinate dehydrogenase and lactate dehydrogenase (LDH) levels remained unaltered. Lysosomal enzymes β-glucuronidase, alkaline phosphatase, acid phosphatase, showed enhanced levels in the serum samples. In proven ten amoebic liver abscess biopsies, the hepatocytes and Kupffer cell preparations showed altered enzyme levels. Hepatocytes showed lowered hexokinase (in 80%), LDH (75%), and higher content of aldolase (in 60%), pyruvate kinase (in 70%), malate dehydrogenase (in 66%), ICDH (in 85%), citrate dehydrogenase (in 70%), phosphogluconate dehydrogenase (66%). Kupffer cells showed higher enzyme levels of β-glucuroronidase (in 80%), leucine aminopeptidase (in 70%), acid phosphatase (in 80%) and aryl sulphatase (in 88%). In these 10 repeat biopsy samples from patients on 2'-nitronidazole clinical recovery, the electron microscopy cytomorphology observations showed swollen bizarre mitochondria, proliferative endoplasmic reticulum, and anisonucleosis. 2'-Nitroimidazole showed reverse effect in favor of liver cell regeneration by recovering hepatic damage. *Conclusion:* The proposed "Hepatocellular Dysfunction Criteria" showed different clinical enzyme activities in these patients as they could distinguish nonspecific amoebic hepatitis from amoebic liver abscess.

Chapter VIII - Aim: To evaluate the cytotoxicity of nitroimidazole in isolated human hepatocytes in cultures by using glucokinase enzyme activity as hepatocyte biomarker and evidence of hormonal dependent glucokinase regulatory behavior. Hypothesis: Hepatocyte hormone dependent glucokinase may be a biomarker of cytotoxicity evaluation of nitroimidazole. Methods and Materials: The selected liver biopsies from 10 patients in ongoing research were processed for isolation and fractionation of hepatocytes. Three groups of liver biopsy heaptocytes were: untreated control (group I); liver biopsy from liver abscess infected (group II); and liver biopsy from nitroimidazole treated liver(group III). Results: The glucokinase enzyme activities showed inhibited enzyme activity by actinomycin D, enhanced activity by insulin with triamicilone, poor glucokinase enzyme activity enhancement by progesterone. Discussion: The glucokinase enzyme synthesis is mainly hormonal dependent and regulated at gene level. Its gene expression control by insulin is significant in beta cells but may be possible in glycogen synthesis in human hepatocytes. The nitroimidazole effect was negative in comparison with actinomycinD. The effect of nitroimidazole is less likely to influence the gene expression of glucokinase unlike the insulin. Conclusion: The nitroimidazole directly affects the nonhormonal human glucokinase enzyme activity and can be used as biomarker. The nitroimidazole effect on hormonal dependent glucokinase synthesis is significant in defining the liver regeneration.

Chapter IX - Estrogens exert powerful effects on physiology by regulating gene expression. Their effects on the transcriptional activities of genes are well described in the literature. However, estrogens are also the hormones that are best known for post-transcriptional gene regulation. With the combination of transcriptional and post-transcriptional regulation, gene expression can be rapidly and powerfully controlled to maximize the utility of genomic information throughout the long lives of vertebrate animals. For some cell responses, up to 50% of the genes with altered expression are the result of changes in the stabilities of the messenger RNAs (mRNAs). For many genes including the estrogen receptor alpha (ER) gene, post-transcriptional regulation is the primary mode of alteration of expression. This indicates that post-transcriptional gene regulation is critical to estrogen actions because the ER protein determines the estrogen-responsiveness of animal tissues to a large extent. Estrogens have been shown to regulate the expression of certain genes by greatly altering the stabilities of mRNAs, including stabilizing ER mRNA. This effect may be ancient as it appears to be conserved from mammals to fish and frogs. Some studies have identified unique proteins that are induced by estrogens to bind and protect specific mRNAs from degradation. Recently, hundreds of microRNAs have been discovered and are estimated to actively regulate about one third of protein-encoding mRNAs. MicroRNAs associate with proteins in complexes on mRNAs, where they usually destabilize the mRNA or block its translation. Estrogens regulate the expression of microRNA genes in responsive tissues during normal physiology and disease processes. Other cell signals alter the expression of certain microRNAs that affect ER gene expression. Elucidation of the molecular mechanisms responsible for these post-transcriptional effects is certain to reveal novel molecular targets for therapeutic control of estrogen actions.

Chapter X - Genomic imprinting, the selective suppression of one of the two parental alleles of various genes, has been proposed to play an important regulatory role in the development of the placenta of eutherian mammals. The "parental conflict hypothesis" views that parents of opposite sex have conflicting interests in allocating resources to their offspring by the mother, proposing that growth-promoting genes are mainly expressed from the paternally inherited genome and are silent in the maternally inherited counterparts. X chromosome inactivation plays a central role in compensating for the double dose of X-linked genes in cells of the female relative to cells of the male.

In the placenta of some species, X inactivation represents a special form of epigenetic imprinting. The paternal X chromosome is preferentially imprinted and silent in mouse trophectoderm, a tissue type from which gestational trophoblastic diseases arise. Growing body of evidence has suggested that the pathogenesis of gestational trophoblastic diseases involves altered genomic imprinting. Abnormal expressions of imprinted genes, such as IGF2 and H19, have been implied in the development of molar pregnancies and gestational choriocarcinoma. Moreover, unique genetic modes have long been established in hydatidiform moles: all complete moles have diandric diploid or tetraploid paternal-only genome and partial moles have triploid diandric and monogynic genome. Consistent with parental imprinting theory, partial mole occurs only with diandroid but digynic troiploidy. Recent studies have found that all human placental site trophoblastic tumors arose from a female conceptus, suggesting that a functional paternal X chromosome is important for the neoplastic transformation, likely through inappropriate expression of paternal X-linked genes.

As epigenetic regulation of genomic imprinting and X chromosome inactivation are important for the genesis of gestational trophoblastic diseases, hydatidiform mole and placental site trophoblastic tumor may provide model systems with which genomic imprinting regulation of placenta development and the proliferative advantage conferred by the paternal X chromosome can be studied.

Chapter XI - Gamma-aminobutyric acid (GABA) and glutamate play critical roles in the mediation and modulation of nociception at peripheral, spinal and supraspinal levels. Supraspinally, these amino acid neurotransmitters, and their receptors, are present in key brain regions involved in the sensory-discriminative, affective and cognitive dimensions of pain perception.

Modulation of central GABAergic and glutamatergic neurotransmission underlies both activation of the endogenous analgesic system and the therapeutic effects of a number of analgesics. Enhancement or suppression of firing of GABAergic and glutamatergic neurons, and associated changes in neurotransmitter release, have been reported in supraspinal sites associated with nociception in animal models of acute, inflammatory and neuropathic pain. Moreover, pharmacological modulation of central GABAergic and glutamatergic signaling results in altered nociceptive behaviour.

Here the author review recent evidence in this area. The author consider how this research has enhanced our understanding of the neurochemical mechanisms underpinning nociception and discuss its implications for the development of novel analgesic agents.

Chapter XII - Currently, femtosecond lasers (femtolasers) are being extensively employed in diverse research and application fields. Femtolasers-mediated multiphoton excitation laser scanning microscopy is one of the most exciting recent developments in biomedical imaging and becomes more and more an inspiring imaging technique in the intact bulk tissue examinations. In this review, this non-linear excitation imaging technique including two-photon auto fluorescence (2PF) and second harmonic generated signal imaging (SHG) was employed to investigate the microstructures of whole-mount corneal, retinal, and scleral tissues in their native environment. Image acquisition was based on intense ultrafast femtosecond near-infrared (NIR) laser pulses, which were emitted from a mode-locked solid-state Ti: sapphire system. By integrating high-numerical aperture diffraction-limited objectives, multiphoton microscopy/tomography of ocular tissues was performed at a high light irradiance order of MW-GW/cm^2, where two or more photons were simultaneously absorbed by endogenous molecules located in the thick tissues. As a result, the cellular and fibrous components of intact scleral and corneal tissues were selectively displayed by in-tandem detection with 2PF and SHG without the assistance of any exogenous dye. High-resolution optical images of keratocytes in cornea, fibroblasts, mature elastic fibers and blood capillaries in sclerae as well as of the retina radial Müller glial cells, ganglion cells, bipolar cells, photoreceptors, and retina pigment epithelial (RPE) cells were acquired. Furthermore, this promising technique has been proved to be an indispensable tool in assisting femtolasers intratissue surgery, especially for *in situ* assessing the obtained microsurgical effects. Most remarkably, the activated keratocytes, also named myofibroblasts during wound repair, were *in vivo* detected using the multiphoton excitation imaging in the treated animals twenty-four hours after the intrastromal surgery. Data show that the in-tandem combination of 2PF and SHG allows for *in situ* co-localization imaging of various microstructural components in the whole-mount ocular tissues. Qualitative and quantitative assessment of microstructures was obtained. The selective displaying merits of tissue components only with the excitation of different wavelengths is the most exciting development for bulk tissue imaging, which allows to selectively studying of three-dimensional (3-D) architecture of cellular microstructures and extracellular matrix arrangement at a substantial depth. Using the laser power within the threshold value, the bulk tissues can be imaged numerous times without visible photodisruption. Intrinsic emission multiphoton microscopy/tomography is consequently confirmed to be an efficient and sensitive non-invasive imaging approach, featured with high contrast and subcellular spatial resolution. The non-linear optical imaging yields vivid insights into biological specimens that may ultimately find its clinical application in optical pathological diagnostics. The author believe that this promising technique will also find more applications in the biological and medical basic research in the near future.

Versions of these chapters were also published in *International Journal of Medical and Biological Frontiers,* Volume 16, Numbers 1-12, edited by Tsisana Shartava, published by Nova Science Publishers, Inc. They were submitted for appropriate modifications in an effort to encourage wider dissemination of research.

In: Biological Aspects of Human Health and Well-Being   ISBN: 978-1-61209-134-1
Editor: Tsisana Shartava   © 2011 Nova Science Publishers, Inc.

Chapter I

CONTROL OF EMERGING INFECTIOUS AGENTS CAUSING NOSOCOMIAL AND COMMUNITY-ACQUIRED CROSS INFECTIONS IN IMMUNOCOMPROMISED HOSTS.

Hossam M. Ashour

Department of Microbiology and Immunology
Faculty of Pharmacy, Cairo University, Egypt.

COMMENTARY

Cross infection is the transmission of an infectious agent from one person to another because of a poor barrier protection as in patients and other immunocompromised hosts. This can be a direct transmission or an indirect transmission through instruments, appliances, and surfaces. The most common are nosocomial cross infections, which are acquired at hospitals or other healthcare facilities such as outpatient clinics. Community-acquired cross infections have also been described.

Many of the microbes causing cross infections are resistant to antimicrobial agents and thus present a challenge in treatment and prevention. Examples include the traditional nosocomial cross-infection with methicillin-resistant *Staphylococcus aureus* (MRSA) and the recently characterized vancomycin- and linezolid- resistant *Staphylococcus aureus* and imipenem-resistant Gram-negative bacteria isolated from hospitalized cancer patients [1, 2]. The misuse of antibiotics might have contributed to the rapid evolution of vancomycin-resistant *Staphylococcus aureus* (VRSA) and linezolid- resistant *Staphylococcus aureus* strains in Egypt. This emphasizes the importance of sparing new effective antimicrobial agents and not using them routinely for the treatment of MRSA. An example of community-acquired cross infections is MRSA isolated from inhalational and intravenous drug abusers, who could be a source or a reservoir of community-acquired-MRSA infection in the non-addict population [3].

Cross Infection Control (CIC) is a term that encompasses all the measures taken to prevent the transfer of pathogenic microorganisms, which can potentially result in infection. Control of cross infections in a hospital, outpatient clinic, or a community setting is crucial, especially for immunocompromised patients, such as cancer patients and drug addicts [1-3]. Unfortunately, the chain of cross infection control is easily broken when simple procedures are not observed. One important measure that is not adequately practiced by many healthcare workers is proper hand washing of healthcare workers, use of gloves, masks, and gowns, and compliance with strict hand hygiene guidelines and appropriate hand disinfection techniques [4]. It is noteworthy that the care of fingernails and the skin of hands are components of hand hygiene [5]. Healthcare practitioners who wear artificial nails are more likely to harbor pathogens on their fingertips than healthcare practitioners who have natural nails [6]. Wearing rings was also reported to increase the frequency of hand contamination with nosocomial pathogens [7].

Another CIC measure is to routinely examine the prevalence of MRSA and VRSA in different institutions and to strive to limit the introduction of antibiotic-resistant microorganisms into the health care institutions. This can be accomplished by efficient eradication of carriers and prompt physical isolation of patients bearing such organisms. A proper understanding of the routes of cross infection is critical for CIC. A review of the literature revealed several modes of cross infection that are not particularly obvious for many health care practitioners. Transmission of antibiotic-resistant microorganisms via the face is one route of cross-infection in hospitals [8]. Thus, hospital personnel must not touch their faces or elevate their hands above the shoulder in the ward, because many antibiotic-resistant bacteria can colonize the face. Pre-operative marking of immunocompromised patients with marker pens should also be very limited although disposable markers can be used, if necessary [9]. This is because marker pens can act as fomites for nosocomial infections. Beds, mattresses, curtains, and sphygmomanometers may be contaminated with micro-organisms. Surprisingly, the patient's own flora (carried intransally or elsewhere) can also result in nosocomial cross-infections of patients or non-patients with MRSA and other antibiotic-resistant pathogens [3].

All new staff members in a health care institution (a hospital or an outpatient facility) should be trained in accordance to the most up-to-date methods and procedures for infection control. The current staff should also be re-trained regularly. Auditing should be continuously performed to ensure effectiveness of the procedures. The infection control protocols should be written and made available to the staff and the public.

A fundamental and motivating factor to CIC is self preservation, which is generated from the fear of becoming infected. Thus, raising the public awareness of the importance of the infection control measures will put an extra pressure on the health care staff members, which will make them more cognizant to the infection control measures. By practicing continuous efficient CIC measures, a health care institution will build up a reputation of safety and care for both patients and staff, which, over a period of time, can result in a large turnout of patients through recommendations. Time pressure due to a heavy workload might make healthcare practitioners neglect some simple cross infection control measures. To avoid this, healthcare institutions should limit the number of patients that one healthcare practitioner could manage per day. Nonetheless, it will always be a personal responsibility for each member of the health care team to implement CIC consistently. Researchers should not only continue to characterize resistant microbes causing cross infections, but should also propose novel therapeutic and preventive approaches that could keep these infectious agents under control.

REFERENCES

[1]   Ashour, HM; A. el-Sharif, Microbial spectrum and antibiotic susceptibility profile of gram-positive aerobic bacteria isolated from cancer patients. *J Clin Oncol*, 2007, 25(36), p. 5763-9.

[2]   Ashour, HM; A. el-Sharif. Species distribution and antimicrobial susceptibility of gram-negative aerobic bacteria in hospitalized cancer patients. *J Transl Med*, 2009, 7(1).14.

[3]   El-Sharif, A; Ashour, HM. Community-acquired methicillin-resistant Staphylococcus aureus (CA-MRSA) colonization and infection in intravenous and inhalational opiate drug abusers. *Exp Biol Med* (*Maywood*), 2008, 233(7), p. 874-80.

[4]   Tanner, J. Double gloving to reduce surgical cross-infection. *J Perioper Pract*, 2006. 16(12): p. 571.

[5]   Boyce, JM; Pittet, D. Guideline for Hand Hygiene in Health-Care Settings. Recommendations of the Healthcare Infection Control Practices Advisory Committee and the HICPAC/SHEA/APIC/IDSA Hand Hygiene Task Force. Society for Healthcare Epidemiology of America/Association for Professionals in Infection Control/Infectious Diseases Society of America. *MMWR Recomm Rep*, 2002, 51(RR-16), p. 1-45, quiz CE1-4.

[6]   McNeil, SA. et al., Effect of hand cleansing with antimicrobial soap or alcohol-based gel on microbial colonization of artificial fingernails worn by health care workers. *Clin Infect Dis*, 2001, 32(3), p. 367-72.

[7]   Trick, WE. et al., Impact of ring wearing on hand contamination and comparison of hand hygiene agents in a hospital. *Clin Infect Dis*, 2003, 36(11), p. 1383-90.

[8]   Kuramoto-Chikamatsu, A., et al., Transmission via the face is one route of methicillin-resistant Staphylococcus aureus cross-infection within a hospital. *Am J Infect Control*, 2007, 35(2), p. 126-30.

[9]   Tadiparthi, S., et al., Using marker pens on patients: a potential source of cross infection with MRSA. *Ann R Coll Surg Engl*, 2007, 89(7), p. 661-4.

In: Biological Aspects of Human Health and Well-Being
Editor: Tsisana Shartava

ISBN: 978-1-61209-134-1
© 2011 Nova Science Publishers, Inc.

Chapter II

GENETIC TOOLS APPLICATIONS TO BIOTECHNOLOGY OF CYANOBACTERIA

Olga A. Koksharova[*]

A.N. Belozersky Institute of Physico-Chemical Biology,
M.V. Lomonosov Moscow State University,
Moscow 119992, Russian Federation

ABSTRACT

Cyanobacteria, structurally Gram-negative prokaryotes and ancient relatives of chloroplasts, can assist analysis of photosynthesis and its regulation more easily than can studies with higher plants. Many genetic tools have been developed for unicellular and filamentous strains of cyanobacteria during the past three decades. These tools provide abundant opportunity for identifying novel genes; for investigating the structure, regulation and evolution of genes; for understanding the ecological roles of cyanobacteria; and for possible practical applications, such as molecular hydrogen photo production; production of phycobiliproteins to form fluorescent antibody reagents; cyanophycin production; polyhydroxybutyrate biosynthesis; osmolytes production; nanoparticles formation; mosquito control; heavy metal removal; biodegradative ability of cyanobacteria; toxins formation by bloom-forming cyanobacteria; use of natural products of cyanobacteria for medicine and others aspects of cyanobacteria applications have been discussed in this chapter.

INTRODUCTION

Cyanobacteria, ancient relatives of chloroplasts, are outer membrane-bearing, chlorophyll*a*-containing, photosynthetic bacteria that carry out photosynthesis much as do plants. Cyanobacteria are believed to have been responsible for introducing oxygen into the atmosphere of primitive Earth. A small fraction of the cells of certain cyanobacteria may differentiate into heterocysts, in which dinitrogen (N_2) fixation can take place in an oxygen-

---

[*] OA-Koksharova@rambler.ru

containing milieu. Cyanobacteria are capable of growth, and in some cases differentiation, when provided with little more than sunlight, air, and water. Their potentialities are being enhanced by the availability of genetic tools and genomic sequences. Many vectors and other genetic tools have been developed for unicellular and filamentous strains of cyanobacteria. Transformation, electroporation, and conjugation are used for gene transfer. Diverse methods of mutagenesis allow the isolation of many sought-for kinds of mutants, including site-directed mutants of specific genes. Reporter genes permit measurement of the level of transcription of particular genes, and assays of transcription within individual colonies or within individual cells in a filament [for review Koksharova & Wolk, 2002]. Complete genomic sequences have been obtained for today for the 40 strains and species of cyanobacteria. Genomic sequence data provide the opportunity for global monitoring of changes in genetic expression at transcriptional and translational levels in response to variations in environmental conditions. The availability of genomic sequences accelerates the identification, study, modification and comparison of cyanobacterial genes, and facilitates analysis of evolutionary relationships, including the relationship of chloroplasts to ancient cyanobacteria. The many available genetic tools enhance the opportunities for possible biotechnological applications of cyanobacteria.

CYANOBACTERIA ARE SOURCE OF A LARGE VARIETY OF BIOCOMPOUNDS

Cyanobacteria can also perform syntheses that are of biotechnological significance. Like many genera of eubacteria, they synthesize polyhydroxyalkanoates (PHAs), a thermoplastic class of biodegradable polyesters that includes polyhydroxybutyrate (PHB). PHAs are carbon- and energy-storage compounds that are deposited in the cytoplasm as inclusions. The presence of PHA in about 50 strains of four different phylogenetic subsections of cyanobacteria has been reviewed by Vincenzini & De Philippis [1999]. Eleven different cyanobacteria were investigated with respect to their capabilities to synthesize poly-3-hydroxybutyrate [poly(3HB)] and the type of poly-β-hydroxyalkanoic acid (PHA) synthase accounting for the synthesis of this polyester by using Southern blot analysis, Western blot analysis and sequence analysis of specific PCR products [Hai et al., 2001]. By using the genomic sequence of Synechocystis, Hein et al. [1998] identified and characterized a gene encoding PHB synthase. Two related genes, encoding a PHA-specific β-ketothiolase and an acetoacetyl-CoA reductase, have been identified and characterized [Taroncher-Oldenburg et al., 2000]. Miyake and colleagues [2000] isolated a Tn5 insertion mutant of Synechococcus sp. MA19 with enhanced accumulation of PHB.

Another such synthesis is that of eicosapentaenoic acid (20:5n-3, EPA), a polyunsaturated fatty acid that is an essential nutrient for marine fish larvae and is important for human health. Yu et al. [2000] introduced the EPA-biosynthetic gene cluster from an EPA-producing bacterium, Shewanella sp. SCRC-2738 into a marine Synechococcus sp., strain NKBG15041c, by conjugation. Transgenic cyanobacteria produced amounts of EPA and its precursor, 20:4n-3 that depended upon the culture conditions used.

A polymer unique to cyanobacteria, called cyanophycin, is a copolymer of arginine and aspartic acid, multi-L-arginyl-poly(aspartic acid), discovered and structurally analyzed by Simon [1971, 1987], that comprises the so-called structured granules within the cells [Lang et *al.*, 1972]. Simon [1973a,b; 1976] presented evidence that the polymer is synthesized non-ribosomally and can be degraded to serve as a cellular nitrogen reserve, and extensively purified an enzyme involved in its biosynthesis. The cyanophycin synthetase from *Anabaena variabilis* ATCC 29413 was isolated, microsequenced, and the partial amino acid sequence used to identify the corresponding gene in the *Synechocystis* PCC 6803 database [Ziegler *et al.*, 1998]. It, in turn, permitted isolation, and then sequencing, of the corresponding gene from *Anabaena variabilis* ATCC 29413, over expression of the corresponding protein, and analysis of the mechanism of synthesis of cyanophycin [Berg *et al.*, 2000]. The one gene evidently sufficed for cyanophycin synthesis in *E. coli*. The cyanophycinase gene of PCC 6803, expressed in *E. coli* and purified, hydrolyzed CGP to an asp-arg dipeptide [Richter *et al.*, 1999]. On the basis of the sequence of those genes, the corresponding genes from *Synechocystis* PCC 6308 were then cloned, leading to heterologous expression exceeding 26 % of cell dry mass [Aboulmagd *et al.*, 2000].

C-phycocyanin and allophycocyanin are phycobiliproteins, pigmented components of the photosystem-II antenna structure, the phycobilisome [Glazer, 1988]. Phycocyanin (PC) is a blue, light-harvesting pigment in cyanobacteria and in the two eukaryote algal genera, *Rhodophyta* and *Cryptophyta*. It is PC that gives many cyanobacteria their bluish color and why these cyanobacteria are also known as blue-green algae. PC and related phycobiliproteins are utilized in a number of applications in foods and cosmetics, biotechnology, diagnostics and medicine. Sekar & Chandramohan [2008] counted existing patents on phycobiliproteins and found 55 patents on phycobiliprotein production, 30 patents on applications in medicine, foods and other areas, and 236 patents on applications utilizing the fluorescence properties of phycobiliproteins. PC is a water soluble, nontoxic fluorescent protein with potent antioxidant, anti-inflammatory and anticancer properties [Benedetti *et al.*, 2004; Sabarinathan & Ganesan, 2008; Subhashini *et al.*, 2004]. Phycocyanin could be purified directly from cyanobacteria [Benedetti *et al.*, 2006; Soni *et al.*, 2008] or could be synthesized in *Escherichiacoli* cells by using one expression vector containing all necessary five genes (*hox1*, *pcyA*, *cpcA*, *cpcE*, and *cpcF*) and a His-tag for convenient purification of recombinant protein [Guang *et al.*, 2007]. Different ways of phycocyanin production has been reviewed recently by N.T. Eriksen [2008].

Phycobiliproteins, coupled to monoclonal and polyclonal antibodies to form fluorescent antibody reagents, are valuable as fluorescent tags in cell sorting, studies of cell surface antigens, and screening of high-density arrays [Sun *et al.*, 2003]. *Spirulina* is a convenient and inexpensive source of allophycocyanin and C-phycocyanin (C-PC) [Jung & Dailey, 1989]. As an alternative approach, genetic engineering of *Anabaena* 7120 has permitted the *in vivo* production of stable phycobiliprotein constructs bearing affinity purification tags, and usable as fluorescent labels without further chemical manipulation [Cai *et al.*, 2001]. Recombinant expression of C-PC and holo-C-PC α–subunits in *Anabaena* sp. and *E. coli* has demonstrated that protein engineering can generate C-PC with improved stability or novel functions. Successful pharmaceutical applications will depend on C-PC produced under well-controlled conditions. Recombinant and heterotrophic production procedures seem more promising for novel C-PC synthesis at industrial scales.

Use of *Spirulina* as a source of protein and vitamins for humans or animals has been reviewed by Ciferri [1983] and Kay [1991]. *Spirulina platensis* and *Spirulina maxima* are thought to have been consumed since ancient times as a food in a part of Africa that is now in the Republic of Chad, and in Mexico, respectively [Ciferri & Tiboni, 1985]. These species have unusually high protein content for photosynthetic organisms, up to 70% of the dry weight. *Nostoc flagelliforme* is considered a delicacy in China [Gao, 1998; see also Takenaka *et al.*, 1998]. Other cyanobacteria are eaten in India and the Philippines [Tiwari, 1978; Martinez, 1988]. The amino acid composition of *S. maxima* [Clément *et al.*, 1967], which can grow on animal wastes [Wu and Pond, 1981], is also among the best, for human nutrition, of a photosynthetic organism. Like other microalgae, *Spirulina* is used as a source of natural colorants in food, and as a dietary supplement [Kay, 1991]. The optimal physiological conditions (temperature and pH) for biomass production and protein biosynthesis were demonstrated recently for new isolate of *Spirulina* sp. that was found from an oil-polluted brackish water environment in the Niger Delta [Ogbonda *et al.*, 2007]. Beginning in the early 1980s, another species, including *Aphanizomenon flos-aquae* have been an accepted source of microalgal biomass for food [Carmichael *et al.*, 2000], as well as *Nostochopsislobatus*, which could be a promising bioresource for enhanced production of nutritionally rich biomass, pigments and antioxidants [Pandey & Pandey, 2008].

Only in the last few years cyanobacteria have been recognized as a potent source for numerous biologically active natural products. To date about 800 molecules of cyanobacterial origin are known, among which are pharmacologically interesting compounds, including anticancer, antimicrobial and hypertension lowering activities. It can be assumed that these organisms hold a huge potential for a abundance of pharmacologically relevant compounds. Cyanobacteria are reported to produce secondary metabolites of which toxic and bioactive peptides are of scientific and public interest. Certain of these toxins and other natural products of cyanobacteria have potential for medicinal uses [Patterson *et al.*, 1991; Boyd *et al.*, 1997; Liang *et al.*, 2005]. Comprehensive review of Řezanka & Dembitsky [2006] present a diverse range of metabolites producing by "biochemical factories" of cyanobacteria belonging to *Nostocaceae.*

Wide spectrum of cyanobacterial toxins is poisoning for animals and dangerous for human health. Cyanobacteria synthesize hepatotoxins (microcystins and nodularins), hepato- and cytotoxins (cylindrospermopsins), neurotoxins (anatoxin-a, anatoxin-a(S), and saxitoxins), dermatotoxins, irritant toxins (lipopolysaccharides) and other marine biotoxins (aplysiatoxins, debromoaplysiatoxins, lyngbyatoxin-a) [Wieg and & Pflugmacher, 2005; Stewart *et al.*, 2006; Sivonen & Börner, 2008].

The cyanobacterial hepatotoxins most frequently found in freshwater blooms are the cyclic heptapeptide microcystins, whereas in brackish waters, the cyclic pentapeptide nodularin is common. More than 60 isoforms of microcystins are currently known. Microcystins have been detected in the cyanobactrial genera *Anabaena*, *Anabaenopsis*, *Hapalosiphon*, *Microcystis*, *Nostoc*, *Planktothrix*, *Phormidium*, and *Synechococcus* [Sivonen & Börner, 2008]. Microcystins are cyclic heptapeptides with an unusual chemical structure and a number of nonproteinogenic amino acids [Sivonen & Jones, 1999]. These peptides are synthesized by the non-ribosomal peptide synthesis pathway. Microcystins typically contain three variable methyl groups, including N-methyl l, O-methyl and C-methyl groups. Methylation is a relatively common modification in biologically active natural peptides and is thought to improve stability against proteolytic degradation [Finking & Marahiel, 2004 ; Sieber & Marahiel, 2005]. They are synthesized on large enzyme complexes consisting of non-ribosomal peptide synthetases (NRPS) [Dittmann *et al.*, 1997] and polyketide synthases (PKS) in a variety of distantly related cyanobacterial genera. Non-ribosomal peptide synthetases possess a highly conserved modular structure with each module consisting of catalytic domains responsible for the adenylation, thioester formation and condensation of specific amino acids [Marahiel *et al.*, 1997]. The arrangement of these domains within the multifunctional enzymes determines the number and order of the amino acid constituents of the peptide product [Sieber & Marahiel, 2005]. Additional domains for the modification of amino acid residues such as epimerization, heterocyclization, oxidation, formylation, reduction or N-methylation may also be included in the module [Lautru & Challis, 2004; Marahiel *et al.*, 1997; Sieber & Marahiel, 2005]. So, all these biosynthetic features result in high diversity of cyanobacterial microcystins. For now the biosynthetic gene clusters have been fully sequenced from *Microcystis*, *Planktothrix*, *Anabaena*, *Nodularia* and *Nostoc* [reviewed by Sivonen & Börner, 2008]. In *Anabaena* this enzyme complex is encoded in a 55 kb gene cluster containing 10 genes (*mcyA–J*) encoding peptide synthetases, polyketide synthases and tailoring enzymes [Rouhiainen *et al.*, 2004]. In all analyzed heterocystous cyanobacteria (*Anabaena*, *Nostoc* and *Nodularia*), the gene order follows the co-linearity rule of peptide synthetases and its products. If genes, which are encoding bioactive compounds, are known, they could be inactivated by directed mutagenesis and corresponding mutants with defective biosynthesis of bioactive compounds could help to search for the functions of these metabolites. In the case of microcystins, such mutants could be generated by insertional disruption or deletion of *mcy* genes in *M. aeruginosa* and *P. agardhii* strains [Dittmann *et al.*, 1997; Nishizawa *et al.*, 2000; Christiansen *et al.*, 2003; Pearson *et al.*, 2004]. Ditmann *et al.* [1997] for the first time show by knock-out mutagenesis that peptide synthetase genes were involved in the production of a cyanobacterial bioactive compound and demonstrated that one gene cluster was responsible for the production of all microcystins variants in the strain *Microcystisaeruginosa* PCC 7806. What microcystins significance for cyanobacterial cells is not clear. The lack of all microcystins in an *mcyB* mutant of *M.aeruginosa* PCC 7806 [Dittmann *et al.*, 1997] had no effect on growth on the mutant cells under different light laboratory conditions as compared to the wild-type cells [Hesse *et al.*, 2001]. However, comparative two-dimensional protein electrophoresis showed that microcystin-related protein, MrpA, was strongly expressed in the wild-type PCC 7806, but was not detectable in the *mcyB* mutant [Dittmann *et al.*, 2001]. Application of modern transcriptomic and proteomic approaches in combination with genetical methods could help to reveal key aspects of cyanobacterial toxin biosynthesis in the future experiments.

CYANOBACTERIA ARE PROMISING PRODUCERS OF MOLECULAR HYDROGEN – FUTURE ECOLOGICALY PURE FUEL

Molecular hydrogen is one of the potential future energy sources as an alternative to the limited fossil fuel resources of today. Its advantages as fuel are numerous: it is ecologically clean, efficient, renewable, and during its production and utilization no CO_2 and at most only small amounts of NO_x are generated [Dutta *et al.*, 2005]. Advances in hydrogen fuel cell technology and the fact that the oxidation of H_2 produces only H_2O increase its attractiveness.

The cyanobacteria form a diverse subdivision of prokaryotic oxygenic phototrophic microorganisms with 2,654 species classified and 43 draft genomes completed or in progress. Genomic DNA sequences are available for 43 different strains and species of cyanobacteria for the moment of this review writing (http://www.ncbi.nlm.nih.gov/sutils/genom_table.cgi). Many, but not all, strains are capable of H_2 production. Hydrogenase encoding genes are found in all five major taxonomic groups; at least 50 genera and about a hundred strains so far were found to metabolize H_2 [Rupprecht et al., 2006]. Liquid suspension cultures or immobilized cells of cyanobacteria offer opportunities for photoproduction of molecular hydrogen [Gisby et al., 1987; Lindblad, 1999; Serebryakova & Tsygankov, 2007].

To optimize hydrogen production by cyanobacteria we have to learn more about a regulation all the genes and proteins that are involved it this process. Molecular genetic analysis of hydrogen metabolism systems is a prerequisite for the use of genetic and genetic engineering methods to create optimized cyanobacterium strains with high rates of hydrogen production. The genetic control of hydrogen metabolism in cyanobacteria has been discussed in the recent review of S.V. Shestakov and L.E. Mikheeva [2006]. Transcriptional analysis of hydrogenases genes in *Anacystis nidulans* and *Anabaena variabilis* has been monitored by RT-PCR [Boison et al., 2000] In Peter Lindblad laboratory transcription and regulation of the bidirectional hydrogenase have been studied in *Nostoc* sp. strain PCC 7120 [Sjöholm et al., 2007].

Comparative analysis of several unicellular and filamentous, nitrogen-fixing and non-nitrogen-fixing cyanobacterial strains on the molecular and the physiological level have been accomplished in order to find the most efficient organisms for photobiological hydrogen production [Schütz et al., 2004]. Among them are symbiotic, marine, and thermophilic cyanobacteria, as well as species, capable of hydrogen production under aerobic conditions. Cyanobacteria possess several enzymes directly involved in hydrogen metabolism: (i) nitrogenase(s), catalyzing the production of H_2 as a side product of reduction of N_2 to NH_3; (ii) an uptake hydrogenase, catalyzing the consumption of H_2 produced by the nitrogenase; and (iii) a bidirectional hydrogenase, which has the capacity both to take up and to produce H_2 [Papen et al., 1986; Schmitz et al., 1995; Tamagnini et al., 2000; 2002; 2007; Vignais & Colbeau, 2004]. The formation of hydrogen from water via the bioconversion of photon energy is a multistage process involving photosystems I and II responsible for electron transfer to NADP along the chain of transporters. This results in the formation of the transmembrane electrochemical gradient of proton transport that is necessary for ATP synthesis. Cyanobacteria use electrons transported via ferredoxin (or flavodoxin) in the nitrogenase reaction of ammonium synthesis and proton reduction yielding molecular hydrogen, which is reutilized by means of uptake hydrogenase rater than released from cells. This FeNi-containing enzyme provides additional energy, is involved in the control of electron flow, and protects nitrogenase from oxygen inactivation. Cells of all nitrogen-fixing cyanobacteria contain membrane-bound uptake hydrogenase. Another enzyme of hydrogen metabolism is cytoplasmic enzyme, bidirectional hydrogenase that catalyzes the reversible reaction $2H^+ + 2e^- \leftrightarrow H_2$. Many but not all cyanobacteria contain bidirectional hydrogenase. Three possible functions are considering for this enzyme: (1) the removal of excess reductants under anaerobic conditions, (2) hydrogen oxidation in the periplasm and electron delivery to the respiratory chain, and (3) a valve for electrons generated in the light reaction of photosynthesis [Tamagnini et al., 2002].

Mutagenesis and genetic engineering methods can be used to create genetically modified strains suitable for commercial use in photobiotechnology. To maximize H_2 production, mutants of *A. variabilis* strain ATCC 29413 defective in H_2-utilization first were isolated after chemical mutagenesis by L. Polukhina-Mikheeva and O. Koksharova at LomonosovM.V. Moscow State University [Mikheeva *et al*, 1994]. Two mutants altered in hydrogen metabolism were characterized [Mikheeva *et al.*, 1995; Sveshnikov *et al.*, 1997] and one of them, PK 84, has been used for hydrogen production in an automated helical tubular photobioreactor [Borodin *et al.*, 2000; Tsygankov *et al.*, 2002]. Later insertional inactivated Δ *hupL* mutants of *Anabaena* sp. PCC 7120 [Masukawa *et al.*, 2002], *Nostocpunctiforme* strain ATCC 2913 [Lindberg *et al.*, 2002, 2004] and *Nostoc* sp. PCC 7422 [Yoshino *et al.*, 2007] have been studied. Masukawa and colleagues [2007] demonstrated a strategy for improving H_2 production activity over that of the parent Δ*hupL* strain derived from *Nostoc* sp. strain PCC 7120, as evidenced by the greater sustained H_2 production and higher nitrogenase activities of the Δ*hupL* Δ*nifV1* mutant culture grown under air. Mutants with blocked uptake hydrogenase or inactivated structural *hup* genes or *hyp* and *hupW* genes controlling nickel metabolism, the assemblage and functions of hydrogenase complexes, and regulation of their activity are promising.

Realization of a semiartificial system for biohydrogen production involves the integration of photosynthetic protein complexes and hydrogenases into a bioelectronic or bioelectrochemical device. Practically, this can be achieved by the immobilization of the protein complexes on conductive supports (e.g. noble-metals, carbon or semiconductors), in which the first step is the efficient electron transfer from light-driven water splitting by PS2 to the conductive surface. To establish a semiartificial device for biohydrogen production utilizing photosynthetic water oxidation, the immobilization of a Photosystem 2 on electrode surfaces has been reported [Badura *et al.*, 2006]. For this purpose, an isolated Photosystem 2 with a genetically introduced His tag from the cyanobacterium *Thermosynechococcuselongatus* was attached onto gold electrodes modified with thiolates bearing terminal Ni (I1)-nitrilotriacetic acid groups. Other artificial system for hydrogen production was presented by engineered a "hard-wired" protein complex consisting of a hydrogenase and photosystem I (hydrogenase-PSI complex) as a direct light-to hydrogen conversion system. The key component was an artificial fusion protein composed of the membrane-bound [NiFe] hydrogenase from the β-proteobacterium *Ralstonia eutropha* H16 and the peripheral PSI subunit PsaE of the cyanobacterium *Thermosynechococcuselongatus*. The hydrogenase-PSI complex displayed light-driven hydrogen production [Ihara *et al.*, 2006]. Immobilized culture of *Gloeocapsaalpicola* CALU 743 placed in a photo-bioreactor (PhBR) operated in a two-stage cyclic regime "photosynthesis- endogenous fermentation" was operated successfully over a period of more then three months, giving stable hydrogen production [Serebryakova & Tsygankov, 2007].

To increase production of hydrogen the strain used, the gas phase composition, irradiance and the medium used during growth and H_2 production stages, as well as the specific growth, CO_2 consumption, could be optimized [Tsygankov *et al.*, 1998, 1999; Gutthann *et al.*, 2007; Berberoğlu *et al.*, 2008; Yoon *et al.*, 2008]. Some heterological systems could be applied for hydrogen production. So, enhanced hydrogen production in *Escherichiacoli* cells expressing the cyanobacterial *Synechocystis* sp. PCC 6803 HoxEFUYH (the reversible or bidirectional hydrogenase) has been obtained by inhibiting hydrogen uptake of both hydrogenase I and hydrogenase 2 [Maeda *et al.*, 2007]. Other example of heterological systems for producing hydrogen is the system in which uptake hydrogenase negative mutants of bloom forming cyanobacteria (*Nostoc* and *Anabaena*) and the fermentative bacteria *Rhodopseudomonaspalustris* P_4 were used together for producing hydrogen within the reverse micelles as microreactor [Pandey *et al.*, 2007].

To increase the H_2 production by heterocyst-forming cyanobacteria different approaches could be undertaken. The main directions are : (i) increasing the efficiency of H_2 production by heterocysts, (ii) increasing the heterocysts number, typically up to 10% of cells [Wolk, 2005; Shestakov & Mikheeva, 2006], (iii) reducing the antenna size and redirecting greater H^+ and e⁻fluxes toward the hydrogenase [Kruse *et al.*, 2005]. (i) The efficiency of H_2 production by heterocysts could be increased by either (a) genetically modifying *Anabaena* nitrogenase to produce primarily or exclusively H_2, as has been done in *Azotobactervinelandii* [Fisher *et al.*, 2000], potentially increasing H_2 production 4-fold, or (b) replacing nitrogenase with a different enzyme, an efficient reversible hydrogenase, potentially producing yet more H_2. In *Anabaena* it could be reversible hydrogenase that is encoded by the *hoxEFUYH* genes [Schmitz *et al.*, 1995]. Other prospective projects are aimed at the creation of strains combining the block of uptake hydrogenase with the substitution of Mo-containing nitrogenase (gene *hif1*) by vanadium-containing nitrogenase (gene *nifDGK*), which more efficiently uses the energy of electrons for reducing protons to molecular hydrogen [Prince & Kheshgi].

The amount of heterocysts can be increased via genetic engineering manipulations with genes controlling the formation of heterocysts (e.g., gene *hetN*, *hetR*, *hetC*, *patA*) and genes involved in the control of nitrogen metabolism. One of these genes is *ntcA*, which encodes the DNA-binding protein interacting with the promoters of uptake hydrogenase genes whose transcription is activated in heterocysts [Axelsson *et al.*, 1999; Herrero *et al.*, 2001; Sjöholm *et al.*, 2007]. H_2-production aside, no biotechnological use has yet been made of the capacity of heterocyst-forming cyanobacteria, while growing in air, to support reactions that require microoxic conditions.

In theory, significant improvements in the light-driven H_2 capacity and stability could be engineered into the cyanobacterial system by reducing the antenna size and by redirecting greater H^+ and e⁻fluxes toward the hydrogenase [Kruse *et al.*, 2005]. For example, a fivefold stimulation in the light-driven H_2 production rate was observed in an engineered strain of *Synechocystis* PCC 6803 by deletion of an assembly gene for the type 1 NADPH dehydrogenase (NDH -1) [Cournac *et al.*, 2004]. This deletion blocks cyclic electron flow from PSI into the PQ pool, thus redirecting the flux of e⁻ into $NADP^+$ reduction. Genetic engineering holds considerable promise for the future because to date, few engineered strains have been reported to test these principles.

There is no doubt that the photobiological production of hydrogen thus represents a potentially valuable renewable energy resource for the future. A prerequisite challenge is to improve current systems at the biochemical level so that they can clearly generate hydrogen at a rate and efficiency that approaches the 10% energy efficiency that has been surpassed in photoelectrical systems.

CYANOBACTERIA AND ENVIRONMENT

Cyanobacteria play an important role in diverse ecological systems. They are common in aqueous environments, including marine, brackish and fresh waters; in soil and rocks, and especially on moist surfaces; and in some habitats that lack eukaryotic life, such as some hot springs and highly alkaline lakes. They form symbioses with algae, bryophytes, ferns, cycads and an angiosperm, and with invertebrates (corals) [Fogg *et al.*, 1973; Wilkinson & Fay, 1979; Adams 2000; Adams 2002; Bergman *et al.*, 2008]. In association with fungi, they form lichens that help to transform rocks to soil in which other forms of life can grow. It is thought that cyanobacteria were the first colonizers of land, providing a physical and chemical substrate for the later growth of eukaryotic plants. Acclimation studies in cyanobacteria offer useful information about adjustment to new environment. Cyanobacteria in particular accumulate a variety of osmolytes depending upon the nature of stress. Trehalose is among the well-known compatible solutes that stabilize membrane and protein during dehydration and is a compound accumulated in cyanobacterial cell during salt or osmotic stress [Potts, 1996; Leslie *et al.*, 1995; Page-Sharp *et. al*, 1999; Asthana *et al.*, 2005, 2008; Higo *et al.*, 2006]. How a cyanobacterium survives extreme desiccation is now open to genetic analysis, thanks to the development of a genetic system for *Chroococcidiopsis*, which dominates microbial communities in the most extremely arid hot and cold deserts [Billi *et al.*, 2001]. Transfer of just the sucrose-6-phosphate synthase gene (*spsA*) from *Synechocystis* to desiccation-sensitive *E. coli* resulted in a 10^4-fold increase in survival compared to wild-type cells following freeze-drying, air drying, or desiccation over phosphorus pentoxide [Billi *et al.*, 2000]. Modern methods, including DNA microarrays in a combination with insertional mutagenesis of potential sensors and transducers, for the identification of stress-inducible genes and regulatory systems, have allowed identification and study of many genes of *Synechocystis* sp. PCC 6803 and have led to significant progress in understanding the mechanisms of reaction to environmental stress in cyanobacteria [Los *et al.*, 2008].

Cyanobacteria are living everywhere on our Planet. The cyanobacterial diversity of microbial mats growing in the benthic environment of Antarctic lakes has been exploited for the discovery of novel antibiotic and antitumour activities [Biondi *et al.*, 2008]. If care is not taken in the disposal of phosphate- and nitrate-containing industrial, agricultural, and human wastes, these wastes can eutrophicate lakes and ponds, resulting in massive growth ("blooms") of cyanobacteria [Atkins *et al.*, 2001]. The surface of the water becomes turbid and light cannot penetrate to lower levels. A portion of the cyanobacteria then dies, producing unpleasant odors; bacteria that decompose the cyanobacteria use up available oxygen; and fish then die for lack of oxygen. Some bloom-forming cyanobacteria produce toxins that may render water unusable [Hitzfeld *et al.*, 2000; Hoeger *et al.*, 2005]. The microcystins and nodularins may accumulate into aquatic organisms and be transferred to higher trophic levels, and eventually affect vector animals and consumers [Kankaanpää *et al.*, 2005].

Naturally occurring aromatic hydrocarbons [Cerniglia & Gibson, 1979; Cerniglia *et al.*, 1980a,b; Ellis, 1977; Narro *et al.*, 1992] and xenobiotics [Megharaj *et al.*, 1987; Kuritz & Wolk, 1995] can both be degraded by cyanobacteria, and cyanobacteria can be genetically engineered to enhance their biodegradative ability [Kuritz & Wolk, 1995]. Microbial mats rich in cyanobacteria facilitate the remediation of oil-polluted waters and desert in the region of the Arabian Gulf by utilizing crude oil and individual n-alkanes as sources of carbon and energy [Sorkhoh *et al.*, 1992, 1995]. Cyanobacterial nitrogen fixation could provide sufficient nitrogen compounds for heterotrophic oil degradation. Free radicals formed during oxygenic photosynthesis could indirectly enhance photo-chemical oil degradation [Nicodem *et al.* 1997]. Currently, there is a growing interest in the application of phototrophic biofilms, for instance, in wastewater treatment [Schumacher *et al.*, 2003] and removal of heavy metals [Mehta & Gaur, 2005; Baptista & Vasconcelos, 2006; De Philipps *et al.*, 2007]. Cyanobacteria are effective biological metal sorbents, representing an important sink for metals in aquatic environment. Parker *et al.* [2000] showed that mucilage sheaths isolated from the cyanobacteria *Microcystis aeruginosa* and *Aphanothece halophytica* exhibit strong affinity for heavy metal ions such as copper, lead, and zinc. In addition to biosorption and bioaccumulation, the elevated pH inside photosyntetically active biofilms may favor removal of metals by precipitation [Liehr *et al.*, 1994]. Cyanobacteria are able to biotransformation of Hg (II) into β-HgS and may be useful for bioremediation of mercury [Lefebvre *et al.*, 2007]. Recently Micheletti and colleagues [2008] discovered that sheathless mutant of *Gloeothece* sp. strain PCC 6909 were even more effective in the removal of the heavy metal (copper ions) than the wild type.

The microorganisms produce extracellular polymeric substances (EPS) that hold the biofilm together [Wimpenny *et al.*, 2000]. EPS produced by algae and cyanobacteria can improve the soil water-holding capacity and prevent erosion [Rao & Burns 1990]. Filamentous nitrogen-fixing cyanobacteria may be natural biofertilizers for the rice fields [Jha *et al.*, 2001; Pereira *et al.*, 2008].

Cyanobacteria, because they inhabit the same ecological niches as mosquito larvae, and are eaten by them [Thiery *et al.*, 1991; Avissar *et al.*, 1994], are attractive candidates for mosquito control. Transgenic mosquitocidal cyanobacteria, filamentous as well as unicellular, have been engineered [Tandeau de Marsac *et al.*, 1987; Angsuthanasombat & Panyim, 1989; Chungjatupornchai, 1990; Murphy & Stevens, 1992; Soltes-Rak *et al.*, 1993, 1995; Xu *et al.*, 1993, 2000; Wu *et al.*, 1997]. A high level of toxicity was observed with recombinant clones of *Anabaena* 7120 bearing two δ-endotoxin genes (*cryIVA* and *cryIVD*) and gene *p20* of *Bacillusthuringiensis* subsp. *israelensis* [Wu *et al.*, 1997]. Outdoor tests indicated that genetically altered *Anabaena* 7120 could keep containers with natural water from being inhabited by *Culex* larvae for over 2 months [Xu *et al.*, 2000]. Although one might expect laboratory strains to have low competitive ability compared with indigenous species that same characteristic may help to prevent unwanted spread of a genetically modified microorganism. Notably, a lyophilized (and presumably non-proliferate) preparation of the recombinant cells retained the same high mosquitocidal activity as the original culture. It is interesting recombinant cyanobacteria, expressing δ-endotoxin proteins of *Bacillusthuringiensis* subsp. israelensis, can protect mosquito larvicidal toxins from UV inactivation [Manasherob *et al.*, 2002].

CYANOBACTERIA AND NANOTECHNOLOGY

Nanotechnology involves the production, manipulation and use of materials ranging in size from less than a micron to that of individual atoms. Synthesis of nanoparticles using biological entities has great interest due to their unusual optical [Krolikowska *et al.*, 2003], chemical [Kumar *et al.* 2003], photoelectrochemical [Chandrasekharan & Kamat, 2000], and electronic [Peto *et al.* 2002] properties. Among different organisms, cyanobacteria also could be useful for nanoparticles synthesis. So, gold nanoparticles could be obtained by using a filamentous cyanobacterium, *Plectonemaboryanum* UTEX 485 [Lengke *et al.*, 2006a]. The mechanisms of gold bioaccumulation by this cyanobacteria from gold (III) - chloride solutions have documented that interaction of cyanobacteria with aqueous gold (III)-chloride initially promoted the precipitation of nanoparticles of amorphous gold (I)-sulfide at the cell walls, and finally deposited metallic gold in the form of octahedral (III) platelets near cell surfaces and in solutions [Lengke *et al.*, 2006b]. In addition, *Plectonemaboryanum* UTEX 485 has been used for biosynthesis of silver and palladium nanoparticles [Lengke *et al.*, 2007a, b].

In order to elucidate and replicate the clean energy generation system cyanobacterial photosystem complexes with the genetically introduced polyhistidine tag and coupled with solid-state electronic devices [Das *et al.*, 2004] and gold nanoparticles have been applied [Terasaki *et al.*, 2006, 2008]. Metal nanoparticles in the combination with the cyanobacterial photosynthetic molecular complexes can be a base for creation of energy-conversion devices and sensors [Govorov & Carmeli, 2007].

CONCLUSION

During the past half century, cyanobacteria have been used increasingly to study, among other topics, photosynthesis and its genetic control; photoregulation of genetic expression; cell differentiation and N_2 fixation; metabolism of nitrogen, carbon, and hydrogen; resistance to environmental stresses; and molecular evolution. The availability of powerful genetic techniques allows the biotechnological application of cyanobacteria to produce specific products, including photosynthetic pigments and molecular hydrogen, to biodegrade organic pollutants in surface waters, to control mosquitoes, to produce nanoparticles and for many other purposes. The combination of genome sequencing with studies of transcriptomes, proteomes, and metabolomes is likely to discover key aspects of the biology of these amazing organisms in the near future and permit to use them even more effectively in biotechnology.

ACKNOWLEDGMENTS

I express my sincere gratitude to Professor S.V. Shestakov and Dr. L.E. Mikheeva at M.V.Lomonosov MoscowStateUniversity for the pleasure to work together in the hydrogen producing mutants project. This review writing was supported by grant from the Russian Foundation for Basic Research (number 08-04-00878).

REFERENCES

Aboulmagd, E., Oppermann-Sanio, F. B. & Steinbüchel, A. 2000. Molecular characterization of the cyanophycin synthetase from *Synechocystis* sp. strain PCC6308. *Arch. Microbiol.174*, 297-306.

Adams, D. G. Symbiotic interactions. In Whitton,B., Potts,M. editors. *Ecology ofCyanobacteria: Their Diversity in Time and Space*. Dordrecht: Kluwer Academic Publishers; 2000; 523-561.

Adams, D. G. Cyanobacteria in symbiosis with hornworts and liverworts. In Rai, A.N., Bergman, B., Rasmussen, U. editors.*Cyanobacteria in symbiosis*. Dordrecht: Kluwer Academic Publishers; 2002; 117-135.

Angsuthanasombat, C. & Panyim, S. 1989. Biosynthesis of 130-kilodalton mosquito larvicide in the cyanobacterium *Agmenellum quadruplicatum* PR-6. *Appl. Environ. Microbiol.55*, 2428-2430.

Asthana, R. K., Nigam, S., Maurya, A., Kayastha, A. M., Singh, S. P. 2008.Trehalose-Producing Enzymes MTSase and MTHase in *Anabaena* 7120 Under NaCl Stress. *Curr. Microbiol.56*, 429–435.

Asthana, R. K., Srivastava, S., Singh, A. P., Kayastha, A. M. & Singh, S. P. 2005. Identification of maltooligosyltrehalose synthase and maltooligosyltrehalose trehalohydrolase enzymes catalysing trehalose biosynthesis in *Anabaen* a 7120 exposed to NaCl stress.*Journal of Plant Physiology.162*, 1030-1037.

Atkins, R., Rose, T., Brown, R. S., Robb, M. 2001. The *Microcystis* cyanobacteria bloom in the SwanRiver--February 2000. *Water Sci. Technol.43*, 107-114.

Avissar, Y. J., Margalit, J., Spielman, A. 1994. Incorporation of body components of diverse microorganisms by larval mosquitoes. *J. Amer. Mosquito Control Assoc.10*, 45-50.

Axelsson, R., Oxelfelt, F., & Lindblad, P. 1999. Transcriptional regulation of *Nostoc* uptake hydrogenase. *FEMS Microbiol. Letts. 170*, 77–81.

Badura, A., Esper, B., Ataka, K., Grunwald, C., Wöll, C., Kuhlmann, J., Heberle, J. & Rögner, M. 2006. Light-Driven Water Splitting for (Bio-)Hydrogen Production: Photosystem 2 as the Central Part of a Bioelectrochemical Device.*Photochemistry and Photobiology.82*, 1385-1390.

Baptista, M. S. & Vasconcelos, M. T. 2006.Cyanobacteria Metal Interactions: Requirements, Toxicity, and Ecological Implications.*Critical Reviews in Microbiology.32*, 127–137.

Benedetti, S., Benvenuti, F., Pagliarani, S., Francogli, S., Scoglio, S., Canestrari, F. 2004. Antioxidant properties of a novel phycocyanin extract from the blue-green alga *Aphanizomenonflos-aquae.Life Sciences.75*, 2353 – 2362.

Benedetti, S., Rinalducci, S., Benvenuti, F., Francogli, S., Pagliarani, S., Giorgi, L., Micheloni, M., D' Amici, G. M, Zolla, L. & Canestrari, F. 2006. Purification and characterization of phycocyanin from the blue-green alga *Aphanizomenonflos-aquae. Journal of ChromatographyB, 833*, 12-18.

Berberoğlu, H., Barra, N., Pilon, L. & Jay, J. 2008.Growth, CO_2 consumption and H_2 production of *Anabaenavariabilis* ATCC 29413-U under different irradiances and CO_2 concentrations. *JournalofAppliedMicrobiology.104*, 105–121.

Berg, H., Ziegler, K., Piotukh, K., Baier, K., Lockau, W., Volkmer-Engert, R. 2000. Biosynthesis of the cyanobacterial reserve polymer multi-L-arginyl-poly-L-aspartic acid (cyanophycin) Mechanism of the cyanophycin synthetase reaction studied with synthetic primers. *Eur. J. Biochem.267*, 5561-5570.

Bergman, B., Ran, L. & Adams, D. G. Cyanobacterial-plant Symbiosies: Signaling and Development. In Herrero,A., Flores,E. editors. *The Cyanobacteria Molecular Biology, Genomics and Evolution*. Norfolk, UK: Caiser Academic Press; 2008; 447-473.

Billi, D., Wright, D. J., Helm, R. F., Prickett, T., Potts, M., Crowe, J. H. 2000. Engineering desiccation tolerance in *Escherichia coli. Appl. Environ. Microbiol.66*, 1680-1684.

Billi, D., Friedmann, E. I., Helm, R. F., Potts, M. 2001. Gene transfer to the desiccation-tolerant cyanobacterium *Chroococcidiopsis. J. Bacteriol.183*, 2298-2305.

Biondi, N., Tredici, M. R., Taton, A., Wilmotte, A., Hodgson, D. A., Losi, D. & Marinelli, F. 2008.Cyanobacteria from benthic mats of Antarctic lakes as a source of new bioactivities. *Journal ofApplied Microbiology.105*, 105–115.

Boison, G., Bothe, H. & Schmitz, O. 2000.Transcriptional analysis of h hydrogenase genes in the cyanobacteria *Anacystis nidulans* and *Anabaena variabilis* monitored by RT-PCR.*Current Microbiology.40*, 315–321.

Borodin, V. B., Tsygankov, A. A., Rao, K. K., Hall, D. O. 2000. Hydrogen production by *Anabaena variabilis* PK84 under simulated outdoor conditions. *Biotechnol. Bioeng.69*, 478-485.

Boyd, M. R., Gustafson, K. R., Mc Mahon, J. B., Shoemaker, R. H., O'Keefe, B. R., Mori, T., Gulakowski, R. J., Wu, L., Rivera, M. I., Laurencot, C. M., Currens, M. J., Cardellina, J. H., Buckheit, R. W., Nara, P. L., Pannell, L. K., Sowder, R. C., Henderson, L. E. 1997. Discovery of cyanovirin-N, a novel human immunodeficiency virus-inactivating protein that binds viral surface envelope glycoprotein gp120: potential applications to microbicide development. *Antimicrob.Agents Chemother.41*, 1521-1530.

Cai, Y. A., Murphy, J. T., Wedemayer, G. J. & Glazer, A. N. 2001. Recombinant phycobiliproteins. *Analyt. Biochem.290*, 186-204.

Carmichael, W. W., Drapeau, C. & Anderson, D. M. 2000. Harvesting of *Aphanizomenon flos-aquae* Ralfs ex Born.& Flah. var.*flos-aquae* (Cyanobacteria) from KlamathLake for human dietary use. *Journal of Applied Phycology.12*, 585–595.

Cerniglia, C. E., Gibson, D. T., van Baalen, C. 1979. Algal oxidation of aromatic hydrocarbons: formation of 1-naphthol from naphthalene by *Agmenellum quadruplicatum,* strain PR-6. *Biochem.Biophys. Res. Commun.88*, 50-58.

Cerniglia, C. E., van Baalen, C., Gibson, D. T. 1980a. Metabolism of naphthalene by the cyanobacterium *Oscillatoria* sp., strain JCM. *J. Gen. Microbiol.116*, 485-494.

Cerniglia, C. E., Gibson, D. T., van Baalen, C. 1980b. Oxidation of naphthalene by cyanobacteria and microalgae. *J. Gen. Microbiol.116*, 495-500.

Chandrasekharan, N., Kamat, P. V. 2000. Improving the photo-electrochemical performance of nanostructured TiO_2 films by adsorption of gold nanoparticles. *J. Phys. Chem. B, 104*, 10851–10857.

Christiansen, G., Fastner, J., Erhard, M., Börner, T. & Dittmann, E. 2003. Microcystin biosynthesis in *Planktothrix*: genes, evolution, and manipulation. *J. Bacteriol.185*, 564-572.

Chungjatupornchai, W. 1990. Expression of the mosquitocidal-protein genes of *Bacillus thuringiensis* subsp. *israelensis* and the herbicide-resistance gene *bar* in *Synechocystis* PCC 6803. *Curr. Microbiol.21*, 283-288.

Ciferri, O. 1983. *Spirulina*, the edible microorganism. *Microbiol. Rev.47*, 551-578.

Ciferri, O. & Tiboni, O. 1985. The biochemistry and industrial potential of *Spirulina. Annu. Rev.Microbiol.39*, 503-526.

Clément, G., Giddey, C., Menzi, R. 1967. Amino acid composition and nutritive value of the alga *Spirulina maxima. J. Sci. Fd. Agric.18*, 497-501.

Das, R., Kiley, P.J., Segal, M., Norville, J., Yu, A.A., Wang, L., Trammell, S.A., Reddick, L.E., Kumar, R., Stellacci, F., Lebedev, N., Schnur, J., Bruce, B.D., Zhang, S. & Baldo, M. 2004.

Integration of photosynthetic protein molecular complexes in solid-state electronic devices. *NanoLetters.4*, 1079-1083.

De Philipps, R., Paperi, R. & Sili, C. 2007. Heavy metal sorption by released polysaccharides and whole cultures of two exopolysaccharide-producing cyanobacteria. *Biodegradation.18*, 181–187.

Dittmann, E., Neilan, B. A., Erhard, M., von Döhren, H. & Börner, T. 1997. Insertional mutagenesis of a peptide synthetase gene that is responsible for hepatotoxin production in the cyanobacterium *Microcystis aeruginosa* PCC 7806. *Mol. Microbiol.26*, 779–787.

Dittmann, E., Erhard, M., Kaebernick, M., Scheler, C., Neilan, B. A., Dohren, H. & Börner, T. 2001. Altered expression of two light-dependent genes in a microcystin-lacking mutant of *Microcystisaeruginosa* PCC 7806.*Microbiology.147*, 3113–3119.

Dutta, D., De, D., Chaudhuri, S. & Bhattacharya, S.K. 2005. Hydrogen production by cyanobacteria. *Microbial. Cell Factories.4*, 36.

Ellis, B. E. 1977. Degradation of phenolic compounds by fresh-water algae. *Plant Sci. Lett.8*, 213-216.

Eriksen, N. T. 2008. Production of phycocyanin – a pigment with applications in biology, biotechnology, foods and medicine. *Appl. Microbiol. Biotechnol.80*, 1-14.

Finking, R. & Marahiel, M. A. 2004. Biosynthesis of nonribosomal peptides. *Annu Rev. Microbiol.58*, 453–488.

Fisher, K., Dilworth, M. J. &Newton, W. E. 2000. Differential effects on N_2 binding and reduction, HD formation, and azide reduction with α-195^{His}- and α-191^{Gln}- substituted MoFe proteins of *Azotobactervinelandii* nitrogenase. *Biochemistry.39*, 15570-15577.

Fogg, G. E, Stewart, W. D. P., Fay, P. & Walsby, A. E. The Blue-Green Algae. London: Academic Press; 1973.

Gao, K. 1998. Chinese studies on the edible blue-green alga *Nostoc flagelliforme*: a review. *J. Appl. Phycol.10*, 37-49.

Gisby, P. F., Rao, K., Hall, D. O. 1987 Entrapment techniques for chloroplasts, cyanobacteria and hydrogenases. *Methods Enzymol.135,* 440-454.

Glazer, A. N. 1988. Phycobiliproteins. *Methods Enzymol.167*, 291-303.

Govorov, A. O. & Carmeli, I. 2007. Hybrid structures composed of photosynthetic system and metal nanoparticles: plasmon enhancement effect. *Nano Lett. 7*, 620-625.

Guang, X., Qin, S., Su, Z., Zhao, F., Ge, B., Li, F. & Tang, X. 2007.Combinational biosynthesis of a fluorescent cyanobacterial holo-α-phycocyanin in *Escherichiacoli* by using one expression vector.*Appl. Biochem. Biotechnol.142*, 52-59.

Gutthann, F., Egert, M., Margues, A. & Appel, J. 2007. Inhibition of respiration and nitrate assimilation enhances photohydrogen evolution under low oxygen concentrations in *Synechocystis* sp. PCC 6803. *Biochimica et Biophysica Acta.1767*, 161-169.

Hai, T., Hein, S. & Steinbüchel, A. 2001.Multiple evidence for wide spread and general occurrence of type-III PHA synthases in cyanobacteria and molecular characterization of the PHA synthases from two thermophilic cyanobacteria: *Chlorogloeopsisfritschii.* PCC 6912 and *Synechococcus* sp. strain MA19.*Microbiology. 147*, 3047–3060.

Hein, S., Tran, H., Steinbüchel, A. 1998. *Synechocystis* sp. PCC6803 possesses a two-component polyhydroxyalkanoic acid synthase similar to that of anoxygenic purple sulfur bacteria. *Arch.Microbiol. 170*, 162-170.

Herrero, A., Muro-Pastor, A. M. &Flores, E. 2001. Nitrogen control in cyanobacteria. *J. Bacteriol., 183*, 411–425.

Hesse , K. & Kohl, J.G. 2001. Effects of light and nutrient supply on growth and microcystin content of different strains of *Microcystisaeruginosa*. In: Chorus,I. editor. *Cyanotoxins – Occurrence, Causes, Consequences*. Berlin: Springer; 2001; 104-114.

Higo, A., Katoh, H., Ohmori, K., Ikeuchi, M., Ohmori, M. 2006. The role of a gene cluster for trehalose metabolism in dehydration tolerance of the filamentous cyanobacterium *Anabaena* sp. PCC 7120. *Microbiology.152*, 979–987.

Hitzfeld, B. C, Hoger, S. J., Dietrich, D. R. 2000. Cyanobacterial toxins: removal during drinking water treatment, and human risk assessment. *Environ. Health Perspect.108*, 113-122.

Hoeger, S. J., Hitzfeld, B. C. & Dietrich, D. R. 2005. Occurrence and elimination of cyanobacterial toxins in drinking water treatment plants.*Toxicology and Applied Pharmacology.203*, 231-242.

Jha, M. N., Prasad, A. N., Sharma, S. G. & Bharati, R. C. 2001. Effects of fertilization rate and crop rotation on diazotrophic cyanobacteria in paddy field. *World Journal of Microbiology &Biotechnology.17*, 463-468.

Ihara, M., Nishihara, H., Yoon, K-S, Lenz, O., Friedrich, B., Nakamoto, H., Kojima, K., Honma, D., Kamachi, T. & Okura, I. 2006.Light-driven hydrogen production by a hybrid complex of a [NiFeI]-hydrogenase and the cyanobacterial photosystem I.*Photochemistry and Photobiology.82*, 676-682

Jung, T. & Dailey, M. 1989. A novel and inexpensive source of allophycocyanin for multicolor flow cytometry. *J. Immunol. Meth.121*, 9-18.

Kankaanpää, H.T., Holliday, J., Schröder, H., Goddard, T. J., von Fister, R. & Carmichael, W.W. 2005.Cyanobacteria and prawn farming in northern New South Wales, Australia - a case study on cyanobacteria diversity and hepatotoxin bioaccumulation. *Toxicology and Applied Pharmacology.203*, 243-256.

Kay, R. A. 1991. Microalgae as food and supplement. *Crit. Rev. Food Sci. Nutr.30*, 555-573.

Koksharova O. A. & Wolk, C. P.2002.Genetic tools forcyanobacteria.*Appl. Microbiol.Biotechnol. 58*, 123-137.

Krolikowska, A., Kudelski, A., Michota, A., Bukowska, J. 2003. SERS studies on the structure of thioglycolic acid monolayers on silver and gold. *Surf. Sci.532*, 227–232.

Kruse, O., Rupprecht, J., Mussgnug, J. H., Dismukes, G. C. & Hankamer, B. 2005. Photosynthesis: a blue print for energy capture and conversion technologies. *Photochem. Photobiol.4*, 957 –970.

Kumar, A., Mandal, S., Selvakannan, P. R., Parischa, R., Mandale, A. B., Sastry, M. 2003. Investigation into the interaction between surface-bound alkylamines and gold nanoparticles. *Langmuir. 19*, 6277–6282.

Kuritz, T., Wolk, C. P. 1995. Use of filamentous cyanobacteria for biodegradation of organic pollutants. *Appl. Env. Microbiol.61*, 234-238.

Lang, N .J., Simon, R. D., Wolk, C. P. 1972. Correspondence of cyanophycin granules with structured granules in *Anabaena cylindrica*. *Arch. Mikrobiol. 83*, 313-320.

Lautru, S. & Challis, G. L. 2004. Substrate recognition by nonribosomal peptide synthetase multi-enzymes. *Microbiology. 150*, 1629–1636.

Lefebvre, D.D., Kelly, D., & Budd, K. 2007. Biotransformation of Hg(II) by cyanobacteria.*Appl. Environ. Microbiol.73*, 243–249.

Lengke, M., Fleet, M. E. & Southam, G. 2006a. Morphology of gold nanoparticles synthesized by filamentous cyanobacteria from gold(I)-thiosulfate and gold(III)-chloride complexes. *Langmuir.22*, 2780–2787.

Lengke, M., Ravel, B., Fleet, M. E., Wanger, G., Gordon, R. A. & Southam, G. 2006b. Mechanisms of gold bioaccumulation by filamentous cyanobacteria from gold(III)-chloride com plex. *Environ. Sci. Technol.40*, 6304–6309.

Lengke, M. V., Fleet, M. E. & Southam, G. 2007a. Biosynthesis of silver nanoparticles by filamentous cyanobacteria. *Langmuir.23*, 2694-2699.

Lengke, M. V., Fleet, M. E. & Southam, G. 2007b. Synthesis of palladium nanoparticles by reaction of filamentous cyanobacterial biomass with a palladium(II) chloride complex. *Langmuir.23*, 8982-8987.

Leslie, S. B., Israeli, E., Lighthart, B., Crow, J. H., Crowe, L. M. 1995. Trehalose and sucrose protect both membranes and proteins in intact bacteria during drying. *Appl. Environ. Microbiol.61*, 3592–3597.

Liang, J., Moore, R. E., Moher, E. D., Munroe, J. E., Al-Awar, R. S., Hay, D. A., Varie, D. L., Zhang, T. Y., Aikins, J.A., Martinelli, M.J., Shih, C., Ray, J.E., Gibson, L.L., Vasudevan, V., Polin, L., White, K., Kushner, J., Simpson, C., Pugh, S. & Corbett, T.H. 2005. Cryptophycins-309, 249 and other cryptophycin analogs: preclinical efficacy studies with mouse and human tumors. *Invest. New Drugs.23*, 213–224.

Liehr, S. K., Chen, H. J. & Lin, S. H. 1994. Metals removal by algal biofilms. *Water Sci. Technol.30*, 59-68.

Lindberg, P., Schutz, K., Happe, T., & Lindblad, P. 2002. A hydrogen-producing, hydrogenase-free mutant strain of *Nostocpunctiforme* ATCC 29133. *Int. J. Hydrogen Energ.* , *27*,1291-1296.

Lindberg, P., Lindblad, P. & Cournac, L. 2004. Gas exchange in the filamentous cyanobacterium *Nostocpunctiforme* strain ATCC 29133 and its hydrogenase-deficient mutant strain NHM5. *Applied and Environmental Microbiology.70*, 2137-2145.

Lindblad, P. 1999. Cyanobacterial H_2-metabolism: knowledge and potential/strategies for a photobiotechnological production of H_2. *Biotecnol. Apl.16*, 141-144.

Los, D. A., Suzuki, I., Zinchenko, V. V. & Murata, N. Stress responses in *Synechocystis*: regulated genes and regulatory systems. In Herrero, A. & Flores, E. editors.*The Cyanobacteria Molecular Biology, Genomics and Evolution*.Norfolk, UK: Caiser Academic Press; 2008; 117-157.

Maeda, T., Vardar G., Self W.T. & Wood, T.K. 2007. Inhibition of hydrogen uptake in *Escherichiacoli* by expressing the hydrogenase from the cyanobacterium *Synechocystis* sp. PCC 6803. *BMC Biotechnology.7*, 25.

Manasherob, R., Ben-Dov, E., Xiaoqiang, W., Boussiba, S. & Zaritsky, A. 2002. Protection from UV-B damage of mosquito larvicidal toxins from *Bacillusthuringiensis* subsp. *israelensis* expressed in *Anabaena* PCC 7120. *Current Microbiology.45*, 217–220.

Marahiel, M. A., Stachelhaus, T. & Mootz, H. D. 1997. Modular peptide synthetases involved in nonribosomal peptide synthesis. *Chem. Rev.97*, 2651–2674.

Marin, K., Zuther, E., Kerstan, T., Kunert, A. & Hagemann, M. 1998. The *ggpS* gene from *Synechocystis* sp. strain PCC 6803 encoding glucosyl-glycerol-phosphate synthase is involved in osmolyte synthesis. *J. Bacteriol.180*, 4843-4849.

Martinez, M. R. 1988. *Nostoc commune* Vauch., a nitrogen-fixing blue-green alga, as source of food in the Philippines. *Philippine Naturalist.71*, 295-307.

Masukawa, H., Mochimaru, M., Sakurai, H. 2002. Disruption of the uptake hydrogenase gene, but not of the bidirectional hydrogenase gene, leads to enhanced photobiological hydrogen production by the nitrogen-fixing cyanobacterium *Anabaena* sp. PCC 7120. *Appl. Microbiol. Biotechnol.58*, 618–624.

Masukawa, H., Inoue, K. & Sakurai, H. 2007. Effects of disruption of homocitrate synthase genes on *Nostoc* sp.strain PCC 7120 photobiological hydrogen production and nitrogenase. *Applied andenvironmental microbiology.73*, 7562–7570.

Megharaj, M., Venkateswarlu, K., Rao, A. S. 1987. Metabolism of monocrotophos and quinalphos by algae isolated from soil. *Bull. Environ. Contam. Toxicol.39*, 251-256.

Mehta, S. K. & Gaur, J. P. 2005. Use of algae for removing heavy metal ions from wastewater: progress and prospects. *Crit. Rev. Biotechnol.25*, 113-152.

Micheletti, E., Pereira, S., Mannelli, F., Moradas-Ferreira P., Tamagnini, P. & De Philipps, R. 2008. Sheathless mutant of cyanobacterium *Gloeothece* sp. strain PCC 6909 with increased capacity to remove copper ions from aqueous solutions. *Appl. Environ. Microbiol.74*, 2797-2804.

Mikheeva, L.E., Koksharova, O.A., & Shestakov, S.V.1994. Hydrogen-Producing Mutant of the Cyanobacterium*Anabaenavariabilis* Strain ATCC 29 413, Vestn. Mosk.Univ., *Ser. Biol.*, 2, 54–57.

Mikheeva, L. E., Schmitz, O., Shestakov, S. V. &Bothe, H. 1995. Mutants of the cyanobacterium *Anabaenavariabilis* altered in hydrogenase activity, *Z. Naturforsch., A: Phys. Sci.*, *50c*, 505–510.

Miyake, M., Takase, K., Narato, M., Khatipov, E., Schnackenberg, J., Shirai, M., Kurane, R. & Asada, Y. 2000. Polyhydroxybutyrate production from carbon dioxide by cyanobacteria. *Appl.Biochem.Biotechnol.84*, 991-1002.

Murphy, R. C., Stevens, S. E. 1992. Cloning and expression of the *cryIVD* gene of *Bacillus thuringiensis* subsp. *israelensis* in the cyanobacterium *Agmenellum quadruplicatum* PR-6 and its resulting larvicidal activity. *Appl. Environ. Microbiol.58*, 1650-1655.

Narro, M. L., Cerniglia, C. E., van Baalen, C., Gibson, D. T. 1992. Metabolism of phenanthrene by the marine cyanobacterium *Agmenellum quadruplicatum* PR-6. *Appl. Environ. Microbiol.58*, 1351-1359.

Nicodem, D. E., Fernandes, M. C. Z., Guedes, C. L. B., Correa, R. J. 1997. Photochemical processes and the environmental impact of petroleum spills. *Biogeochemistry.39*, 121-138.

Nishizawa, T., Ueda, A., Asayama, M., Fujii, K., Harada, K.-I., Ochi, K. & Shirai, M. 2000. Polyketide synthase gene coupled to the peptide synthetase module involved in the biosynthesis of the cyclic heptapeptide microcystin. *J. Biochem.*, *127*, 779-789.

Ogbonda, K. H., Aminigo, R. E. & Abu, G. O. 2007. Influence of temperature and pH on biomass production and protein biosynthesis in a putative *Spirulina* sp. *Bioresource Technology.98*, 2207-2211.

Page-Sharp, M., Behm, C. A. & Smith G. D. 1999.Involvement of the compatible solutes trehalose and sucrose in the response to salt stress of a cyanobacterial *Scytonema* species isolated from desert soils.*Biochimica et Biophysica Acta.1472*, 519-528.

Pandey, A., Pandey, A., Srivastava, P. & Pandey, A. 2007. Using reverse micelles as microreactor for hydrogen production by coupled systems of *Nostoc/R. palustris* and *Anabaena/R. palustris*. *World J. Microbiol. Biotechnol.23*, 269-274.

Papen, H., Kentemich, T., Schmülling, T., Bothe, H. 1986. Hydrogenase activities in cyanobacteria. *Biochimie.68*, 121-132.

Parker, D. L., Mihalick, J. E., Plude, J .L., Plude, M. J., Clark, T. P., Egan, L., Flom, J. J., Rai, L. C., Kumar, H. D. 2000. Sorption of metals by extracellular polymers from the cyanobacterium *Microcystis aeruginosa f. flos-aquae* strain C3-40. *J. Appl. Phycol.12*, 219 – 224.

Patterson, M. L. G., Baldwin, C. L., Bolis, C. M., Caplan, F. R., Karuso, H., Larsen, L. K., Levine, I. A., Moore, R. E., Nelson, C. S., Tschappat, K. D., Tuang, G. D., Furusawa, E., Furusawa, S., Norton, T. R., Raybourne, R. B. 1991. Antineoplastic activity of cultured blue-green algae (Cyanophyta). *J. Phycol.27*, 530-536.

Pearson, L. A., Hisbergues, M., Börner, T., Dittmann, E. & Neilan, B.A. 2004. Inactivation of an ABC transporter gene, *mcyH*, results in loss of microcystin production in the cyanobacterium *Microcystisaeruginosa* PCC 7806. *Appl. Environ. Microbiol.70*, 6370-6378.

Pereira, I., Ortega, R., Barrientos, L., Moya, M., Reyes, G. & Kramm, V. 2009. Development of a biofertilizer based on filamentous nitrogen-fixing cyanobacteria for rice crops in Chile. *J. Appl. Phycol.21*, 135-144

Peto, G., Molnar, G. L., Paszti, Z., Geszti, O., Beck, A., Guczi, L. 2002. Electronic structure of gold nanoparticles deposited on SiOx/Si. *Mater Sci. Eng. C.19*, 95–99.

Potts, M. 1996. The anhydrobiotic cyanobacterial cell. *Physiol. Plant.* 97,788–794.

Prince, R. C. & Kheshgi, H. S. 2005. The photobiological production of hydrogen: potential efficiency and effectiveness as a renewable fuel. *Crit. Rev. Microbiol. 31*, 19–31.

Rao, D. L. N. & Burns, R. G. 1990. The effect of surface growth of blue-green-algae and bryophytes on some microbiological, biochemical, and physical soil properties. *Biol. Fertil. Soils.9*, 239-244.

Řezanka, T. & Dembitsky, V. M. 2006. Metabolites produced by cyanobacteria belonging to several species of the family *Nostocaceae*. *Folia Microbiol. 51*, 159-182.

Richter, R., Hejazi, M., Kraft, R., Ziegler, K., Lockau, W. 1999. Cyanophycinase, a peptidase degrading the cyanobacterial reserve material multi-L-arginyl-poly-L-aspartic acid (cyanophycin). Molecular cloning of the gene of *Synechocystis* sp. PCC 6803, expression in *Escherichia coli*, and biochemical characterization of the purified enzyme. *Eur. J. Biochem.263*, 163-169.

Rouhiainen, L., Vakkilainen, T., Siemer, B.L., Buikema, W., Haselkorn, R. & Sivonen, K. 2004. Genes coding for hepatotoxic heptapeptides (microcystins) in the cyanobacterium *Anabaena* strain 90. *Appl. Environ. Microbiol.70*, 686-692.

Rupprecht, J., Hankamer, B., Mussgnung, J.H., Ananyev, G., Dismukes, C. & Kruse O. 2006. Perspectives and advances of biological H2 production in microorganisms. *Appl. Microbiol.Biotechnol.72*, 442– 449.

Sabarinathan, K. G. & Ganesan, G. 2008. Antibacterial and toxicity evaluation of C-phycocyanin and cell extract of filamentous freshwater cyanobacterium- *Westiellopsis* sps. *Eur. Rev. Med.Pharmacol. Sci.12*, 79-82.

Schmitz, O., Boison, G., Hilscher, R., Hundeshagen, B., Zimmer, W., Lottspeich, F. & Bothe, H. 1995. Molecular biological analysis of a bidirectional hydrogenase from cyanobacteria. *Eur. J. Biochem.233,* 266-276.

Schumacher, G., BlumeT. & SekoulovI. 2003. Bacteria reduction and nutrient removal in small wastewater treatment plants by an algal biofilm. *Water Sci. Technol.47*, 195– 202.

Schütz, K., Happe, T., Troshina, O., Lindblad, P., Leitão, E., Oliveira, P. & Tamagnini, P. 2004. Cyanobacterial H2 production – a comparative analysis. *Planta.218*, 350-359.

Schwartz, S. H., Black, T. A., Jäger, K., Panoff, J-M, Wolk, C. P. 1998. Regulation of an osmoticum-responsive gene in *Anabaena* sp. strain PCC 7120. *J. Bacteriol.180*, 6332-6337.

Shestakov, S. V & Mikheeva, L. E. 2006. Genetic control of hydrogen metabolism in cyanobacteria. *Russian Journal of Genetics.42*, 1272-1284.

Sekar, S. & Chandramohan, M. 2008. Phycobiliproteins as a commodity: trends in applied research, patents and commercialization. *J. Appl. Phycol.20*, 113-136.

Serebryakova, L. T. & Tsygankov, A. A., 2007.Two-stage system for hydrogen production by immobilized cyanobacterium *Gloeocapsaalpicola* CALU 743. *Biotechnol. Prog. 23*, 1106-1110.

Sieber, S. A. & Marahiel, M. A. 2005. Molecular mechanisms underlying non-ribosomal peptide synthesis: approaches to new antibiotics. *Chem. Rev.105*, 715-738.

Simon, R. D. 1971. Cyanophycin granules from the blue-green alga *Anabaena cylindrica*: a reserve material consisting of copolymers of aspartic acid and arginine. *Proc. Natl. Acad. Sci. U.S.A.68*, 265-267.

Simon, R. D. 1973a. The effect of chloramphenicol on the production of cyanophycin granule polypeptide in the blue-green alga *Anabaena cylindrica. Arch. Mikrobiol.92*, 115-123.

Simon, R. D. 1973b. Measurement of the cyanophycin granule polypeptide contained in the blue-green alga *Anabaena cylindrica. J. Bacteriol.114*, 1213-1216.

Simon, R. D. 1976. The biosynthesis of multi-L-arginyl-poly(L-aspartic acid) in the filamentous cyanobacterium *Anabaena cylindrica. Biochim. Biophys. Acta.422*, 407-418.

Simon, R. D. Inclusion bodies in the cyanobacteria: cyanophycin, polyphosphate, polyhedral bodies. In: FayP,Van BaalenC. editors. *The Cyanobacteria*. Amsterdam: Elsevier Science Publishers B.V.; 1987; 199-225.

Sivonen, K. & Jones, G. Cyanobacterial toxins. In: Chorus, I., Bartram, J. editors. *Toxic Cyanobacteria in Water*. London: E&FN Spon; 1999; 41-111.

Sivonen, K. & Börner, T. Bioactive compounds produced by cyanobacteria. In: Herrero,A.,Flores, E. editors. *The Cyanobacteria Molecular Biology, Genomics and Evolution*. Norfolk, UK: Caiser Academic Press; 2008; 158-197.

Sjöholm, J., Oliveira, P. & Lindblad, P. 2007.Transcription and regulation of the bidirectional hydrogenase in the cyanobacterium *Nostoc* sp. strain PCC 7120.*Appl. Environ. Microbiol.73, 5435–5446*.

Soltes-Rak, E., Kushner, D. J., Williams, D. D, Coleman, J. R. 1993. Effect of promoter modification on mosquitocidal *cryIVB* gene expression in *Synechococcus* sp. strain PCC 7942. *Appl. Environ. Microbiol.59*, 2404-2410.

Soltes-Rak, E., Kushner, D. J., Williams, D.D., Coleman, J. R. 1995. Factors regulating *cryIVB* gene expression in the cyanobacterium *Synechococcus* PCC 7942. *Mol. Gen. Genet.246*, 301-308.

Soni, B., Trivedi, U. & Madamwar, D. 2008. A novel method of single step hydrophobic interaction chromatography for the purification of phycocyanin from *Phormidium fragile* and its characterization for antioxidant property. *Bioresource Technology.99*, 188-194.

Sorkhoh, N., Al-Hasan, R., Radwan, S. & Höpner, T. 1992. Self-cleaning of the Gulf. *Nature. 359*, 109.

Sorkhoh, N. A., Al-Hasan, R. H., Khanafer, M. & Radwan, S. S. 1995. Establishment of oil-degrading bacteria associated with cyanobacteria in oil-polluted soil. *J. Appl. Bacteriol.78*, 194-199.

Stewart, I., Schluter, P. J. & Shaw, G. R. 2006. Cyanobacterial lipopolysaccharides and human health – a review.*Environmental Health: A Global Access Science Source.5*, 7.

Subhashini, J., Mahipal, S. V., Reddy, M. C., Mallikarjuna Reddy, M. Rachamallu, A. & Reddanna, P. 2004. Molecular mechanisms in C-Phycocyanin induced apoptosis in human chronic myeloid leukemia cell line-K562. *Biochemical Pharmacology.68*, 453-462.

Sun, L., Wang, S., Chen, L. & Gong, X. 2003. Promising fluorescent probes from phycobiliproteins. *IEEE J. Sel. Top Quantum Electron.9*, 177-188.

Suzuki, I., Kanesaki, Y., Mikami, K., Kanehisa, M., Murata, N. 2001. Cold-regulated genes under control of the cold sensor Hik33 in *Synechocystis*. *Mol. Microbiol.40*, 235-244.

Sveshnikov, D. A., Sveshnikova, N. V., Rao, K. K., Hall, D. O. 1997. Hydrogen metabolism of mutant forms of *Anabaena variabilis* in continuous cultures and under nutritional stress. *FEMSMicrobiol. Lett.147*, 297-301.

Takenaka, H., Yamaguchi, Y., Sakaki, S., Watarai, K., Tanaka, N., Hori, M., Seki, H., Tsuchida, M., Yamada, A., Nishimori, T. & Morinaga, T. 1998. Safety evaluation of *Nostoc flagelliforme* (nostocales [sic], Cyanophyceae) as a potential food. *Food Chem. Toxicol.36*, 1073-1077.

Tamagnini, P., Costa, J.-L., Almeida, L., Olivera, M.-J., Salema, R.& Lindblad, P. 2000. Diversity of cyanobacterial hydrogenases, a molecular approach. *Curr. Microbiol. ,40*, 356-361.

Tamagnini, P., Axelsson, R., Lindberg, P., Oxelfelt, F., Wünschiers, R. & Lindblad, P. 2002. Hydrogenases and hydrogen metabolism of cyanobacteria.*Microbiol. Mol. Biol. Rev. 66*, 1–20.

Tamagnini, P., Leitão, E., Oliveira, P., Ferreira, D., Pinto, F., Harris, D. J., Heidorn, T.& Lindblad, P. 2007. Cyanobacterial hydrogenases: diversity, regulation and applications.*FEMS Microbiol. Rev.31*, 692–720.

Tandeau de Marsac, N., de la Torre, F. & Szulmajster, J. 1987. Expression of the larvicidal gene of *Bacillus sphaericus* 1593M in the cyanobacterium *Anacystis nidulans* R2. *Mol. Gen. Genet.209*, 396-398.

Taroncher-Oldenburg, G., Nishina, K. & Stephanopoulos, G. 2000. Identification and analysis of the polyhydroxyalkanoate-specific β-ketothiolase and acetoacetyl coenzyme A reductase genes in the cyanobacterium *Synechocystis* sp. strain PCC6803. *Appl. Environ. Microbiol.66*, 4440-4448.

Terasaki, N., Yamamoto, N., Hiraga, T., Sato, I. Inoue, Y. & Yamada, S. 2006. Fabrication of novel photosystem I-gold nanoparticle hybrids and their photocurrent enhancement. *Thin. SolidFilms.299*, 153-156.

Terasaki, N., Iwai, M., Yamamoto, N., Hiraga, T., Yamada, S. & Inoue, Y. 2008. Photocurrent generation properties of Histag-photosystem II immobolized on nanostructured gold electrode. *Thin SolidFilms.516*, 2553-2557.

Thiery, I., Nicolas, L., Rippka, R., Tandeau de Marsac, N. 1991. Selection of cyanobacteria isolated from mosquito breeding sites as a potential food source for mosquito larvae. *Appl. Environ.Microbiol.57*, 1354-1359.

Tiwari, D. N. 1978. The heterocysts of the blue-green alga *Nostochopsis lobatus*: effects of cultural conditions. *New Phytol.81*, 853-856.

Tsygankov, A. A., Borodin, V. B., Rao, K. K. & Hall, D. O.1999. H_2photoproduction by batch culture of *Anabaenavariabilis* ATCC 29413 and its mutant PK84 in a photo-bioreactor. *Biotechnol. Bioeng. 64*, 709–715.

Tsygankov, A. A., Serebryakova, L. T., Rao, K. K., & Hall, D. O. 1998. Acetylene reduction and hydrogen photo-production by wild type and mutant strains of *Anabaena* at different CO_2 and O_2 concentrations. *FEMS Microbiol. Lett. 167*, 13–17.

Tsygankov, A. A., Fedorov, A. S., Kosourov, S. N. & Rao, K. 2002. Hydrogen production by cyanobacteria in an automated outdoor photobioreactor under aerobic conditions.*Biotechnology and Bioengineering.80*, 777-783.

Vignais, P.M. & Colbeau, A. 2004. Molecular biology of microbial hydrogenases. *Curr. Issues Mol. Biol. 6*,159-188.

Vincenzini, M. & De Philippis, R. Polyhydroxyalkanoates. In Cohen,Z., editor. *Chemicals fromMicroalgae.*London: Taylor & Francis; 1999; 292-312.

Vinnemeier, J. & Hagemann, M. 1999. Identification of salt-regulated genes in the genome of the cyanobacterium *Synechocystis* sp. strain PCC 6803 by subtractive RNA hybridization. *Arch. Microbiol.172*, 377-386.

Wiegand, C. & Pflugmacher, S. 2005. Ecotoxicological effects of selected cyanobacterial secondary metabolites a short review. *Toxicology and Applied Pharmacology.203*, 201 – 218.

Wilkinson, C. R. & Fay, P. 1979. Nitrogen fixation in coral reef sponges with symbiotic cyanobacteria. *Nature.279*, 527-529.

Wimpenny, J., Manz, W. & Szewzyk, U. 2000. Heterogeneity in biofilms.*FEMS Microbiol. Rev.24*, 661-671.

Wolk, C. P. Developmental biology of nitrogen-fixing cyanobacteria.In: Bird K., editor, *MSU-DOE Plant Research Laboratory, Fortieth Annual Report*; Howard Printing Cmpany, Kalamazoo, MI, 2005; 81-90.

Wu, J. & Pond, W. 1981. Amino acid composition and microbial contamination of *Spirulina maxima*, a blue-green alga, grown on the effluent of different fermented animal wastes. *Bull.Environ. Contam. Toxicol.27*, 151-159.

Wu, X., Vennison, S. J., Liu, H., Ben-Dov, E., Zaritsky, A. & Boussiba, S. 1997. Mosquito larvicidal activity of transgenic *Anabaena* strain PCC 7120 expressing combinations of genes from *Bacillus thuringiensis* subsp. *israelensis. Appl. Environ. Microbiol.63*, 4971-4975.

Xu, X., Kong, R. & Hu, Y 1993. High larvicidal activity of intact recombinant cyanobacterium *Anabaena* sp. PCC 7120 expressing gene 51 and gene 42 of *Bacillus sphaericus* sp. 2297. *FEMSMicrobiol. Lett.107*, 247-250.

Xu, X., Yan, G., Kong, R., Liu, X., Yu, L. 2000. Analysis of expression of the binary toxin genes from *Bacillussphaericus* in *Anabaena* and the potential in mosquito control. *Curr. Microbiol.41*, 352-356.

Yoon, J. H., Shin, J-H. & Park T.H. 2008. Characterization of factors influencing the growth of *Anabaena variabilis* in a bubble column reactor. *Bioresource Technology.99*, 1204-1210.

Yoshino, F., Ikeda, H., Masukawa, H. & Sakurai, H. 2007. High photobiological hydrogen production activity of a *Nostoc* sp. PCC 7422 uptake hydrogenase-deficient mutant with high nitrogenase activity. *Marine Biotechnology.* 9, 101–112.

Yu, R., Yamada, A., Watanabe, K., Yazawa, K., Takeyama, H., Matsunaga, T. & Kurane, R. 2000. Production of eicosapentaenoic acid by a recombinant marine cyanobacterium, *Synechococcus* sp. *Lipids.35*, 1061-1064.

Ziegler, K., Diener, A., Herpin, C., Richter, R., Deutzmann, R. & Lockau, W. 1998. Molecular characterization of cyanophycin synthetase, the enzyme catalyzing the biosynthesis of the cyanobacterial reserve material multi-L-arginyl-poly-L-aspartate (cyanophycin). *Eur. J. Bch.254*, 154-159.

In: Biological Aspects of Human Health and Well-Being
Editor: Tsisana Shartava

ISBN: 978-1-61209-134-1
© 2011 Nova Science Publishers, Inc.

Chapter III

WHY GLUCOSE IS THE PRINCIPAL SOURCE OF ENERGY FOR LIVING BEINGS? AND THE EXPLANATION OF HUMAN DISEASES

"The natural forces within us are the true healers of disease."

Hippocrates (460 - 377 BC)

Alberto Halabe Bucay[*]
Hospital Angeles Lomas
Av. Vialidad de la Barranca s/n, Huixquilucan, 52763, Mexico

DEDICATION

It is with great pride that I dedicate this essay to the memory of my grandfather, Alberto Halabe Rayek, from whom I learned that the only people that fail are those that fear failure.

INTRODUCTION

Man has attempted to explain the appearance of life on Earth in a very complex manner, therefore the understanding of the diseases that affect human beings has been equally complicated, and thus the treatment of many diseases has had to be very aggressive; it being sufficient to mention the current treatments for cancer, autoimmune diseases, and mitochondrial diseases.

But in reality life is much more straightforward; it's all a question of understanding the origin of multi-cellular organisms.

[*] Tel/Fax: 5255-52469555; Email: doctorhalabe@hotmail.com

Charles Darwin's theory of evolution, despite its critics over the last 150 years [6], consolidated the whole of a fundamental fact in the understanding of the presence of life on our planet: time, adaptation to the environment, the need for survival, and the maintenance of every species through life, are facts that are conclusive and which explain our existence today.

There is Biblical evidence of evolution, the pages of the Book of Genesis explain the continuity of events that occurred in the creation of the planet Earth up until the appearance of man, and then woman, although metaphorically speaking these events are described as occurring over one week [7].

Just as Darwin's theory and the Book of Genesis explain it, the appearance of multi-cellular organisms was a successive process, and, not intending to suggest a reductionist focus on this point, the key moment in evolution, after which multi-celled organisms appeared, including invertebrates, reptiles, and mammals, is very simple:

There already were living cells in the ocean at this time, cells with vital functions and independence from their environment, and also there were bacteria in the ocean; it is calculated that this type of life has been around approximately 1.5 billion years ago [8], and at a certain moment, one or more bacteria incorporated into one of these primitive cells, most probably for the purpose of invading a system different from theirs, or perhaps to protect themselves from the marine environment; but the fact is that this invasion, or something very similar, occurred [9].

All living multi-cellular organisms appeared after this fusion of two different living systems, and then continue evolving up to the present day.

Because of all of this, I believe this interaction, or struggle, that appeared in this new biological system, between the cells and the invading bacteria and which now are called mitochondria, and which I refer to as biological competition [2], could explain the great majority of human diseases, which are more than diseases, they are consequences.

This reasoning will explain the appearance of life on Earth and its consequences, including diseases.

THE HYPOTHESIS OF THE BIOLOGICAL COMPETITION AND ITS CONSEQUENCES

I initially explained the content of the hypothesis I refer to as biological competition in the essay I published in the Journal Medical Hypotheses, published by Elsevier, titled: "The biological significance of cancer" [2], which consists of understanding that all multi-cellular organisms that inhabit our planet originated as a result of the fusion between one or more bacteria and a primitive cell [8,9]; All the consequences of this phenomenon I shall explain in this chapter are completely hypothetical, but I have tried to remain steadfast to the same rational system under which I initially wrote the hypothesis on biological competition [2]. And it should be mentioned that the succession of evolutionary events I describe do not necessarily appear in chronological order, and several of these events could have occurred simultaneously.

To continue, in order to understand the appearance of multi-cellular organisms on Earth, imagine a scenario where one cell, with its vital functions already established, was invaded by one or more bacteria, this occurred in the ocean approximately 1.5 billion years ago [1].

Both the recently invaded cell and the bacteria in question were at risk of losing their existence, however, when this invasion occurred both systems remained stable; and a new and different living being was created; the cell did not suffer great damage as it did not disappear, and the invading bacteria found a favorable environment to subsist, isolated from the dangers of the exterior environment by an already existing membrane.

Some time passed and both living systems, the cells and the bacteria, which now comprised a single organism, began to "realize" that they had to dominate these living surroundings in order to preserve their biological identity, which they had managed to establish in Nature.

For its part, the cell was obliged to protect its genetic material, from which it guided its vital functions within the system, which is why the nuclear membrane appeared [10], which insulated itself from the cytoplasm so as not to be vulnerable to the action of the bacteria living there, and thus be able to maintain its independence.

This is also how I believe eukaryote organisms appeared, from protozoa [10].

In the cytoplasm of these cells the bacteria began to divide and attempted to dominate the whole system; so, the cell had to balance this domain, which is why the Golgi Apparatus was created by the cell, it is a system also made up of membranes similar to that of the bacteria, but with basic functions directed at benefiting the cell, and not the bacteria; this is why the Golgi Apparatus has a whole regulation system for the life of these bacteria, now mitochondria, based on the production and control of lysosomes, whose functions include destroying mitochondria when these are no longer necessary [11].

These bacteria, which were already living in the cytoplasm of the cell, also attempted to dominate these biological system for their benefit; the most basic thing they did was to maintain their identity as bacteria: they maintain their double membrane system, they have their own DNA and RNA, their ribosomes just as the bacteria, and a personal transcription system that has not been changed by evolution [2,12]; and with this genetic regulation system that had been present since these bacteria were independent, perhaps a little more complex than that of the cell, the bacteria began to exchange genetic material with the cell nucleus; there is scientific evidence of this fact as gene codifier proteins have been discovered that can be transferred from mitochondria to the nucleus [13]; this is how I believe oncogenes appeared, such as the C-myc oncogene, which is found in the chromosomes of the cell nucleus and when activated, activates glycolysis [14]; this genetic activation, as will be discussed below, benefits the mitochondria, as these benefit from glycolysis, and the same thing happens with cancer, which is the uncontrollable splitting of a cell where these mitochondria reside. This splitting mechanism is partially regulated by oncogenes, which is why I explained the biological significance of cancer based on the hypothesis of biological competition where the bacteria, now mitochondria, dominate in order to maintain their own identity indefinitely within a safe living environment [2].

Oncogenes were strategically introduced into the cell nucleus, directed by the mitochondria, to achieve these purpose of dominating the cell.

Analyzing this occurrence of the transfer of genes from the mitochondria to the cell nucleus [13], I also proposed that part of this genetic material could have escaped during the transfer to the exterior of the cell, and thus the first viruses appeared; this explains why there are oncogenic viruses with oncogenes inside [3].

It is also highly likely that as part of the biological competition to dominate the living system, the cell nucleus has also transferred genetic material to the mitochondria, carrying information to encode enzymes and proteins in order to dominate and control the mitochondria; these enzymes and proteins could have been responsible for the regulation of the life and programmed death of the mitochondria, which somehow results in the process of apoptosis [3].

Another of the most significant occurrences retain by these bacteria that already resided in the cell was the capacity of using glucose as a source of energy, property that these bacteria had developed when they were living independently in the ocean, and I believe that because of the same biological competition, the cells acquired the capacity to metabolize lipids in an alternate manner, by beta oxidation, which occurs in the cytoplasm of cells; glycolysis also occurs in the cytoplasm, but the end product of the glycolysis, the pyruvate, is that which the mitochondria use to their benefit to start the Krebs cycle and generate energy; the pyruvate is converted into acetyl coenzyme A within the mitochondria by one of most complex enzymes in nature, which the mitochondria conserved since they were independent bacteria: the pyruvate dehydrogenase enzyme [2].

By using the products of the reaction that occurs with the mitochondrial pyruvate dehydrogenase enzyme and the whole of the energy system generated from this reaction, including the Krebs cycle, the mitochondria perform one of their basic functions since they were bacteria 1.5 billion years ago: They use glucose as their vital substrate.

The use of glucose as a principal source of energy has been maintained throughout the whole of evolution, and this was a consequence of the domination of the cell by bacteria as part of the biological competition.

As we all know, lipids provide much more energy than glucose, but as the mitochondria dominated, in this context, the decision that evolution made was to use glucose primarily; however, the cell for its part, to compete with this, as more complex organisms were created, decided to create adipocytes at the systemic level; adipocytes are specialized cells that perform a wide variety of functions, which is why they are distributed strategically throughout the body [15], they are cells that are limited in their reproduction and when their uncontrolled reproduction is activated, the result is generally benign, in the form of lipomas [16].

Another type of cell that also presents a low rate of malignity is muscle cells, both flat and striated muscles [17,18], and continuing with the rational structure of the hypothesis of the biological competition, these muscle cells were designed by the cell machinery to dominate the mitochondria; these muscle cells acquired the capacity to use the mitochondria without the risk of malignity, just as the adipocytes, they also have a limited reproduction and various of these cells contain many cell nuclei, as evidence of their dominance over the mitochondria. Muscle cells can also use lactic acid as a source of energy, which we know as the Cori cycle; the lactic acid comes from the pyruvate, which no longer converts into acetyl coenzyme A and is no longer used by the mitochondria, so that the pyruvate dehydrogenase enzyme described is not activated.

Myoglobin appeared under this same context, this being the protein that transports oxygen in muscle cells, as a counterpart to hemoglobin, which transports red blood cells; the hemoglobin was designed by the mitochondria at the systemic level to deliver the oxygen necessary for their ancestral bacterial functions; this oxygen is used in the respiratory chain of the mitochondria and is also of bacterial origin.

Hemoglobin contains iron just like the cytochromes of the respiratory chain; and the porphyrins, which form part of the hemoglobin synthesis, are synthesized from the metabolites from the Krebs cycle, which occurs within the mitochondria [19]; it would appear that the mitochondria have great control of their domination in the production of red blood cells; another interesting fact regarding this is that there are various types of anemias secondary to antibiotics use, such as cloramphenicol [20] and penicillin [21]; what occurs is that these antibiotics attack the bacteria, and also the mitochondria, which is why their control over the production of red blood cells is spoiled.

Some scientific evidence that is worth analyzing regarding this is the relationship of gastric cancer and other types of cancer with the red blood cell blood types [22], the mitochondria dominate the red blood cells and also they are responsible for the development of cancer [2]. This is a topic that opens a whole new horizon in research: the reason for blood types ... which could be explained by the biological competition hypothesis.

Another scientific area I am interested in considering in terms of this point of erythrocyte domination by mitochondria is the appearance of the metastasis of malignant tumors, which presents principally through the blood stream, somehow, the red blood cells transport the malignant cells that detach from tumors through the blood stream. One of the conditions the mitochondria dictated in their contract for evolutionary dominance was: if they are at risk of losing their existence, the whole organism in which they reside must respond, there must be malignant cells all over the body that maintain their survival; one of the evolutionary functions of red blood cells was to facilitate the distribution of the malignant cells activated on the decision of the mitochondria.

And what occurred with the biological competition in order for the erythrocyte control to not dominate the system was that the cell modified the red blood cell precursor cells, the pluripotential stems cells in the bone marrow; these cells acquired the capacity to produce other cells, of which, some white blood cells have the capacity to destroy cells that become malignant, such as natural killer cells [23], which, it would appear, were created by the cell machinery to dominate the mitochondria when these decided to make a cell malignant.

Platelets, also derived from these stem cells, are also dominated by the cell, as they participate at the systemic level to block metastasis processes by forming blood clots and preventing the migration of malignant cells, and it has also been proven that platelets block the angiogenesis processes necessary for tumor growth [24], lastly, they dominate the mitochondria.

To return to the role of white blood cells in the biological competition as part of cell dominance, it is reasonable to analyze the appearance of antibodies (immunoglobulins), produced by B lymphocytes, which also belong to the cell band, and are capable of destroying "external" bacteria, but at a certain moment, they can also recognize the mitochondria as foreign and attack, and this is what I have proposed as the origin of autoimmune diseases [3].

Many cytokines have played an important role in the biological competition. It is interesting to recapitulate on the origin of tumor necrosis factor (TNF), which is produced precisely by the white blood cells that were specialized in engulfing bacteria and destroying these: the macrophages; and as has already been mentioned, white blood cells, in general, belong to the cell band in the biological competition; this TNF also participates in tumor destruction, beside the cells, attacking the mitochondria.

And the role of TNF in diabetes mellitus type I is also very interesting, as this is one of the reasons why the beta cells of the islets of Langerhans in the pancreas are destroyed in this type of diabetes mellitus, these beta cells responsible for producing insulin [25].

I believe insulin to be the ball in the game of the biological competition, it is a hormone that acts for the two bands, the cell and the mitochondria: insulin participates in the metabolism of lipids, increasing lipogenesis, which controls cells according to my hypothesis of the biological competition, and also acts in the metabolism of glucose, as it directly activates the pyruvate dehydrogenase enzyme, which converts the pyruvate from glucose into acetyl coenzyme A; this reaction within the mitochondria starts the energy process necessary for their vital functions [26].

Continuing with the metabolism of carbohydrates and lipids as part of the biological competition, a relevant fact is gluconeogenesis, meaning the obtaining of glucose from other sources, such as amino acids, which occurs in the liver and is activated to supply glucose in case of emergency and also supplies the glucose the cells that become cancerous require. The mitochondria designed gluconeogenesis to obtain the glucose they require at all times, and this could also explain the etiology of the fatty liver: in this case the cell dominates and controls the mitochondria to prevent gluconeogenesis from occurring and the liver fills with lipids to block the use of glucose by the mitochondria.

Another fact that supports this hypothesis on the metabolic control of the metabolism of carbohydrates by mitochondria is that an emergency glycolysis activator appeared designed by the mitochondria, which guarantees the mitochondria will be able to continue using glucose as a source of energy regardless of any circumstance; fructose 2,6 biphosphate [27], which is part of the same glycolysis.

One consequence of the biological competition that is also very interesting is mitochondrial inheritance. Only the ovum has mitochondria when it is fertilized to form a new human being [28]; regarding this I believe that in order to dominate the mitochondria, the cell system established that sperm do not supply mitochondria on fertilizing the ovum so that the mitochondria necessary to form a new human being would not be so high, numerically speaking, to ensure a competitive equilibrium.

This same logic could explain the appearance of other physiological systems in the organism based on the hypothesis of the biological competition between the cell and the mitochondria, such as is the case of the function of various hormones, like thyroid hormones, which act in the metabolism and the growth of the organism, somehow activating the mitochondria [29]; other hormones that also activate mitochondria, and could have appeared based on their domination of the cell, are the growth hormone [30], steroid hormones [31], including estrogen [32], and of course the erythropoietin hormone of the kidneys, which in addition to its function of increasing the red blood cells for mitochondrial dominance, also acts in the mitochondria of neurons [33].

Other more complex physiological functions can also be analyzed from the evolutionary perspective of the biological competition, such as the respiratory function, whose principal function is to obtain oxygen in the lungs caught by the hemoglobin and delivered to all the mitochondria in the body.

The function of the exocrine pancreatic and other parts of the digestive system, whose principal purpose is to digest food, including carbohydrates, benefit the mitochondria, in order to make the carbohydrates be absorbed from the intestine into the blood, to finally arrive at the mitochondria throughout the organism in the form of glucose.

I invite readers to open their minds to this hypothetical concept known as biological competition in order to better understand the origin of multi-cellular organisms and all the vital functions has based on evolutionary consequences of this competition between the cell and the mitochondria.

EXPLANATION OF HUMAN DISEASES BASED ON THE BIOLOGICAL COMPETITION

Throughout this chapter I have explained the elements of the biological competition between the cell machinery and the mitochondria, which appeared when bacteria invaded a primitive cell 1.5 billion years ago, and the consequences of this biological competition, which has persisted throughout evolution; these were the origins for the appearance of multi-cellular organisms and their physiological functions. Also, the origin of a large number of human diseases can be explained as a consequence of the biological competition.

Cancer is a good example; the mitochondria dominate the cell and make the cell replicate uncontrollably in order for the mitochondria to remain in an isolated cell environment indefinitely, as a preservation system for the bacterial species [2]. This is why some antibiotics, like actinomycin, are effective for treating cancer, as they attack both the bacteria and the mitochondria, which preserve many bacterial qualities [2].

And even when cancer presents in an organism, the biological competition continues, which is why mitochondrial alterations present in various types of malignant tumors [2], the cell attempts to attack the mitochondria to the end; and one of the manifestations of some types of cancer is lactic acidosis, the cell machinery is attempting to make the pyruvate from the glycolysis convert into acetyl coenzyme A, preventing the pyruvate dehydrogenase enzyme in the mitochondria from activating, as described in the previous section, just as the muscle cells do.

Autoimmune diseases are another example of a pathological situation that appeared as a consequence of the biological competition, the cell machinery attempts to dominate and attack the mitochondria through the immune system it designed, as it recognizes the mitochondria as foreign, as though they were bacteria [3].

Another interesting scientific fact regarding this, and which provides further support for my hypothesis of the biological competition, are mitochondrial diseases, such as Leigh, Kerans-Sayre, Pearson, and MELAS syndromes; one of the principal clinical characteristics of these diseases is muscular disorder; what occurs is that the cell machinery is literally attacking the mitochondria so that these present dysfunction, and the muscle cells, where the cell is more dominant over the mitochondria, is where this attack phenomenon is seen more [4].

Neurodegenerative diseases, such as Alzheimer's and Parkinson's among others, are also explained by this hypothesis of the biological competition, the mitochondria also attempt to dominate the cell machinery in these diseases, however as with other cases where neurons are involved, and these cannot replicate to generate cancer, the mitochondrial domination manifests clinically as neurodegenerative diseases, which is why mitochondrial alterations have been found in these diseases and why neurodegenerative diseases have been associated with cancer [34]; and the role of insulin in these diseases is very interesting, insulin, as has been explained, is the ball in the game between the cell and the mitochondria and it has been proven that insulin can increase neuronal damage in neurodegenerative diseases [35], what occurs is that the insulin activates the pyruvate dehydrogenase enzyme in the mitochondria and, somehow, benefits its dominance, and thus increases the damage to the neurons. Another fact that supports the findings presented throughout this chapter is the damage suffered by the Golgi apparatus of the neurons affected by neurodegenerative diseases [36]; it would appear that the mitochondria attempted to damage these organelles generally known as Golgi apparatus as part of the biological competition, because as has already been described the Golgi apparatus was created by the cell to compete against the mitochondria.

Diabetes mellitus type I, which presents as a result of the destruction of the beta cells of the islets of Langerhans in the pancreas, largely due to TNF as has been explained, is also a consequence of the biological competition; the cell machinery attempted to dominate the mitochondria, eliminating the insulin they require to activate their pyruvate dehydrogenase enzyme 4], to prevent the dominance of the mitochondria and the conversion of a cell into cancer, or neurodegenerative diseases in any part of the organism; it is therefore logical that diabetes mellitus type I has been associated with some mitochondrial diseases, such as Kearns-Sayre syndrome [37,38], or MELAS syndrome [39]; it would appear that in these cases the dominance by the cell is of such magnitude that both the mitochondria and the production of insulin are affected.

Diabetes mellitus type II can be explained by the same mechanism of cell dominance over mitochondria, but in this case, to dominate the mitochondria it was only necessary that resistance to insulin be created, not to eliminate it completely as with diabetes mellitus type I; the result is the same, the mitochondria fail to activate their pyruvate dehydrogenase enzyme with insulin, and in the end, the cell dominates.

Many of the inherent errors of metabolism can be understood with this same reasoning of the biological competition, principally in terms of those that affect the metabolism of the carbohydrates, such as galactosemia, fructosinemia, and glucose 6 phosphate dehydrogenase deficiency, as what these diseases represent is the dominance of the cell to prevent the mitochondria from using glucose as their principal source of energy.

And examining further other inherent errors of metabolism, such as alterations of amino acids, including phenylketonuria or alkaptonuria, to recapitulate, could lead us to the conclusion that these also present as a consequence of the biological competition, as the evolutionary result of these alterations is to the prevention of substrates in the form of amino acids to produce the glucose through gluconeogenesis; gluconeogenesis is the metabolic pathway that mitochondria need to continue obtaining glucose under extreme conditions and in cancer.

Storage diseases, especially those that affect glycogen, are also explained by the biological competition. The cell dominated to block another resource by which mitochondria obtain glucose to be used as a source of energy: glycogen.

Some types of anemia are also a consequence of the biological competition, the cell dominates to reduce the number of red blood cells necessary to transport the oxygen the mitochondria require, which is why there are autoimmune anemias, the cell instructed its immune system to destroy some red blood cells and limit the supply of oxygen to the mitochondria.

The majority of alterations in coagulation associated with bleeding could have been established by the mitochondria to achieve an active blood flow, without risk of thrombosis, for which, among other circumstances, malignant cells could have disseminated in the form of metastasis and thus the mitochondria could then continue to dominate the whole of the organism.

Inherited immunodeficiencies are also examples of the dominance of mitochondria, as their purpose is to block the immune system created by the cell so that it does not recognize the mitochondria as foreign and does not attack them.

Cystic fibrosis is more difficult to explain with the biological competition, but what occurs with this disease is that the pancreatic enzymes in the digestive system are being blocked so as to not obtain glucose for the mitochondria, on missing these enzymes necessary for the absorption of carbohydrates, it is the mitochondria that suffer, the reason of cystic fibrosis is cell dominance.

Following this line of thinking, some types of hypoglycemia may also have appeared as a result of cell dominance, the final result is the same, preventing the supply of glucose to the mitochondria; and, in the other hand, insulinomas and nesidioblastosis appeared, which are dominated by mitochondria, guaranteeing the supply of insulin to the mitochondria in order to activate their pyruvate dehydrogenase enzyme.

Hypertriglyceridemia and hypercholesterolemia could appear by the domination of the cells, lipids are attempted to be used instead of glucose as a principal source of energy, to be used by the cells; and the mitochondria will not be able to use glucose.

Gout is interesting because of the fact that it is more prominent in men: would this have anything to do with the fact that men do not contribute with mitochondria in inheritance? It seems that gout appeared as a result of the dominance of cells in order for debilitate the mitochondria and consequently, many neurodegenerative diseases could not appear, like Parkinson's disease [40], which as has been explained, results from the mitochondrial dominance over neurons. And this could also explain other diseases that are exclusive to men, including diseases associated with the X chromosome, such as Duchenne disease.

Psychiatric disorders have also been associated with mitochondrial alterations [41], could it also be in these cases that the cell tries to dominate the mitochondria? And does the subsequent damage to the mitochondria by the cell cause emotional alterations?

Lastly, and this is much more complex, some forms of resistance to antibiotics that bacteria have acquired when these are in the human body could be explained based on the biological competition, could it be that the mitochondria are somehow trying to defend these bacteria, which were of the same species a long time ago?

DISCUSSION

I have liked to end a scientific essay with questions, but I believe these to be justified due to the type of content in this chapter.

CONCLUSION

The clinical applications that could result from the hypothesis of the biological competition I suggest in this chapter have yet to be fully determined. However, some answers have already been given.

Firstly, I have proposed that citric acid may be effective for treating cancer, which is the internal regulator from the Krebs cycle that blocks the pyruvate dehydrogenase enzyme of the mitochondria, which is activated in cancer, as the mitochondria are dominating at that time and using glucose as their principal source of energy [2].

In the case of autoimmune diseases, including rheumatoid arthritis, where it is the cell that dominates on activating the immune system to attack the mitochondria, I have proposed provoking a mild infection to distract the immune system and instead of attacking the mitochondria in the body itself, it attacks the infection provoked [42], or somehow try to develop immunological tolerance to their own mitochondria of patients suffering from autoimmune diseases [3].

And in the case of mitochondrial diseases, where the cell dominates and even damages the mitochondria, I have proposed the administering of insulin, as this activates the pyruvate dehydrogenase enzyme of the mitochondria affected [4].

We may begin to consider that mental and psychiatric disorders could be treated with lipids, such as omega 3 [43], trying to use lipids instead of glucose as the source of energy for neurons, or some how reduce the action of insulin in the neurons.

However, the most important point regarding the hypothesis of the biological competition between the cell and the mitochondria is that it may help us to understand more about our existence on Earth.

REFERENCES

[1]   van der Giezen M, Tovar J. Degenerative mitochondria. *EMBO Rep.* 2005; 6(6):525-530

[2]   Halabe Bucay A. The biological significance of cancer: Mitochondria as a cause of cancer and the inhibition of glycolysis as a cancer treatment. *Med. hypotheses.* 2007; 69(4):826-828

[3]   Halabe Bucay A. The origin of autoimmune diseases: An evolutionary point of view. In: *New Research on Innate Immunity.* Editors: M. Durand, C. V. Morel. Chapter 5: 93:108. Nova Science Publishers, New York, 2008.

[4]   Halabe Bucay A. Treatment of mitochondrial diseases strengthening mitochondria by stimulating glycolysis with insulin. *Med. Hypotheses.* 2008; 71(6):979-980

[5]   Betley S, Alberti KG, Agius L. Regulation of fatty acid and carbohydrate metabolism by insulin, growth hormone and tri-iodothyronine in hepatocyte cultures from normal and hypophysectomized rats. *Biochem. J.* 1989; 258(2):547-552

[6]   The complete work of Charles Darwin. Book published by University of Cambridge. (www.darwin-online.org.uk)

[7]   Book of Genesis, Holly Bible. Chapters 1 and 2

[8]   Kachane AN, Timmis KN, Martis Dos SantosVA. Dynamics of reductive genome evolution in mitochondria and obligate intracellular microbes. *FEBS lett.* 2001; 501(1):11-18

[9]   Sogin ML. Early evolution and the origin of eukaryotes. *Curr. Opin. Genet. Dev.* 1991; 1(4):457-463

[10]   Seravin LN. [The origin of the eukaryotic cell. II. A critical analysis of the symbiotic (exogenous) concept]. Article in Russian. *Tsitologlia.* 1986; 28(7):659-669

[11]   Locke M, SykesAK. The role of the Golgi complex in the isolation and digestion of organelles. *Tissue cell.* 1975; 7(1):143-158

[12]   Mootha VK, Bunkenborg J, Olsen JV, Hjerrild M, Wisniewski JR, Stahl E, et al. Integrated analysis of protein composition, tissue diversity, and gene regulation in mouse mitochondria. *Cell.* 2003; 115(5):629-640

[13]   von Heijne G. Why mitochondria need a genome. *FEBS lett.* 1986; 198(1):1-4

[14]   Dang CV, Lewis BC, Dolde C, Dang G, Shim H. Oncogenes in tumor metabolism, tumorigenesis and apoptosis. *J. Bioenerg. Biomembr.* 1997; 29(4):345-354

[15]   Halabe Bucay A. The role of lipotropins as hematopoietic factors and their potential therapeutic use. *Exp. Hematol.* 2008; 36(6):752-754

[16]   Springfield D. Liposarcoma. *Clin. Orthop. Relat. Res.* 1993; 289:50-57

[17]   Bettocchi S, Siristatidis C, Pontrelli G, Di Spiezo Sardo A, Ceci O, Nappi L, et al. The density of myomas: should we treat small submucous myomas in women of reproductive age? *Fertil. Steril.* 2008; 90(4):905-910

[18]   Willis J, Abdul-Karim FW. Di Sant'Ágnese PA. Extracardiac rhabdomyomas. *Semin. Diagn. Pathol.* 1994; 11(1):15-25

[19]   Atamna H. Heme, iron, and the mitochondrial decay of ageing. *Ageing Res. Rev.* 2004; 3(3):303-318

[20]   Petitpierre-Gabathuler MP, Beck EA. Effects of chloramphenicol on heme synthesis. *Acta Haematol.* 1972; 47(5):257-263

[21]   Kerr RO, Cardamone J, Dalmasso AP, Kaplan ME. Two mechanisms of erythrocyte destruction in penicillin-induced hemolytic anemia.*N. Eng. J. Med.* 1972; 287(26):1322-1325

[22]   Eklund AE. Studies on the relation between ABO blood groups and gastric carcinoma. I. Relation of blood groups to different types of tumour. *Acta Chir. Scand.* 1965; 129:211-218

[23]   Andoniou CE, Coudert JD, Degil-Esposti MA. Killers and beyond: NK-cell-mediated control of immune responses. *Eur. J. Immunol.* 2008; 38(11):2938-2942

[24]   Klement GL, Yip TT, Cassiola F, Kikuchi L, Cervi D, Podust V, et al. Platelets actively sequester angiogenesis regulators. *Blood.* 2008, Nov 25. In Press. DOI 10.1182/blood-2008-06-159541

[25]   CampbellIL, Iscaro A, Harrison LC. IFN-gamma and tumor necrosis factor alpha: cytotoxicity to murine islets of Langerhans. *J. Immunol.* 1988; 141(7):2325-2329

[26]   Czech MP. Insulin action. *Am. J. Med.* 1981; 70(1):142-150

[27]   Depré C, Veitch K, Hue L. Role of fructose 2,6-biphosphate in the control of glycolysis. Stimulation of glycogen synthesis by lactate in the isolated working rat heart. *Acta Cardiol.* 1993; 48(1):147-164

[28]   Hoekstra RF. Evolutionary origin and consequences of uniparental mitochondrial inheritance. *Human Reprod.* 2000; 15 Suppl 2:102-111

[29] Fernández-Vizarra E, Enriquez JA, Pérez-Martos A, Montoya J, Fernández-Silva P. Mitochondrial gene expression is regulated at multiple levels and differentially in the heart and liver by thyroid hormones. *Curr. Genet.* 2008; 54(1):13-22

[30] Short KR, Moller N, Bigelow ML, Coenen-Schimke J, NairKS. Enhancement of muscle mitochondrial function by growth hormone. *J. Clin. Endocrinol. Metab.* 2008; 93(2):597-604

[31] Psarra AM, Sekeris CE. Steroid and thyroid hormone receptors in mitochondria. *IUBMB Life.* 2008; 60(4):210-223

[32] Chen JQ, Brown TR, Yager JD. Mechanisms of hormone carcinogenesis: evolution of views, role of mitochondria. *Adv. Exp. Med. Biol.* 2008; 630:1-18

[33] Xiong Y, Chopp M, Lee CP. Erythropoietin improves brain mitochondrial function in rats after traumatic brain injury. Neurol Res 2008, Dec 18. In Press. DOI 10.1179/174313208X353703

[34] Staropoli JF. Tumorigenesis and neurodegeneration: Two sides of the same coin? *Bioessays.* 2008; 30(8):19-27

[35] Cohen E, Dillin A. The insulin paradox: aging, proteotoxicity and neurodegeneration. *Nat. Rev. Neurosci.* 2008; 9(10):759-767

[36] Fan J, Hu Z, Zeng L, Lu W, Tang X, Zhang J, et al. Golgi apparatus and neurodegenerative diseases. *Int. J. Dev. Neurosci.* 2008; 26(6):523-534

[37] Poulton J, O'Rahilly S, Morten KJ, Clark A. Mitochondrial DNA, diabetes and pancreatic pathology in Kearns-Sayre syndrome. *Diabetologia.* 1995; 38(7):868-871

[38] Laloi-Michelin M, Virally M, Jardel C, Meas T, Ingster-Moati I, Lombés A, et al. Kearns Sayre syndrome: an unusual form of mitochondrial diabetes. *Diabetes Metab.* 2006; 32(2):182-186

[39] Huang CN, Jee SH, Hwang JJ, Kuo YF, Chuang LM. Autoimmune IDDM in a sporadic MELAS patient with mitochondrial tRNA (Leu (UUR)) mutation. *Clin. Endocrinol.* (Oxf) 1998; 49(2): 265-270

[40] Alonso A, Rodriguez LA, Logroscino G, Hernán MA. Gout and risk of Parkinson disease: a prospective study. *Neurology.* 2007; 69(17):1696-1700

[41] Shao L, Martin MV, Watson SJ, Schatzberg A, Akil H, Myers RM, et al. Mitochondrial involvement in psychiatric disorders. *Ann. Med. 2008*; 40(4):281-295

[42] Halabe Bucay A. Treatment of rheumatoid arthritis and other related conditions by provoking a mild infection to be controlled by the immune system itself. *Med. Hypotheses.* 2007; 69(1):27-29

[43] Kapoor S. Omega-3 fatty acids and their rapidly evolving role in the management of psychiatric disorders. *Prog. Neuropsychopharmacol. Biol. Psychiatry.* 2008; 32(8):2014-2015

In: Biological Aspects of Human Health and Well-Being   ISBN: 978-1-61209-134-1
Editor: Tsisana Shartava   © 2011 Nova Science Publishers, Inc.

Chapter IV

THE EVOLUTION BIOLOGY OF HEALTH AND DISEASE CLINICAL MEDICINE AS SEEN FROM A DARWINIAN PERSPECTIVE

*Gerhard Mertens**

Department of Clinical Biology, University of Antwerp,
Antwerp, Belgium

ABSTRACT

In its population genetic sense, evolution is defined as the ongoing change of gene frequencies in populations due to one or several of the driving forces of evolution: selection, drift, mutation and migration. Evolution's role is central in the sub-discipline of biology that addresses health and disease in humans and training in evolutionary thinking can both help biomedical researchers and clinicians ask useful questions they might not otherwise pose.

The co-evolution of man and his environment of pathogenic micro-organisms, the rapidly shifting antibiotic resistance of these pathogens and our persistent vulnerability to chronic diseases should all be seen from an evolutionary perspective. These subjects form the core of "evolutionary medicine", which will be illustrated by a number of thought inspiring examples.

The hypothesis that allergy can be viewed like cough and pain as a defence mechanism evolved by natural selection, is gaining support from toxicological studies measuring lower levels of carcinogens in allergic individuals.

Recent research, combining the effects of genes and environment, has provided surprising clues to the cause of atherosclerosis, a major public health problem.

In medical microbiology, the combination of the short generation time of bacteria, the exchange of resistance genes between species and the swift transfer of bacteria from animals to humans and between humans, forms a life threatening cocktail with a critical role for evolutionary mechanisms.

The HLA system which encodes proteins of the immune response, shows the most extensive polymorphism of the whole human genome. The global distribution of HLA

---

* Gerhard.Mertens@uza.be

alleles illustrates evolution by migration, while the polymorphism itself is promoted by natural selection, operating through pre- and post-conceptual mechanisms.

An example of "recent" evolution in *Homo sapiens* by natural selection and a genetic bottleneck, comes from the relation between *Yersinia pestis* and hemochromatosis. The geographical distribution of the hemochromatosis gene correlates strictly with the area of the 14th century bubonic plague that raged through Europe, which can be explained by a protective mechanism of the hemochromatosis gene against bacterial infection.

The examples above make a strong case for recognizing evolution biology as a basic science for medicine.

INTRODUCTION

"Nothing in biology makes sense except in the light of evolution", is the often quoted title of a 1973 article by the great evolutionary biologist Theodosius Dobzhansky [1]. He writes: "Seen in the light of evolution, biology is - perhaps - intellectually the most satisfying and inspiring science. But without this light, it becomes a pile of sundry facts, some of them interesting or curious but making no meaningful picture as a whole."

Evolution's role is also of paramount importance in the sub-discipline of biology that addresses health and disease in humans.

The co-evolution of man and his pathogenic micro-organisms, the rapidly shifting antibiotic resistance of these pathogens and our persistent vulnerability to chronic diseases should all be viewed in the context of continuing evolution. These subjects form the core of "evolutionary medicine", also known as "Darwinian medicine" [2].

Evolutionary medicine is not a form of "alternative medicine". Actually, it is the opposite, complementing traditional science-based medicine by not only seeking the immediate causes of disease - the "how" - but also the longer-term reasons for the existence of these diseases - the "why".

While the practice of medicine by itself aims at counteracting natural selection, the following pledge for evolutionary medicine is not paradoxical. Indeed, it is by improving the knowledge on evolutionary biology of clinicians, learning them to see health and disease from an evolutionary perspective, that more effective treatment and preventive strategies can be developed.

AUGMENTING MEDICINE WITH EVOLUTIONARY INSIGHTS

If evolutionary medical research is carried out superficially, it may degenerates into adaptationism, telling stories of "it is so because it is so", explaining why in nature "tout est pour le mieux dans le meilleur des mondes" without much objective proof. When the evidence is evaluated critically, however, the evolutionary perspective provides truly fundamental insights into the nature of the disease and the best way to treat it.

Lack of evolutionary understanding leads to "the clinician's illusion", in which natural defensive reactions are regarded as defects to be treated. In fact, some reactions, like fever and cough, are useful in certain circumstances. Others, such as pain and fatigue, serve as warning signals. Blocking such symptoms with drugs is like responding to a car's oil light by clipping the wires rather than by addressing the underlying cause. Fortunately, the body's defences are redundant, but medicine should distinguish between the defence reaction and the primary disease [3].

WHY NATURAL SELECTION HAS LEFT
US VULNERABLE TO DISEASE

Knowledge of evolution provides new insight into our continued vulnerability to disease. Natural selection does not only explain why some things work well, but also why sometimes things work badly.

The evolutionary explanations for disease can be classified into four categories.

The first category reflects the intrinsic inertia in humans of natural selection, in response to a changing environment or evolving bacteria and viruses. Myopia, for example, is clearly a heritable disease, but which manifests itself only in cultures where children learn to read at school. Most chronic diseases, such as diabetes and high blood pressure, result from a discrepancy between our evolutionary determined anatomy and physiology, which is still largely as in the days of hunters and gatherers, and our modern way of life. In responding to micro-organisms, we are handicapped by the slow pace of our evolution. We have difficulty outpacing the pathogens because they evolve so much faster than man. The limited impact on reproduction of many diseases also limits the effect of natural selection on the population gene pool.

The second category of medical-evolutionary insights recognizes that adjustments are not for free and come with trade-offs. This sometimes results in physiology that may look surprisingly maladaptive. Normal blood contains limited amounts of uric acid and bilirubin. An excess however, leads to gout and jaundice respectively. But uric acid as well as bilirubin are powerful endogenous antioxidants, such as vitamin C and vitamin E, and are therefore indispensable [4]. The haemoglobin S gene and HFE mutation make up other examples of evolutionary compromises. These genes protect the heterozygous carrier against malaria and bubonic plague respectively, but cause sickle cell anaemia and hemochromatosis in the homozygote. According to the laws of population genetics, the homozygous genotype is significantly less frequent than the heterozygous carrier, explaining the net positive effect of the existence of these genes. Finally, we are limited by our evolutionary tree. *Homo sapiens* is still living with the imperfect retina arrangement that arose 150 million years ago, while, in a different phylogenetic branch, the eye of the octopus shows that a better structure is feasible [5].

The third category is not about flaws in natural selection, but reflects our own misunderstanding about it. Natural selection shapes organisms for their reproductive success, not necessarily for their health. The higher mortality for men is an example. For every woman who dies in early adulthood, three men die.

Fourth, we should keep in mind that diarrhoea, cough, fever and pain are not the disease itself, but defences evolved by natural selection. The clinician shall distinguish the primary disease from its associated signs and symptoms in order to install a causal treatment.

ALLERGY AND CANCER: A COMPLEX RELATIONSHIP

An example of a defence reaction somewhat less obvious than diarrhoea, cough and pain, is given by the relation between allergy and cancer. Although this relation is complex, correlation studies have shown a possible inverse relation between allergy and certain cancers [6]. It thus seems that some forms of cancer are less frequent in allergic patients than in non-allergic people. The protective effect is most pronounced for cancers of organs exposed to the external environment - such as skin and lung cancer - suggesting that the allergic symptoms protect against environmental carcinogens [7, 8]. The hypothesis is supported by the observation that allergic people have lower blood and tissue levels of carcinogenic substances, such as dioxins and PCBs [9].

GENE/ENVIRONMENT INTERACTION
IN ATHEROSCLEROSIS

Atherosclerosis is an example of the contribution of Darwinian insights into the cause of a major chronic disease. Recent research combining the effects of genes and environment provides surprising clues to the cause of this important public health problem.

Genetic factors have an undeniable role in atherosclerosis. The primary genetic risk factor is the epsilon4 allele of the gene encoding apolipoprotein E [10], which transports triglycerides and cholesterol in the blood. This allele is an evolutionary relic from the pre-agricultural history of *Homo sapiens* and is not adapted to a modern, highly nutrient-rich culture. Through the process of natural selection, other, better adapted alleles are replacing the epsilon4 allele, generation after generation. Still, the allele has now, 8000 years after the discovery of agriculture, still a high frequency. The assumption that natural selection is eliminating the epsilon4 allele starting with the introduction of agriculture, is supported by the fact that in contemporary populations with a longer tradition of agriculture, the allele is less frequent than in populations that have shifted more recently to agriculture, and even much less frequent than in "primitive" populations with a hunter-and-gatherer lifestyle today.

Aside from genetic factors, there are also environmental – other than dietary – factors to consider in atherosclerosis. First, histology shows, contrasting with what was long assumed, that fat and cholesterol do not simply accumulate on the inner surface of the artery, but within the arterial wall itself. In atherosclerotic lesions - somewhat surprisingly - bacteria have been demonstrated, especially *Chlamydia pneumoniae*.

The gene product of the epsilon4 allele has been shown to function as a receptor protein permitting *Chlamydia pneumoniae* to penetrate the cells of the arterial wall. Thus, all pieces of the puzzle come together and we get close to proof that *Chlamydia pneumoniae* is causally involved in atherosclerosis [11, 12, 13].

When an infection is viewed as the primary cause of atherosclerosis, the paradox of "passive smoking" [14] is resolved. Traditionally, the risk of atherosclerosis by passive smoking is attributed to the direct effect of toxic compounds in the smoke. Still, the link between second-hand exposure to smoke and the increased atherosclerosis risk is not correct quantitatively. Compared with a non-smoker, a heavy smoker increases his risk by 100%. Living with a smoker increases the risk by 30%, although the exposure to smoke is 100 times smaller in the passive smoker. The paradox goes up in smoke if the primary cause of atherosclerosis is infection, spread by airway transmission. Smoking indeed increases the susceptibility to respiratory infections and the cohabitation with a smoker increases the chance of acquiring the same infection.

The insight that a frequent and potentially mortal disease could be caused by a bacterium opens new doors for treatment and prevention.

ECOLOGY OF ANTIBIOTIC RESISTANCE

An alarming application of biological evolution and medicine is the growing resistance of bacteria to antibiotics.

It is often stated that antibiotic resistance is inevitable because of the biological reality of mutation and natural selection. Nonetheless, the ubiquity of resistance genes should not by itself be a problem. The real problem is created by the huge selective pressure on bacteria by the widespread use of antibiotics [15].

Resistance genes can arise spontaneously by mutation within a bacterial strain. In addition, bacteria have evolved a range of mechanisms to exchange genes, including antibiotic resistance genes, between strains of the same species and between different species. Thus, resistance is not stationary but can evolve rapidly.

The selective pressure for resistance is enormous. In the European Union, more than 3 billion euros was spent in 2006 on antibiotics in human medicine [16]. The figure for use of antibiotics in agriculture is not known exactly but is estimated to be even higher. There are two major purposes for veterinary antibiotic use: the prevention of infectious diseases in livestock in overcrowded stables and the promotion of animal growth. In human medicine, both these indications are considered as complete abuse of antibiotics.

Antibiotics differ from other drugs because of their impact not only on the patient taking the drug. Antibiotic treatment of the individual patient also affects his family and eventually even the entire community [17].

Hospitals have long been recognized as a breeding ground for resistant bacteria because of their extensive antibiotic use. However, multi-resistant strains are now arising elsewhere. Resistance may also appear first in bacteria different from the intended target of the antibiotic. Studies in outpatients have shown that skin flora developed resistance to the antibiotic the patient had taken for his bronchitis. Then his family and later his other contacts had a similar change of their skin flora. When these skin commensals "meet" with pathogens, antibiotic resistance is transferred and spreads further.

Evidence of rapid evolution of antibiotic resistance outside the hospital is also found in agriculture. Bywater [18] compared a chicken farm where the animals were fed low doses of antibiotics and an antibiotic free farm. In the first farm, resistance appeared in two weeks. More disturbingly, after 12 weeks all workers of the first farm showed resistant bacteria, although none had taken antibiotics. In the antibiotic free farm none of the chickens, nor the people, developed resistant bacteria during the study period.

It is evident that the widespread use of antibiotics should be halted. The combination of the short generation time of bacteria, the exchange of resistance genes between species and the swift transfer of bacteria from animals to humans and between humans, is a life threatening cocktail for man with a critical role for evolutionary mechanisms.

EXPERIMENTAL EVOLUTION

Evolutionary research of non-pathogenic micro-organisms may also yield insights on infectious diseases. Experiments with bacteriophage phiX174 have led to mathematical models for the development of micro-organisms which can also be applied to human pathogens. In the longest experiment, 13000 subsequent generations of bacteriophages were studied, accumulating a genetic difference comparable to that between humans and chimpanzees [19]. In a typical experiment, a phage was taken that was well adapted to one environment or host cell and was exposed to a novel environment. As the viruses evolved to become better adapted to their new situation, their fitness, as measured by the doubling rate, often increased dramatically in a single mutation.

These experiments have demonstrated microbial evolution by large fitness jumps [20]. In pathogens, such fitness jumps may be host switching or becoming drug resistant.

Thus, new insights are gained in evolutionary biology with possible application in infectious disease management, complementing the classic "genetic landscape" model of Fisher [21]. In this model, fitness landscapes are used to visualise the relationship between genotypes and reproductive success, every genotype having a defined replication rate or "fitness", corresponding to the "height" in the "landscape". According to Fisher, adaptation is dominated by small-fitness changes toward optimal fitness.

THE POLYMORPHIC MAJOR HISTOCOMPATIBILITY COMPLEX

Another medical evolutionary topic can be found in the alleles of the Major Histocompatibility Complex, coined "HLA" in humans. The pattern of global distribution of the alleles of the HLA system correlates with the great prehistoric journey of *Homo sapiens*, starting from East Africa [22]. This illustrates evolution by migration.

Furthermore, the HLA system is the most polymorphic region of the human genome. For the HLA-B gene, for example, 830 distinct alleles have been identified [23]. This genetic diversity reflects the evolutionary pressure favouring immune responses against a wide variety of foreign proteins. The polymorphism is enhanced through pre- and post-conceptual mechanisms, both resulting in fewer offspring of people with the same HLA alleles. First, there is pure sexual selection, women feeling more attracted to men with a different HLA genotype than their own. This attraction – interestingly – is mediated through olfactory signals [24]. The post-conceptual mechanism operates through the more frequent spontaneous early abortion of HLA homozygous foetuses. Indeed, couples who happen to share HLA alleles need significantly longer intervals before a successful pregnancy ensues [25].

"RECENT" EVOLUTION IN *HOMO SAPIENS*

While most of the anatomy and physiology of *Homo sapiens* was shaped by evolution over the last few hundred thousand of years, one can wonder whether there are any examples of more recent and rapid evolution. Well, the geographic distribution of the hemochromatosis gene – a cysteine-to-tyrosine substitution in the gene called HFE - correlates with the area of the 14th century bubonic plague that raged through Europe. It was shown by Moalem et al. [26] that heterozygote carriers of the mutated HFE gene have a lowered iron level in their macrophages. This confers protection against bacterial pathogens, including *Yersinia pestis*. The 14th century plague epidemics, wiping out 30-60% of the European population while selecting in favour of carriers of a mutated HFE gene, created a population bottleneck. As a result of this bottleneck, six centuries later 13% of the European population are carriers of a mutated HFE gene [27]. The evolutionary price Europeans have paid is the incidence of hemochromatosis in one quarter of the offspring of a couple of HFE mutation carriers.

CONCLUSION

The examples above witness the progress made at the intersection of medicine and evolution, of both humans and pathogenic micro-organisms. Still much research is needed to understand diseases more profoundly, in order to optimally treat and prevent. A major problem for evolutionary medicine is the difficulty to prove evolutionary hypotheses in prospective experiments. First, the long generation time of humans makes it difficult to observe evolution "in real time" and then certain experimental interventions are impossible for ethical reasons. The key questions are centered around which aspects of modern environment and lifestyle are pathogenic. That is what we ultimately need to know and it should be clear to all: Medicine Needs Evolution [28].

REFERENCES

[1]   Dobzhansky T. Ethics and values in biological and cultural evolution. *Zygon.* 1973;8:261-81.
[2]   Nesse RM. What evolutionary biology offers public health. *Bull. World Health Organ.* 2008;86:83.
[3]   Nesse RM. Maladaptation and natural selection. *Q. Rev. Biol.* 2005;80:62-70.
[4]   Sedlak TW, Snyder SH. Bilirubin benefits: cellular protection by a biliverdin reductase antioxidant cycle. *Pediatrics.* 2004;113:1776-82.
[5]   Young JZ. Light has many meanings for cephalopods. *Vis. Neurosci.* 1991;7:1-12.
[6]   Zacharia BE, Sharman P. Atopy, helminths, and cancer. *Med. Hypotheses.* 2003;60:1-5.
[7]   Leung DY, Diaz LA, DeLeo V, Soter NA. Allergic and immunologic skin disorders. *JAMA.* 1997;278:1914-23.
[8]   Wong HR, Wispé JR. The stress response and the lung. *Am. J. Physiol.* 1997;273:L1-9.
[9]   Mochida Y, Fukata H, Matsuno Y, Mori C. Reduction of dioxins. 2007;98:106-13.

[10] Davignon J, Gregg RE, Sing CF. Apolipoprotein E polymorphism and atherosclerosis. *Arteriosclerosis.* 1988;8:1-21.

[11] Ewald PW, Cochran GM. Chlamydia pneumoniae and cardiovascular disease: an evolutionary perspective on infectious causation and antibiotic treatment. *J. Infect. Dis.* 2000;181 Suppl. 3:S394-401.

[12] Watson C, Alp NJ. Role of Chlamydia pneumoniae in atherosclerosis. *Clin. Sci.* 2008;114:509-31.

[13] Hoymans VY, Bosmans JM, Ieven MM, Vrints CJ. Chlamydia pneumoniae-based atherosclerosis: a smoking gun. *Acta Cardiol.* 2007;62:565-71.

[14] Barnoya J, Glantz SA. Cardiovascular effects of secondhand smoke: nearly as large as smoking. *Circulation.* 2005;111:2684-98.

[15] Levy SB, Marshall B. Antibacterial resistance worldwide: causes, challenges and responses. *Nature Med.* 2004;10:S122-9.

[16] Coenen S, Ferech M, Haaijer-Ruskamp FM, Butler CC, Vander Stichele RH, Verheij TJ, Monnet DL, Little P, Goossens H. European Surveillance of Antimicrobial Consumption (ESAC): quality indicators for outpatient antibiotic use in Europe. *Qual. Saf. Health Care.* 2007;16:440-5.

[17] Levy SB. The 2000 Garrod lecture. Factors impacting on the problem of antibiotic resistance. *J. Antimicrob. Chemother.* 2002;49:25-30.

[18] Bywater RJ. Veterinary use of antimicrobials and emergence of resistance in zoonotic and sentinel bacteria in the EU. *J. Vet. Med. B. Infect. Dis. Vet. Public Health.* 2004;51:361-3.

[19] Pepin KM, Wichman HA. Experimental evolution. 2008;8:85.

[20] Pepin KM, Wichman HA. Variable epistatic effects between mutations at host recognition sites in phiX174 bacteriophage. *Evolution.* 2007;61:1710-24.

[21] Fisher RA. Gene frequencies*Biometrics.* 1950;6:353-61.

[22] Meyer D, Single RM, Mack SJ, Erlich HA, Thomson G. Signatures of demographic history and natural selection in the human major histocompatibility complex. *Genetics.* 2006;173:2121-42.

[23] Single RM, Meyer D, Mack SJ, Lancaster A, Erlich HA, Thomson G. 14th International HLA and Immunogenetics Workshop: report of progress in methodology, data collection, and analyses. *Tissue Antigens.* 2007;69 Suppl. 1:185-7.

[24] Wedekind C, Seebeck T, Bettens F, Paepke A. MHC-dependent mate preferences in humans. *Proc. Royal Soc. London Ser.* 1995;260:245-9.

[25] Ober C, Elias S, Kostyu DD, Haucks WW. Decreased fecundability in Hutterite couples sharing HLA-DR. *Am. J. Hum. Gen.* 1992;50:6-14.

[26] Moalem S, Percy ME, Kruck TP, Gelbart RR. Epidemic pathogenic selection: an explanation for hereditary hemochromatosis? *Med. Hypotheses.* 2002;59:325-9.

[27] Cayley WE. Haemochromatosis. *BMJ.* 2008;336:506.

[28] Nesse RM, Stearns SC, Omenn GS. Medicine needs evolution. *Science.* 2006;311:1071.

In: Biological Aspects of Human Health and Well-Being   ISBN: 978-1-61209-134-1
Editor: Tsisana Shartava   © 2011 Nova Science Publishers, Inc.

Chapter V

A QUANTITATIVE STRUCTURE-ACTIVITY RELATIONSHIP FOR THE GASTROPROTECTIVE EFFECT OF FLAVONOIDS EVALUATED IN HUMAN COLON ADENOCARCINOMA HT-29 CELLS

*Jingli Zhang** and *Margot A. Skinner*

The New Zealand Institute for Plant and Food Research Ltd,
Auckland, New Zealand

ABSTRACT

Flavonoids are widely distributed in fruit and vegetables and form part of the human diet. These compounds are thought to be a contributing factor to the health benefits of fruit and vegetables in part because of their antioxidant activities. Despite the extensive use of chemical antioxidant assays to assess the activity of flavonoids and other natural products that are safe to consume, their ability to predict an *in vivo* health benefit is debateable. Some are carried out at non-physiological pH and temperature, most take no account of partitioning between hydrophilic and lipophilic environments, and none of them takes into account bioavailability, uptake and metabolism of antioxidant compounds and the biological component that is targeted for protection. However, biological systems are far more complex and dietary antioxidants may function via multiple mechanisms. It is critical to consider moving from using 'the test tube' to employing cell-based assays for screening foods, phytochemicals and other consumed natural products for their potential biological activity. The question then remains as to which cell models to use. Human immortalized cell lines derived from many different cell types from a wide range of anatomical sites are available and are established well-characterized models.

The cytoprotection assay was developed to be a more biologically relevant measurement than the chemically defined antioxidant activity assay because it uses human cells as a substrate and therefore accounts for some aspects of uptake, metabolismand location of flavonoids within cells. Knowledge of structure-activity relationships in the cytoprotection assay may be helpful in assessing potential *in vivo*

---

* Telephone: ++64 9 9257100, email: Jingli.Zhang@plantandfood.co.nz.

cellular protective effects of flavonoids. This study will discuss the cytoprotective properties of flavonoids and focuses on the relationship between their cytoprotective activity, physicochemical properties such as lipophilicity (log P) and bond dissociation enthalpies (BDE), and their chemical structures. The factors underlying the influence the different classes of flavonoids have in modulating their ability to protect human gut cells are discussed and support the contention that the partition coefficients of flavonoids as well as their rate of reaction with the relevant radicals define the protective abilities in cellular environments. By comparing the geometries of several flavonoids, we were able to explain the structural dependency of the antioxidant action of these flavonoids.

INTRODUCTION

The flavonoids are among the most numerous and widespread natural products found in plants and have many diverse applications and properties. Over the years, a wide range of beneficial properties related to human health, including cancer (Colic and Pavelic, 2000; Eastwood, 1999; Middleton et al., 2000); cardiovascular diseases (Riemersma et al., 2001), including coronary heart disease (Eastwood, 1999; Giugliano, 2000; Middleton et al., 2000) and atherosclerosis (Wedworth and Lynch, 1995); anti-inflammatory effects (Manthey, 2000; Middleton et al., 2000); and other diseases in which an increase in oxidative stress have been implicated (Diplock et al., 1998; Harborne and Williams, 2000; Packer et al., 1999). A number of studies have shown that consumption of fruits and vegetables can reduce the risk of cardiovascular diseases and cancer, potentially through the biological actions of the phenolic components such as flavonoids.

The precise mechanisms by which flavonoids may protect different cell populations from oxidative insults are currently unclear. However, potential mechanisms that involve their classical antioxidant properties, interactions with mitochondria, modulation of intracellular signalling cascades, and stimulation of adaptive responses have been proposed. The effects of a flavonoid in a cellular environment may well extend beyond conventional antioxidant actions. In the cellular environment, the coexistence of other factors such as the bioavailability of the compound, the effectiveness of the compound within the cell, and the effectiveness of the compound in the body must also be considered. Therefore, using a cell-based assay format, these compounds react with cells and provide information regarding the cellular response, taking into account some aspects of uptake, metabolism, location of antioxidant compounds within cells and intracellular effects on signalling pathways and enzyme activity. These effects are likely to be the result of differential modulation of cellular activities such as signalling pathways, enzyme activity, transport and bioavailability, rather than simply a result of free radical scavenging. Furthermore, as cells from different anatomical sites respond differently to both stressors and treatments, it is important to use the appropriate cell types to test a particular cellular response, rather than a chemically-defined system for the antioxidant activity of flavonoids. Here, we employed and established cell-based assays using gut-derived cultured human cell lines. The rationale to use cultured human cell lines over primary cells is that human primary cells are not readily available, but human immortalized cell lines derived from many different cell types from a wide range of anatomical sites are available and are established well-characterized models.

The multiple biological activities of flavonoids as well as their structural diversity make this class of compounds a rich source for modelling lead compounds with targeted biological properties. Different classes of flavonoids are not equally physiologically active, presumably because they are structurally different. Despite the enormous interest in flavonoids and other polyphenolic compounds as potential protective agents against the development of human diseases, the real contribution of such compounds to health maintenance and the mechanisms through which they act are still unclear. Structure activity relationships (SARs) represent an attempt to correlate physicochemical or structural descriptors of a set of structurally related compounds with their biological activities or physical properties. Molecular descriptors usually include parameters accounting for electronic properties, hydrophobicity, topology, and steric effects. Activities include chemical and biological measurements. Once developed, SARs provide predictive models of biological activity and allow the identification of those molecular parameters responsible for the biological and physicochemical properties. These may shed light on the mechanism of action.

SARs of flavonoids have been previously reported for scavenging of peroxynitrite, hydroxyl radical and superoxide, and protection against lipid peroxidation(Chen *et al.*, 2002; Choi *et al.*, 2002; Cos *et al.*, 1998), inhibition of LDL oxidation(van Acker *et al.*, 1996; Vaya *et al.*, 2003), and the influence of flavonoid structure on biological systems has also been investigated, e.g., induction of DNA degradation; growth and proliferation of certain malignant cells; acute toxicity in isolated rat hepatocytes, and inhibition of gastric H^+, K^+-ATPase (Agullo *et al.*, 1997; Moridani *et al.*, 2002; Murakami *et al.*, 1999; Sugihara *et al.*, 2003).

CHEMICAL STRUCTURE OF FLAVONOIDS

The flavonoids are a group of phenolic compounds that share common structural features and physicochemical properties, which are important in determining their biological effects. Phenylpropanoid metabolism, which encompasses natural product metabolic pathways unique to plants, transforms phenylalanine into a variety of plant secondary products, including lignins, sinapate esters, stilbenoids and flavonoids. Amongst these phenylpropanoids, flavonoids (C_6-C_3-C_6) have received significant attention in the past few decades because they appear to have diverse functions in plant defence systems and effects on human health such as antiallergic, anti-inflammatory, antithrombotic, anticancer, and antioxidant effects. Flavonoids are plant secondary metabolites that are synthesized from phenylalanine (Figure 1) (Havsteen, 2002; Ververidis *et al.*, 2007; Winkel-Shirley, 2001). Flavonoids constitute a relatively diverse family of aromatic molecules that are derived from phenylalanine via a *p*-coumaric acid (C_6-C_3) intermediate step (Figure 1). They account for a variety of colours in flowers, berries and fruits, from yellow to red and dark purple. The term "flavonoids" is generally used to describe a broad collection of natural products that include a C_6-C_3-C_6 carbon framework, which possess phenylbenzopyran functionality. Chalcones and dihydrochalcones are considered to be the primary C_6-C_3-C_6 precursors and constitute important intermediates in the synthesis of flavonoids. The nomenclature of flavonoids is with respect to the aromatic ring A condensed to the heterocyclic ring C and the aromatic ring B most often attached at the 2-position of the C-ring. The various attached substituents are listed first for the C-ring and A-ring and, as primed numbers, for the B-ring (Figure 1).

Flavonoids differ in the arrangements of hydroxyl, methoxy, and glycosidic side groups, and in the configuration of the C-ring that joins the A- and B-rings. These give rise to a multitude of different compounds (Middleton *et al.*, 2000). In plants, the majority of the flavonoids are found as glycosides with different sugar groups linked to one or more of the hydroxyl groups. They are mainly found in the outer parts of the plants, such as leaves, flowers and fruits, whereas the content in stalks and roots is usually very limited.

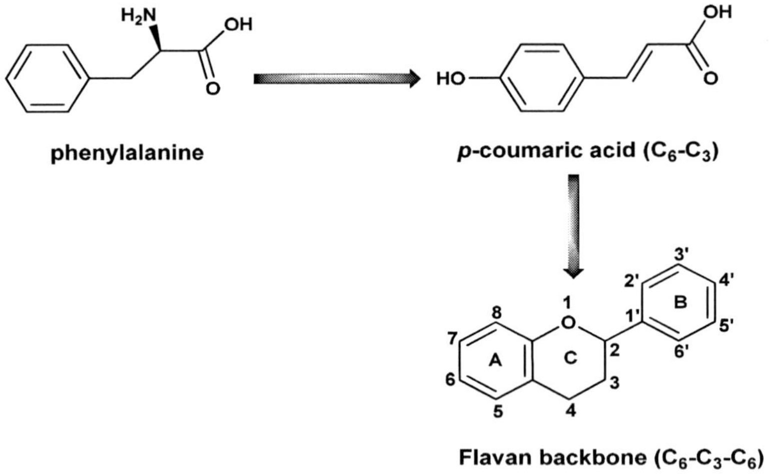

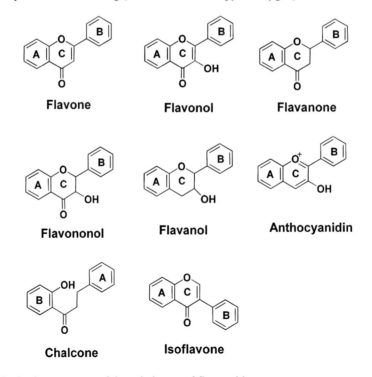

Figure 1. Diagram of biosynthetic formation of flavonoid backbone from phenylalanine. The basic flavonoid structure consists of the fused A and C-rings, with the phenyl B-ring attached through its 1'-position to the 2-position of the C-ring (numbered from the pyran oxygen).

Figure 2. The basic ring structure of the subclasses of flavonoids.

The flavonoids located in the upper surface of the leaf or in the epidermal cells have a role to play in the physiological survival of plants. They contribute to the disease resistance of the plant, either as constitutive antifungal agents or as induced phytoalexins (Harborne and Williams, 2000). Multiple combinations of hydroxyl groups, sugars, oxygen atoms, and methyl groups attached to the basic ring structural skeleton create the various classes of flavonoids. According to the configuration of the C-ring, flavonoid can be classified as flavonol, flavonone, flavone, flavanol, anthocyanidin, chalcone, and isoflavone (Herrmann, 1976; Herrmann, 1989) as illustrated in Figure 2. It should be noted that chalcones contain an opened C-ring (Dziezak, 1986) and the numbering system for chalcones is reversed. Flavonoids comprise a large group of secondary plant metabolites. Presently more than 7000 individual compounds are known, which are based on very few core structures (Fossen and Andersen, 2006; Stack, 1997). Within each class, individual flavonoids may vary in the number and distribution of hydroxyl groupsas well as in their degree of alkylation or glycosylation.

PHYSICOCHEMICAL PROPERTIES OF FLAVONOIDS

Theoretical parameters employed to characterize radical scavenging activity of a flavonoid can be roughly grouped into the following classes: (1) indices reflecting O–H bond dissociation enthalpy (BDE), where a relatively low BDE value facilitates the H-abstraction reaction between antioxidant and radical (Dewar et al., 1985; van Acker et al., 1993; Zhang et al., 2003b); (2) parameters representing electron-donating ability, such as ionization potential (IP) or relative adiabatic ionization potential (van Acker et al., 1993), enthalpy of single electron transfer (also defined as activation energy of the intermediate cation) (Vedernikova et al., 1999); (3) factors stabilizing the corresponding radical after hydrogen-abstraction (Vedernikova et al., 1999); (4) electrochemical properties, such as redox potentials (van Acker et al., 1996; Vedernikova et al., 1999); and (5) solubility, which controls the mobility of the antioxidant between lipid membranes (Gotoh et al., 1996; Noguchi et al., 1997), e.g. lipophilicity (logarithm of octanol/water partition coefficient).

Bond Dissociation Energy (BDE)

BDE is the measure of the energy change on bond making or bond breaking and is defined as the amount of energy required to break a given bond to produce two radical fragments when the molecule is in the gas phase at 25°C (298.15 °K) (McMurry, 1992).

$$A \text{:} B \xrightarrow{\text{BDE}} A^{\bullet} + B^{\bullet} \tag{1}$$

Bond dissociation energies have long been considered to provide the best quantitative measure of the stabilities of the radicals formed (Bordwell and Zhang, 1993). Since the rate constants of this reaction depend largely on the strength of the ArO-H bond (ArOH represents flavonoid molecule), the BDE of flavonoids is defined by the following equation:

$$A_rOH \xrightarrow{\text{BDE}} A_rO^\bullet + H^\bullet \tag{2}$$

BDE (ArO-H) can be obtained as:

$$BDE = H_{fr} + H_{fh} - H_{fp} \tag{3}$$

where H_{fr} is the enthalpy for radicals generated after H abstraction, H_{fh} is the enthalpy for the hydrogen atom, -0.49792 hartrees, and H_{fp} is the enthalpy of the parent molecule (Zhang *et al.*, 2003a).

The properties of the A_rO-H bond appear to be essential to understanding the chemical and biochemical behaviour of flavonoids. The A_rO-H bond must be broken to generate the truly active species, i.e. the phenoxy radical, in order to exhibit its antioxidant activity. There are a number of studies, using a diversity of modern experimental and computational tools, on the determinations of the BDEs of phenolic derivatives (Bordwell and Zhang, 1993; Lucarini and Pedulli, 1994; van Acker *et al.*, 1996). Their aim was to understand how the strength of the phenolic bond is affected by nature, position, and number of substituents. BDE can be experimentally determined in the gas phase using approaches such as radical kinetics, gas-phase acidity cycles and photoionization mass spectrometry (Berkowitz *et al.*, 1994) and in solutions using techniques such as photoacoustic calorimetry (PAC) (Mulder *et al.*, 1988), electrochemical (EC) (Wayner and Parker, 1993), and other measurements (Mahoney and DaRooge, 1975). It should be noted that neither the EC technique nor the PAC method for measuring BDEs is a stand-alone method. Both techniques are dependent upon at least one gas-phase measurement (Wayner *et al.*, 1995). However, all the above-mentioned methods are limited especially for the larger organic compounds since most of them are not stable in the gas phase. Moreover, these measurements require very sophisticated instruments. For these reasons, the number of experimentally known BDEs for flavonoids is very small (Denisov and Khydyakov, 1987).

Besides these experimental studies, a number of theoretical investigations of varying degrees of sophistication have also been reported in order to understand the structural factors determining the stability of the O-H phenolic bond. Both experimental and theoretical results indicate that the change of the O-H bond strength due to a given substituent is approximately constant in the variously substituted phenols and that, for each substituent in the *ortho, meta,* and *para* positions, an additive contribution may be derived that can be used to estimate the bond BDE of polysubstituted phenols for which experimental data are lacking (Wright *et al.*, 1997; Wright *et al.*, 2001).

As a fundamental chemical parameter (Borges dos Santos and Simoes, 1998), there have been several types of theoretical methods to estimate O-H BDE (Chipman *et al.*, 1994). The first is through the additive rule (Wright *et al.*, 2001). Although this is convenient to estimate the O-H BDEs for monophenols (Brigati *et al.*, 2002; Lucarini *et al.*, 1996), it has not been demonstrated as generally effective for catechols (Zhang *et al.*, 2003a). The second is through semi-empirical quantum chemical calculations by means of intermediate neglect of differential overlap (INDO) (Pople *et al.*, 1968), modified neglect of diatomic overlap (MNDO) (Dewar and Thiel, 1977), the Austin Model 1 (AM1) (Dewar *et al.*, 1985), and the parameterization method 3 (PM3) (Stewart, 1989). The third is through density functional theory (DFT) (Qin and Wheeler, 1995; Ziegler, 1991) or *ab initio* molecular dynamics (Bakalbassis *et al.*, 2001; Car, 2002) calculations. The parameterization method 3 (PM3) uses nearly the same equations as the AM1 method along with an improved set of parameters. PM3 predicts energies and bond lengths more accurately than AM1 (Yong, 2001). Several computer programs, such as Gaussian, Hyperchem, and MOPAC have been developed based on these theories. All these programs can be employed to perform the calculation of BDEs. In this chapter, in order to investigate the cytoprotective mechanism, the BDEs of selected flavonoids were calculated using the PM3 method using the MOPAC2002 program through Chem3D Ultra 2008 (http://www.camsoft.com/). In this chapter, the difference (ΔH_f) between the heat of formation of a parent molecule (H_{fp}) and that of its phenoxyl radical (H_{fr}) is used to represent the BDE. As stated above, the heat of formation of the H atom (H_{fh}) is treated as a constant in order to simplify the calculations (Zhang, 1998; Zhang et al., 1999) and can be ignored. Thus, BDE can be expressed:

$$BDE \approx \Delta H_f = H_{fr} - H_{fp} \tag{4}$$

Therefore, ΔH_f was used in this chapter to approximate BDE and these two terms are interchangeable.

Lipophilicity (log P)

Lipophilicity of compounds of bioactive interest is an important parameter in the understanding of transport processes across biological barriers (Lipinski et al., 1997). The lipophilic behaviour of an antioxidant is determined by its partition between phases differing in polarity. The forces of interaction between molecules that result from attraction of different functional groups can lead to different partition behaviour (Schwarz et al., 1996). It is possible to quantify the degree to which an antioxidant's action is moderated by its ability to enter the locus of autoxidation (Castle and Perkins, 1986; Porter et al., 1989). Uptake of most organic chemicals to the site of action is by passive diffusion and is best modelled by lipophilicity (MacFarland, 1970). Lipophilicity characterizes the tendency of molecules (or parts of molecules) to escape contact with water and to move into a lipophilic environment.

Since Hansch et al. (Hansch et al., 1968) recognized that the partition coefficient of a molecule in the n-octanol/water solvent system mimics molecule transport across biological membranes, the basic quantity to measure lipophilicity has been the logarithm of the partition coefficient, log P. The partition coefficient (P) is defined according to the Nernst Partition Law as the ratio of the equilibrium concentrations (C) of a dissolved substance in a two-phase system consisting of two largely-immiscible solvents, e.g. n-octanol and water (Eadsforth and Moser, 1983). The partition coefficient is therefore dimensionless, being the quotient of two concentrations, and it is customary to express them in logarithmic form to base ten, i.e., as log P because Pvalues commonly range over many orders of magnitude (Fujita et al., 1964). The logarithm of the partition coefficient, log P, has been successfully used as a hydrophobic parameter (Leo, 1991).

Pioneering work by Leo and Hansch (Leo et al., 1971) has led to the use of log P in quantitative structure-activity relation methods (QSAR), as a general description of cell permeability. In the field of drug development, log P has become a standard property determined for potential drug molecules (Lipinski et al., 1997). The lipophilicity of the flavonoids is an important parameter in chemical toxicology as it can indicate metabolic fate, biological transport properties and intrinsic biological activity (Hansch *et al.*, 2000). Lipophilicity is of central importance for biological potency as it plays a role in the interaction of flavonoids with many of the targets in a biological system. Log P probably can be considered the most informative and successful physicochemical property in biochemistry and medicinal chemistry (Leo, 1991).

Since log P is an additive, constitutive molecular property, it is possible to estimate the log Pvalue of a molecule from the sum of its component molecular fragment values (Masuda *et al.*, 1997). Many programs developed to do this are based on substructure approaches such as ClogP (Leo, 1991; Leo, 1993), KOWWIN (Meylan and Howard, 1995), AB/LogP (Japertas et al., 2002), ACD/LogP (Buchwald and Bodor, 1998; Osterberg and Norinder, 2001), and KLOGP (Klopman and Zhu, 2001). The substructure methods usually require a long calculation time because a large number of structural parameters need to be taken into account (Mannhold and Petrauskas, 2003). An alternative approach for the computation of log Pis based on additive atomic contributions. The Ghose-Crippen approach is the most widely used atom-based method (Ghose and Crippen, 1987). The parameters used in the calculation of log Pcan be obtained by first classifying atoms into different types according to their topological environments, which contribute differently to the global log Pvalue. Several computer programs are developed based on atomic contribution techniques such as XLOG P (Wang *et al.*, 1997) and SMILOGP (Convard *et al.*, 1994).

Since the whole is more than the sum of its parts, any method of calculating log Pof a molecule from its parts has limitations. Thus, other methods have been proposed based on calculated molecular properties. Fewer programs are based on a whole-molecule approach compared with a substructure approach. The most widely available one is SciLogP Ultra (Bodor *et al.*, 1989).

From a theoretical perspective, it is difficult to judge the validity of any particular method since it depends on the methodology used in data analysis and algorithm derivation.

CYTOPROTECTION ASSAY

Oxidative stress refers to the cytopathological consequences of an imbalance between the production of free radicals and the ability of the cell to neutralize them. Reactive oxygen species (ROS) have been suggested to be a major cause of neurodegenerative disorders, such as Alzheimer's disease, Parkinson's disease and Huntington's disease (Simonian and Coyle, 1996). Hydrogen peroxide (H_2O_2) can traverse membranes and exerts cytotoxic effects on cells in the proximity of those responsible for its production (Halliwell, 1992). Although H_2O_2 is not a free radical and has a limited reactivity, it is thought to be the major precursor of the highly reactive hydroxyl radical (HO$^{\bullet}$). Recent studies have shown a close association between H_2O_2 and neurodegenerative disease, and it has been suggested that H_2O_2 levels are increased during pathological conditions such as ischemia (Behl *et al.*, 1994; Hyslop *et al.*, 1995). ROS such as H_2O_2 and HO$^{\bullet}$ readily damage biological molecules that can eventually lead to apoptotic or necrotic cell death (Gardner *et al.*, 1997). Exposure of cells to oxidative stress induces a range of cellular events that can result in apoptosis or necrosis (Davies, 1999). Apoptotic cells can be evaluated based on the measurement of the loss of plasma membrane asymmetry (van Engeland *et al.*, 1998). Under normal physiological conditions, a cell maintains a strictly asymmetric distribution of phospholipids in the two leaflets of the cellular membranes with phosphatidylserine (PS) facing the cytosolic side (Devaux, 1991). However, during early apoptosis this membrane asymmetry is rapidly lost without concomitant loss of membrane integrity (van Engeland *et al.*, 1998). Cell surface exposure of PS, which precedes the loss of membrane integrity, can be detected by fluorescein isothiocyanate (FITC)-labelled annexin V, a reagent that has high affinity for PS residues in the presence of millimolar concentrations of calcium (Ca^{2+}) (Andree *et al.*, 1990). By simultaneous probing of membrane integrity by means of exclusion of the nuclear dye propidium iodide (PI), apoptotic cells can be discriminated from necrotic cells (Darzynkiewicz *et al.*, 1997). The importance of apoptosis in the regulation of cellular homeostasis has mandated the development of accurate assays capable of measuring this process. Apoptosis assays based on flow cytometry have proven particularly useful; they are rapid, quantitative, and provide an individual cell-based mode of analysis (rather than a bulk population).

In this study, the relationship between physicochemical properties, chemical structures and cytoprotective capacity of twenty-four different flavonoids was established using human colon adenocarcinoma (HT-29) cells.

MATERIALS AND METHODS

Materials

4-[3-(4-iodophenyl)-2-(4-nitrophenyl)-2H-5-tetrazolio]-1, 3-benzene disulphonate (WST-1 reagent) was obtained from Roche (Basel, Switzerland). Hydrogen peroxide was obtained from BDH Chemicals (Poole, England). Annexin V-FITC and binding buffer were obtained from BD Biosciences (San Diego, CA). All other chemicals were obtained from Sigma (St. Louis, MO). All solvents were of HPLC grade. Deionized water (MilliQ) was used in all experiments. All cell culture media were obtained from Invitrogen-Life Technologies (Carlsbad, CA). Cultured human colon adenocarcinoma HT-29) were obtained from the ATCC (American Type Culture Collection; Manassas, VA).

Assessment of Cell Viability

Cell viability was assessed using the WST-1 assay. The WST-1 assay is based on the cleavage of the tetrazolium salt WST-1 by mitochondrial dehydrogenases to form dark red formazan, which absorbs at 450 nm. Cultured human cells were seeded in 96-well plates at a density of 5×10^5 cells/ml with various concentrations (0.25-20 μM) of testing compounds. Tested compounds were either dissolved in DMSO or deionised water depending on their solubility (the amount of DMSO in cell culture was limited to 0.1%). Equivalent amounts of the DMSO vehicle had no effect compared with results in control cells. After 24 h of incubation in a humidified 5% CO_2, 95% air atmosphere at 37°C, 10 μl WST-1 tetrazolium salt was added to each well and the cells were incubated for 2 h to allow the reaction between the mitochondrial dehydrogenase released from viable cells and the tetrazolium salt of the WST-1 reagent. The absorbance was measured at 450 nm with a reference at 690 nm using a microplate reader (Synergy HT, BioTEK Instruments, Winooski, VT). The level of absorbance directly correlates to viable cell numbers. Each assay was performed in triplicate and the cell viability was expressed as a percentage of the absorbance of cells exposed to test samples compared with that of controls (cells only).

Culture and Treatment of HT-29 Cells

HT-29 cells were grown in McCoy's 5A medium (modified) supplemented with 10% fetal bovine serum (FBS) in the presence of 100 U/ml penicillin and 0.1 g/l streptomycin at 37°C in humidified air with 5% CO2. Cultured HT-29 cells were plated in 24-well plates at a concentration of 5×10^5 cells/ml. A range of non-toxic concentrations (0-20 μM) of testing compounds were added together with 150 μM H_2O_2. Cells were incubated for 24 h at 37°C in humidified air with 5% CO2.

Annexin V Staining and Flow Cytometric Analysis

Annexin V coupled with fluorescein isothiocyanate (FITC) is typically used in conjunction with a vital dye such as propidium iodide (PI) to identify different stages of apoptotic and necrotic cells using flow cytometry. This assay was performed according to the method described by Vermes and co-workers (Vermes *et al.*, 2000) with slight modifications. After 24 h of incubation, cells were harvested and stained with both annexin V and propidium iodide to identify different stages of apoptotic and necrotic cell death using flow cytometry.

Briefly, the washed cells were resuspended in 100 μL of 1X binding buffer containing Annexin V-FITC (5 μl per test according to the manufacturer's instruction) and incubated in the dark for 20 min. Then, another 400 μL binding buffer containing PI (5 μl per test from 1 mg/ml stock solution) was added and incubated for a further 10 min. Flow cytometric analysis was performed within 1 h using a Cytomics FC500 MPL (Beckman Coulter, Miami, FL). The total cell count was set to 35,000 cells per sample.

CALCULATION OF RESULTS

Cell Death Index (CDI)

The percentages of viable, early, late apoptotic and necrotic cells were determined as illustrated in the cytogram (Figure 3). The viable cells are located in the lower left corner (negative in both annexin V-FITC and propidium iodide) (A3). Early apoptotic cells are in the lower right corner (annexin V-FITC positive only) (A4). Late apoptotic cells that show progressive cellular membrane and nuclear damage are in the upper right corner (both annexin V-FITC and PI positive) (A2). Necrotic cells are located in the upper left corner (PI positive only) (A1). The total percentage of damaged cells (both apoptotic and necrotic) was considered as (A1+A2+A4). The cell death index (CDI) was calculated based on the cytogram by the following equation (equation 5):

$$CDI = \frac{(A1+A2+A4)}{A3} \times 100$$

(5)

The CDI is the ratio of total damaged cells to viable cells and is used to remove inter-experimental variations in cell density. The net cell damage (ΔCDI) is derived by subtracting the CDI of incubated control cells (Figure 3B) from that of treated cells (Figure 3C) (equation 6).

$$\Delta CDI = (CDI_{Treated\ cells} - CDI_{Incubated\ control\ cells})$$

(6)

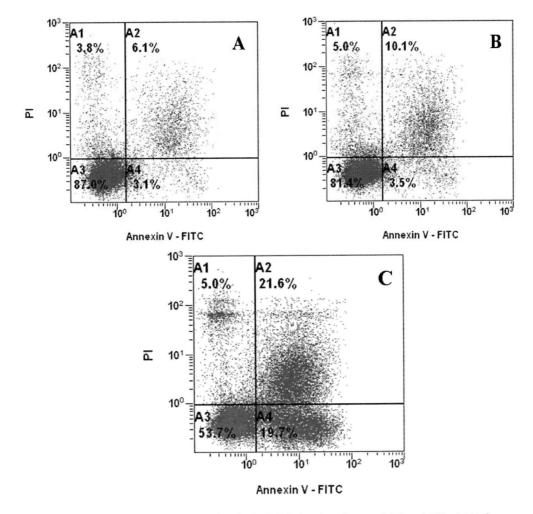

Figure 3. Cytograms of control (prior to incubation) (A), incubated control (B) and 150 μM H$_2$O$_2$ treated (C) human colon adenocarcinoma (HT-29) cells.

Calculation of 50% Reduction in Cell Death (EC50)

The cytoprotective effects of test compounds were measured by the inhibition of the cytotoxic effects of H_2O_2 using both apoptosis and necrosis endpoints (approximately caused 50% total cell death). The percentage of inhibition of cell death was calculated by equation 7:

$$\% \text{ Inhibition of cell death} = \frac{\Delta CDI_{HP} - \Delta CDI_{Sample}}{\Delta CDI_{HP}} \times 100$$

(7)

where ΔCDI_{HP} and ΔCDI_{Sample} are the net cell damage caused by H_2O_2 and test sample, respectively. The EC_{50} values were calculated from the dose-response relationship between the concentrations of antioxidant and % inhibition of cell death.

QUANTUM CHEMICAL CALCULATIONS

Calculation of Heat of Formation

All geometry calculations of flavonoid were performed by using PM3 of the MOPAC2002 molecular package through Chem3D Ultra 2008 interface. The procedures were as follows. The molecular geometries were optimized by MM2 and then by the semiempirical quantum chemical method (PM3), and energies were minimized by using the EF algorithm. After the calculation of the heat of formation of the parent molecules (H_{fp}), the phenolic H was removed to get its free radical and a restricted Hartree-Fock optimization was performed on the phenoxyl radical. The differences in heat of formation was calculated by calculated by $\Delta H_f = H_{fr} - H_{fp}$.

Calculation of Log P

The log P of flavonoids was obtained by using ClogP in the Chem3D Ultra 2008 molecular package.

RESULTS AND DISCUSSION

Cytotoxic effects of H_2O_2

Hydrogen peroxide is known to be able to induce both apoptosis and necrosis in cells (Antunes and Cadenas, 2001; Barbouti et al., 2002; Kim et al., 2000), with the required concentrations and exposure time dependent on the cell type being investigated. The response of cultured human cells to H_2O_2 in terms of both concentration and exposure time was determined to calculate the dosage required to kill approximately half the cells. The CDI increased with increasing concentrations of H_2O_2 on HT-29 cells (Figure 4).

A concentration of 150 μM H_2O_2 was selected for the cytoprotection assay using HT-29 cells (CDI of 63.5 ± 2.7) with a 24-hour exposure to H_2O_2. These conditions were used in this assay to investigate the protective effects of antioxidants against H_2O_2-induced total cell death. Although H_2O_2 itself is a relatively unreactive species and easily scavenged by cellular catalase (Gille and Joenje, 1992), it can cause membrane damage by increasing the release of arachidonic acid from the cell membrane, which may account for the prolonged damage caused by H_2O_2 even after being scavenged (Cantoni *et al.*, 1989).

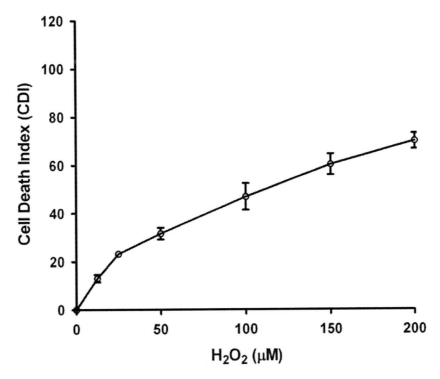

Figure 4. Cell death responses of cultured human HT-29 cells exposed to increasing concentrations of hydrogen peroxide (H_2O_2). HT-29 cells (5 x 10^5 cells per ml) were exposed to different concentrations of H_2O_2 and incubated at 37°C in humidified air with 5% CO_2 for 24 h. Bars indicate standard deviation from the mean of two separate determinations.

Thus, even at low concentrations, H_2O_2 can cause damage to cultured cells. These facts demonstrate that the cytoprotection assay can be used to screen for and compare the protective effects of flavonoids in a biologically relevant cellular environment.

Cytotoxicity of Flavonoids

The cytotoxicities of Trolox, catechol, pyrogallol and selected flavonoids were tested on cultured human HT-29 cells at different concentrations for 24 h using the WST-1 assay (data not shown). None of the compounds tested affected the viability of HT-29 cells within the concentration range used (0–20.0 μM) in this study.

The Influence of Trolox on Cytotoxic Effects of H_2O_2

As described above, Trolox is a common compound used as a standard for most antioxidant assays. Trolox was co-incubated with cultured HT-29 cells at doses of 0, 0.25, 0.5, 1.0, 5.0, 10.0, 20.0 μM immediately prior to the addition of the H_2O_2. Trolox protected HT-29 cells against H_2O_2-induced cell death in a dose dependent manner (Figure 5). The EC_{50} value (7.91 ± 0.22 μM) of Trolox was calculated from its dose-response curve. Trolox could thus be used as a standard antioxidant in this cytoprotection assay for comparison with other antioxidant compounds.

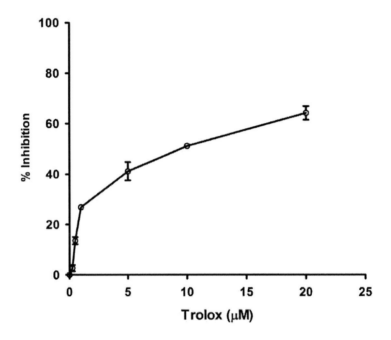

Figure 5. The concentration-response curves of Trolox for protection of HT-29 cells from 150 μM hydrogen peroxide. HT-29 cells (5 x 10^5 cells per well) were incubated at 37°C in humidified air with 5% CO_2 for 24 h. After incubation, cells were stained with both Annexin V-FITC and PI and analyzed by flow cytometry. Bars indicate standard deviation from the mean of two separate determinations.

STRUCTURAL RELATED CYTOPROTECTIVE ACTIVITY OF FLAVONOIDS

In the present Chapter, the cytoprotective activities (EC_{50}) of a range of structurally diverse flavonoids were measured (Table 1). Structural variations within the rings subdivide the flavonoids into several families (Figure 2). Measurement of the potential health effects of dietary-derived phenolic compounds needs to be undertaken at concentration ranges that are relevant to levels that might be achieved *in vivo*. Maximum plasma concentrations attained after a polyphenol-rich meal are thought to be in the range of 0.1–10 μM (Kroon *et al.*, 2004). In this work gut-derived cells were used but general bioavailability was taken into account. Hence, the 24 flavonoids, catechol and pyrogallol were added to HT-29 cells at doses of 0, 0.25, 0.5, 1.0, 5.0, 10.0, 20.0 μM immediately prior to the addition of the H_2O_2.

Effects of Hydroxyl Groups in the B Ring

The manipulation of the hydroxyl substitutions in the B-ring in flavones (with the 2,3-double bond and 4-keto function in the C-ring, but no 3-OH group) allows the observation of the contribution of these hydroxyl groups to their cytoprotective activities. With the $3',4'$-dihydroxyl group, the EC_{50} value of luteolin is 4.05 μM. Dehydroxylation at the $3'$-position as in apigenin increases the value to 12.87 μM, making the cytoprotective activity of apigenin only one-third of luteolin (Figure 6). The EC_{50} value of chrysin is further increased because of the lack of any hydroxyl group in its B-ring. The cytoprotective activity of chrysin can be reasonably attributed to the 5,7-*meta*-dihydroxyl groups of its A-ring.

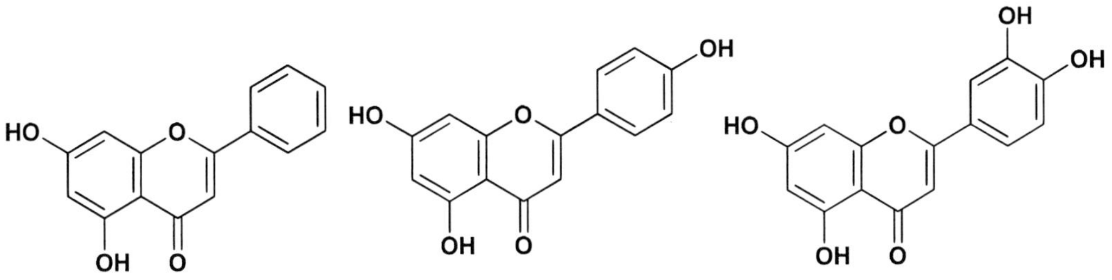

15.42 ± 1.99 μM	**12.87 ± 0.94 μM**	**4.05 ± 0.48 μM**
Chrysin	**Apigenin**	**Luteolin**

Figure 6. The influences of hydroxylation in the B ring on the cytoprotective activity of the flavones.

As a group of flavonols, galangin, quercetin, morin and myricetin have the same structures on the A and C-rings but the number of hydroxyl groups in the B-ring increases from none to three (Figure 7).

As the number of hydroxyl groups increases the EC_{50} value decreases except for quercetin, which is more active than myricetin. With morin, the dihydroxyl groups in the B-ring are arranged *meta* to each other. This significantly reduces its cytoprotective activity compared with quercetin (the EC_{50} value of morin was four times higher than that of quercetin).

This result confirms that the presence of two adjacent hydroxyl groups in the B-ring plays a significant role in the high cytoprotective activity of flavonoids. Possibly, the two adjacent hydroxyl groups at position $3'$ and $4'$ in quercetin are more vulnerable to loss of a proton than the two hydroxyl groups at position $3'$ and $5'$ in morin. Myricetin, which possesses *ortho*-trihydroxyl (pyrogallol) groups in the B-ring, is much less active than quercetin. This suggests that the additional $5'$-hydroxyl group has a negative impact on its cytoprotective activity. However, the cytoprotective activity of pyrogallol is much more active than that of catechol as illustrated in Figure 8.

A fairly stable *ortho*-semiquinone radical can be formed by oxidation of a flavonoid on the B ring, when the $3'4'$-catechol structure is present facilitating electron delocalization (Arora *et al.*, 1998; Mora *et al.*, 1990). The formation of flavonoid aroxyl radicals is an essential step after initial scavenging of an oxidizing radical (Bors *et al.*, 1990).

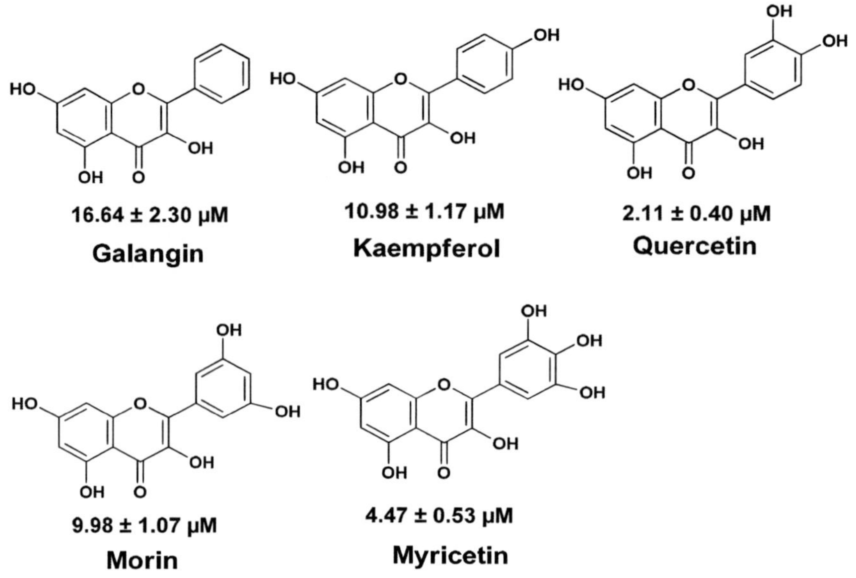

16.64 ± 2.30 µM

Galangin

10.98 ± 1.17 µM

Kaempferol

2.11 ± 0.40 µM

Quercetin

9.98 ± 1.07 µM

Morin

4.47 ± 0.53 µM

Myricetin

Figure 7. The influences of hydroxylation in the B ring on the cytoprotective activity of the flavanols.

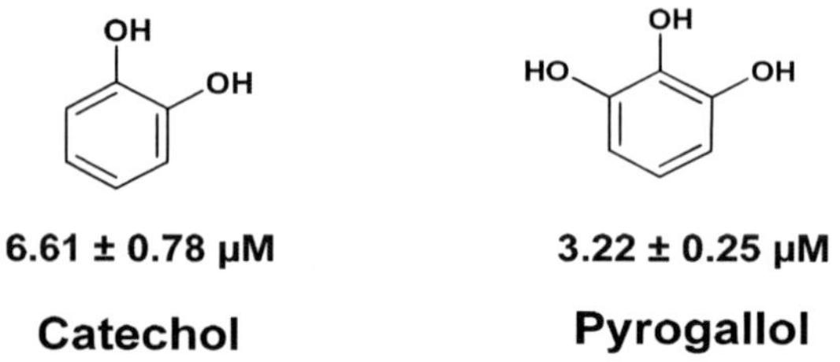

6.61 ± 0.78 µM

Catechol

3.22 ± 0.25 µM

Pyrogallol

Figure 8. The cytoprotective activity of catechol and pyrogallol. This may be to the result of the rest of the quercetin structure (C- and A-rings) stabilizing the oxidation product (*o*-quninoe) as shown in Figure 9.

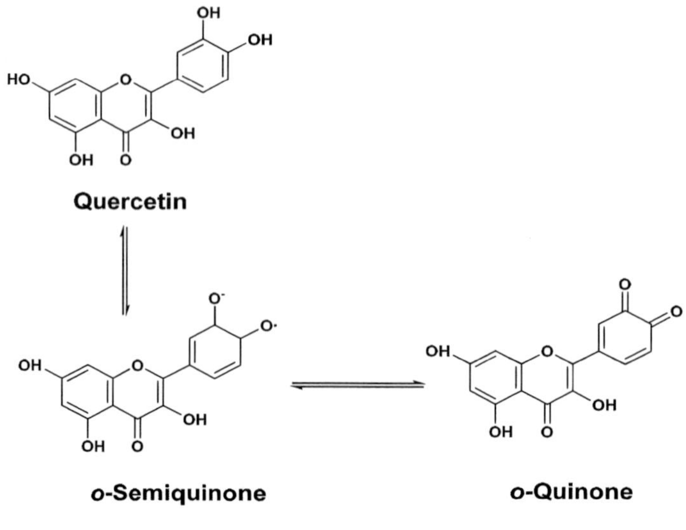

Quercetin

o-Semiquinone

o-Quinone

Figure 9. Quercetin oxidation and its possible consequences.

The stability of aroxyl radicals strongly depends on their bimolecular disproportionation reaction and electron delocalization. For instance the oxidation of quercetin can form an *o*-semiquinone radical and then an *o*-quinone radical (Awad *et al.*, 2001; Awad *et al.*, 2003; Boersma *et al.*, 2000; Metodiewa *et al.*, 1999) as illustrated in Figure 9. However, with only one hydroxyl group in the B-ring the EC$_{50}$ value of kaempferol was significantly increased. Without any hydroxyl substitutions in its B-ring, the cytoprotective activity of galangin was almost negligible like that of chrysin.

The flavanone, naringenin, with only a single 4'-OH group in the B-ring has a EC$_{50}$ value twice than that of hesperitin, which has an identical structure to naringenin except for the 3'-OH, 4'-methoxy substitution in the B-ring (Figure 10).

This finding suggested that methoxylation does not destroy its cytoprotective activity. Repeated studies have shown that flavonoids having greater numbers of hydroxyl groups, or hydroxyl groups localized *ortho* to one another, are more effective antioxidants. The B-ring of most flavonoids is usually the initial target of oxidants, as it is more electron-rich than the A- and C-rings, whose electron densities are somewhat drained away by the carbonyl group.

These properties are consistent with the expected mechanisms of oxidation of phenols; electron-donating substitutes, such as hydroxyl groups, should lower the oxidation potential for a compound, and *ortho* hydroxylation should stabilize phenoxyl radicals.

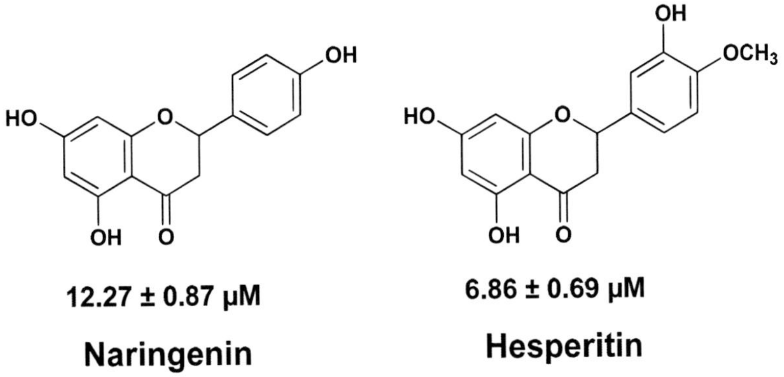

12.27 ± 0.87 µM

Naringenin

6.86 ± 0.69 µM

Hesperitin

Figure 10. The influences of hydroxylation and methoxylation in the B-ring on the cytoprotective activity of the flavanones.

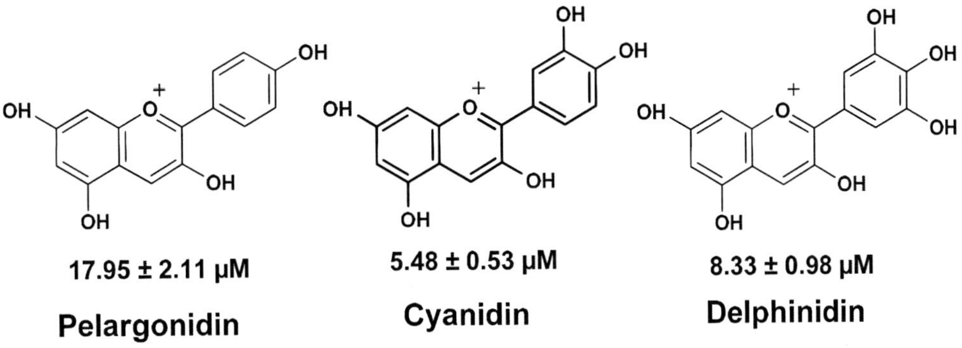

17.95 ± 2.11 µM

Pelargonidin

5.48 ± 0.53 µM

Cyanidin

8.33 ± 0.98 µM

Delphinidin

Figure 11. The influences of hydroxylation in the B-ring on the cytoprotective activity of the anthocyanins.

The cytoprotective activity pattern of pelargonidin, cyanidin and delphinidin (Figure 11) shows a similar trend to that revealed by kaempferol, quercetin and myricetin (Figure 7). The EC_{50} value of cyanidin is much lower than that of delphinidin, and the cytoprotective activity of pelargonidin, which has a lone 4'-OH, is almost negligible. With the anthocyanin C-ring, the cytoprotective activity of cyanidin is only half of that of quercetin. The same trend also applies to delphinidin and myricetin. The presence of a third OH group in the B-ring does not enhance the effectiveness against H_2O_2-induced cell death. This is also supported by the findings that myricetin was less active in protecting liposome oxidation (Zhang *et al.*, 2006). In acidic or neutral media, four anthocyanin structures exist in equilibrium (Figure 12): the flavylium cation, the quinonoidal base, the carbinal pseudobase, and the chalcone (Borkowski *et al.*, 2005; Brouillard, 1983). The equilibrium among the four different structural conformations of anthocyanin is illustrated in Figure 12 (using cyanidin as an example).

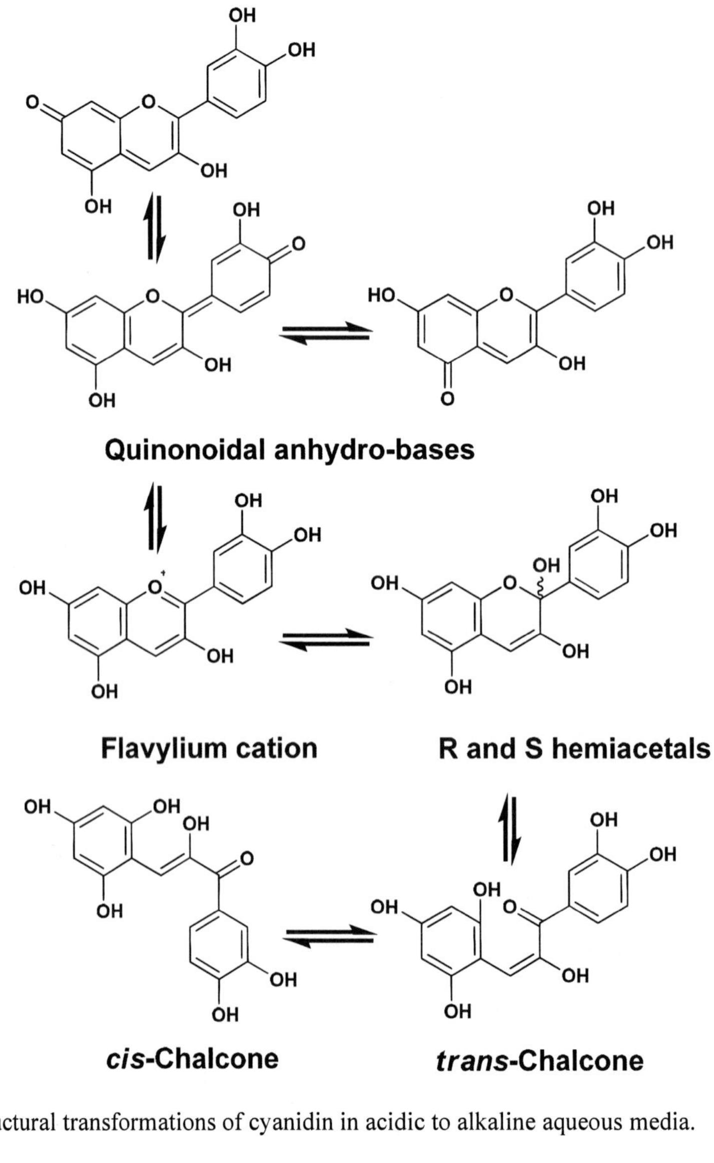

Quinonoidal anhydro-bases

Flavylium cation   **R and S hemiacetals**

***cis*-Chalcone**   ***trans*-Chalcone**

Figure 12. Structural transformations of cyanidin in acidic to alkaline aqueous media.

At pH less than 2, the anthocyanin exists primarily in the form of the red (with a 3-O-sugar substitute) or yellow (with a 3-OH) flavylium cation. As the pH is raised, there is a rapid proton loss to yield the red or blue quinonoidal forms. At higher pH, hydration of the flavylium cation occurs to give the colorless carbinol or pseudobase. The relative amounts of flavylium cation, quinoodal forms, pseudobases and chalcones at equilibrium vary with both pH and the structure of the anthocyanin. At pH 3.5-4.5, a mixture of the flavylium ion and the neutral quinonoidal anhydro-base is found. At pH 4.5-6.0, the concentration of the flavylium ion becomes vanishingly small, the quinonoidal anhydro-base increasingly predominates and there is a mixture of both the neutral and the ionized (blue anionic) quinonoidal anhydro-base forms present at pH 7.0 (around neutrality) (Brouillard and Dubois, 1977). As their quinonoidal anhydro-base or as their flavylium cations, anthocyanins could be strongly stabilized by neutral salts such as magnesium chloride and sodium chloride in concentrated aqueous solutions.

The anthocyanidin structural transformation path is very sensitive to the substitution pattern of the pyrilium ring, especially the C_3 position. The 3-OH substituted anthocyanidins are significantly shifted towards colorless pseudobase forms causing color instability (Timberlake and Bridle, 1967). In addition, an increase in the number of hydroxyl groups tends to deepen the color to a more bluish shade. The hydroxyl groups at C_5, C_7 on the A-ring and $C_{4'}$ on the B-ring of the flavylium cation can lose a proton at pH values close to equilibrium.

It is must be emphasized that the interpretation of the cytoprotective properties of anthocyanins is complicated by the relatively complex pathway of reversible structural transformations of anthocyanins in aqueous solution (Brouillard and Delaporte, 1977; Brouillard and Dubois, 1977), which not only includes proton transfer between coloured forms but also water addition to the pyrilium ring leading to colourless hemiacetal and chalcone forms. Hence, the EC_{50} values of anthocyanins measured here are actually a reflection of the cytoprotective properties of the transformed products, i.e. the quinonoidal anhydro-base (Hoshino, 1991; Hoshino and Goto, 1990; Hoshino et al., 1981).

Effect of the 3-OH Group, 2,3-Double Bond and 4-Keto Group

Without the 3-OH group in the C-ring, the EC_{50} values of apigenin and luteolin are increased compared with those of kaempferol and quercetin, respectively (Figure 13). However, the EC_{50} value is reduced for chrysin compared with that of galangin. The results presented in Figure 13 demonstrate that when the 3-hydroxyl group is absent, its contribution to electron dislocation is substantially reduced and so consequently is the flavonoid cytoprotective activity, although this reduction is smaller when the catechol structure is absent in the B ring. This fact indicated that 3-OH is required to stabilize the catechol structure in the B ring. A distinguishing feature among the flavonoid structural classes is the presence or absence of an unsaturated 2,3-double bond in conjugation with a 4-keto group.

Comparison of naringenin with apigenin shows that the 2,3-double bond in the C-ring has a slightly negative influence on the cytoprotective activity (Figure 13). On the other hand, the introduction of a 2,3-double bound and 4-keto group to catechin with the existing 3-hydroxyl group decreases the EC_{50} value as in quercetin.

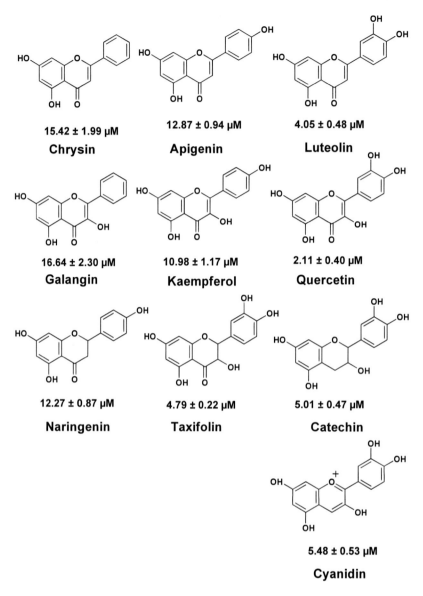

Figure 13. Structure-cytoprotective activity comparisons of the 3-OH, 2,3-double bond and 4-keto group of flavonoids.

This fact indicates that the presence of the 3-hydroxyl group is an important factor in neutralizing the negative impact of the 2,3-double bond on the cytoprotective activity. This may also indicate that the combined effect of the 2,3-double bond in the C-ring and the *ortho*-hydroxyl groups in the B-ring have positive effect on cytoprotective activity as demonstrated by the comparison of apigenin and luteolin.

However, the presence of a 2,3-double bond when the 3-hydroxyl group is absent (apigenin and luteolin) does not significantly change the cytoprotective activity of flavonoids relative to those that do not contain this double bond (naringenin and taxifolin). When the 3-hydroxyl group is present (quercetin), it significantly enhances cytoprotective activity compared with those that do not contain this double bond (taxifolin). The loss of the 4-keto group at the C-ring and introduction of a positive charge decreases cytoprotective activities as seen in cyanidin and quercetin. As shown in Figure 13, quercetin, catechin and cyanidin have identical A- and B-rings, but quercetin is more than twice as cytoprotective as catechin and cyanidin. This observation indicates the important contribution of the 2,3-double bond and 4-keto group to the cytoprotective activity.

Effect of the Carbohydrate Moieties

Blocking the 3-hydroxyl group in the C-ring of quercetin as a glycoside (while retaining the 3′,4′-dihydroxy structure in the B-ring) as in isoquercetin (quercetin-3-glucoside) decreases the cytoprotective activities. Replacement of the hydroxyl group at the C_3 position of quercetin by the disaccharide rutinose in rutin further decreases cytoprotective activity (Figure 14).

The presence of the 3-OH group on the C-ring double bond undoubtedly contributes to attack by free radicals. If the 3-OH is replaced by an O-sugar group (as in the glycoside rutin or isoquercetin, for example), reactivity is decreased by about a factor of 2-3 (Briviba et al., 1993; Tournaire et al., 1993). The results shown in Figure 14 also indicate that when the 3-hydroxyl group is substituted, the reduction in cytoprotective activity of the flavonoids depends on the nature of the substituted sugar group. This reduction is smaller when this hydroxyl group is substituted (isoquercetin or rutin) than when it is just absent (luteolin).

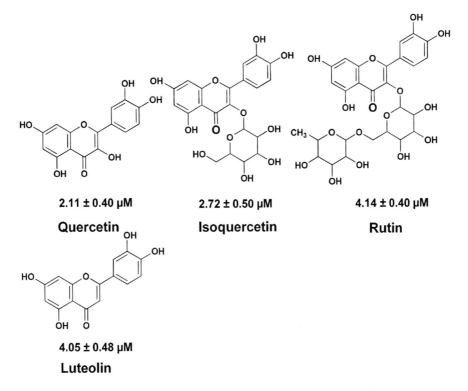

| 2.11 ± 0.40 µM | 2.72 ± 0.50 µM | 4.14 ± 0.40 µM |
| **Quercetin** | **Isoquercetin** | **Rutin** |

4.05 ± 0.48 µM

Luteolin

Figure 14. Influences of glycosylation of flavanols on their cytoprotective activity.

Similar effects are observed when cyanidin is compared with its 3-glucoside, idaein and its 3-rutinoside, keracyanin, and when pelargonidin is compared with its 3-glucoside, callistephin (Figure 15). Comparison of naringenin with naringin shows that glycosylation of the 7-hydroxyl group in a structure with a saturated heterocyclic C-ring and with a single hydroxyl group on the B-ring has a significant negative impact on the EC_{50} values. Similar trends are observed with hesperitin when a 4'-hydroxyl group in the B-ring is replaced by a methoxy and 3'-hydroxyl group, in contrast to naringenin, compared with its rhamnoside, hesperidin, which has a glycosylated 7-hydroxyl group. However, hesperidin is much more cytoprotective than that of naringin because of its B-ring configuration (Figure 16).

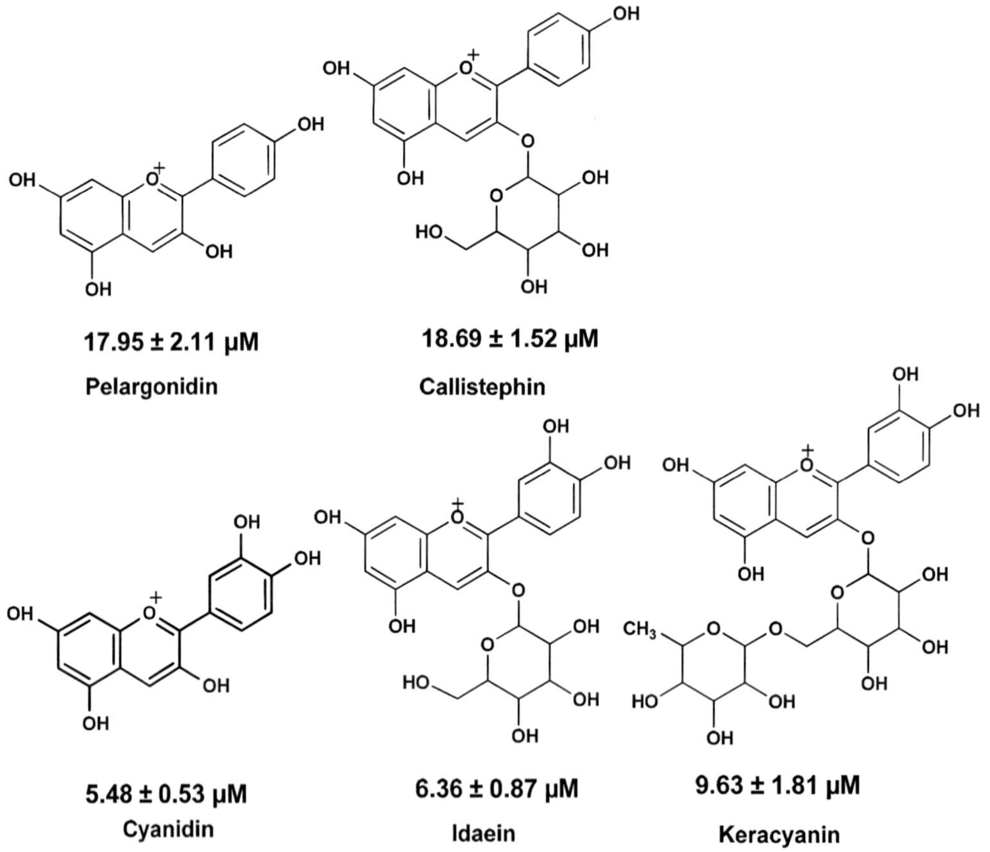

17.95 ± 2.11 µM

Pelargonidin

18.69 ± 1.52 µM

Callistephin

5.48 ± 0.53 µM

Cyanidin

6.36 ± 0.87 µM

Idaein

9.63 ± 1.81 µM

Keracyanin

Figure 15. Influences of glycosylation of anthocyandins on their cytoprotective activity.

The results presented in Figure 16 and 17 demonstrate that the presence of both 3- and 5-hydroxyl groups is also necessary to maximize cytoprotective activity of flavonoids.

The sugar moiety is reported to have a negative effect on the oxidizability of flavonoid glycosides (Hedrickson *et al.*, 1994). The oxidation rate of compounds decreased as the substituent at the 3-position became a poorer leaving group. Disaccharides are a poorer leaving group than monosaccharides, thus rutin is less oxidizable than isoquercetin (Hopia and Heinonen, 1999). This observation may explain why rutin displays a lower cytoprotective activity than quercetin and isoquercetin.

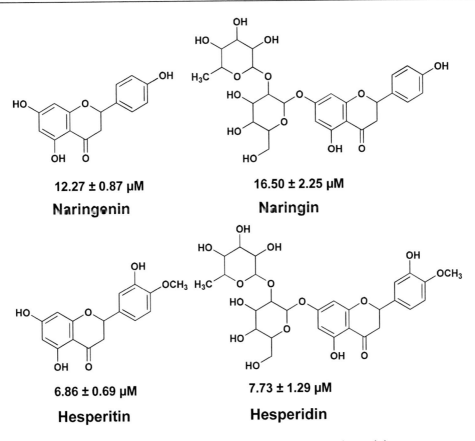

Figure 16. Influences of glycosylation of flavanones on their cytoprotective activity.

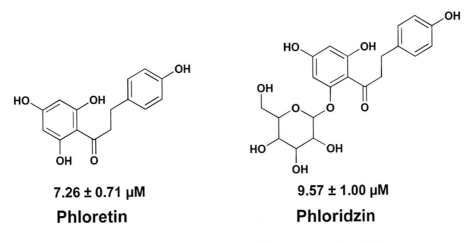

Figure 17. Influences of glycosylation of chalcones on their cytoprotective activity.

The structural criteria for the very high cytoprotective activity by flavonoids can be summarized as: 1) the *o*-dihydroxy (catechol) structure in the B-ring; 2) the 2,3-double bond in conjugation with the 4-keto group in the C-ring; and 3) the 3-hydroxyl group in the C-ring. Thus, quercetin, for example, satisfies all the above-mentioned determinants and has the highest cytoprotective activities among 24 flavonoids tested.

CYTOPROTECTIVE AND PHYSICOCHEMICAL PROPERTIES OF FLAVONOIDS

Correlation between O-H Bond Dissociation Enthalpy (BDE) and Cytoprotective Activity (EC50) of Flavonoids

Possible explanations of the cytoprotective capacity of flavonoids obtained from cell-based assays could be derived by calculating the heat of formation differences (ΔH_f) between radicals and their parent molecules (bond dissociation energy approximation) of flavonoids. Quantum chemical calculations of the geometry of the flavonoids and their corresponding radicals give their heat of formation. The ΔH_f calculated between each flavonoid and its corresponding radicals provides an estimation of the ease with which radicals may be formed (Lien et al., 1999). The ΔH_f of a given compound represents the difference between the parent compound and the appropriate radical, which was constructed by an abstraction of a hydrogen atom from assigned hydroxyl moiety (Zhang, 1998).

This value may represent the relative stability of a radical with respect to its parent compound, and it enables a comparison to be made between the stabilization achieved by hydrogen abstraction (toward radical formation) (Sun et al., 2002; Zhang, 1998; Zhang et al., 2002; Zhang and Wang, 2002). Generally speaking , the smaller the ΔH_f, the more stable the phenoxyl radical and the weaker the O-H bond in the molecule, so the more active is the flavonoid (van Acker et al., 1993).

A summary of calculated ΔH_f for the H-abstraction from hydroxyl groups (in the B ring and 3-OH) in all the flavonoids tested is shown in Table 1. All heat of formations were calculated or selected by the PM3 semi-empirical method, for energy-optimized species as described in the method section.

The ΔH_f of chrysin and galangin were calculated from the hydroxyl groups in their A-ring because there are no hydroxyl groups in their B-ring. The calculated ΔH_f shows that the least energy required for abstracting a hydrogen atom is from the 3-OH, when the C-ring contains the 2,3-double bond and the 4-keto group (flavonols).

In the absence of flavonol structure, the most favored position for donating a hydrogen atom is from the two adjacent hydroxyls in the B-ring, with 3'-OH preferred over 4'-OH. In myricetin and delphinidin, in which 4'-OH is adjacent to two hydroxyl groups (3'-OH and 5'-OH), the donation of a hydrogen atom from 4'-OH is favored over 3'-OH or 5'-OH. The calculated ΔH_f of 5'-OH is larger than that of 3'-OH (Table 1). However, the 3-hydroxyl group is not the determining factor for the cytoprotective activity of flavonoids and this is better demonstrated by galangin, which showed a very weak cytoprotective activity. The ΔH_f of flavonols (galangin, kaempferol, quercetin, morin and myricetin) is almost identical regardless of their cytoprotective activities. Therefore, the least ΔH_f of flavonoids was obtained from their hydroxyl groups in the B ring, and then the A ring.

As shown in previous work, flavonoids with a catechol group in the B ring are the most active free radical scavengers (Zhang et al., 2006). It appears that the rest of the hydroxyl groups of the flavonoid are of little importance to the antioxidant activity, except for quercetin and its derivatives, in which the combination of the catechol moiety with a 2,3-double bond at the C-ring and a 3-hydroxyl group results in an extremely active scavenger.

Therefore, the ΔH_f values were calculated from the O-H bond in the B-ring, and only the most stable phenoxyl radical is considered (lowest ΔH_f) to derive the correlation with the cytoprotective activity of flavonoids (Figure 18).

Table 1. Calculated differences of heat of formation (ΔH_f) between the parent flavonoid, catechol, and pyrogallol and each possible corresponding relative radical, the lipophilities (log P) and their cytoprotective activities

Compounds	EC$_{50}$ (µM)	Least OH ΔH_f (KJ/mol)	3-OH ΔH_f (KJ/mol)	3'-OH ΔH_f (KJ/mol)	4'-OH ΔH_f (KJ/mol)	5'-OH ΔH_f (KJ/mol)	Log P
Chrysin	15.42 ± 1.99	162.83					3.56
Apigenin	12.87 ± 0.94	149.84			149.84		2.91
Luteolin	4.05 ± 0.48	127.03		127.03	137.14		2.31
Galangin	16.64 ± 2.30	167.08	105.67				2.76
Kaempherol	10.98 ± 1.17	154.25	104.87		154.25		2.10
Quercetin	2.11 ± 0.40	125.80	103.06	125.80	138.53		1.50
Isoquercetin	2.72 ± 0.50	110.17		124.61	110.17		-0.34
Rutin	4.14 ± 0.40	109.91		109.91	137.32		-2.68
Morin	9.98 ± 1.07	154.18	104.29	154.18		158.51	1.43
Myricetin	4.47 ± 0.53	123.43	104.06	133.34	123.43	140.68	0.84
Naringenin	12.27 ± 0.87	149.91			149.91		2.44
Naringin	16.50 ± 2.25	158.11			158.11		-0.09
Hesperitin	6.86 ± 0.69	135.31		135.31			2.29
Hesperidin	7.73 ± 1.29	124.38		124.38			-0.29
Taxifolin	4.79 ± 0.22	124.86	210.59	139.16	124.86		0.77
Catechin	5.01 ± 0.47	123.12	211.65	124.65	123.12		0.53
Phloretin	7.26 ± 0.71	141.91	163.29		141.91		2.22
Phloridzin	9.57 ± 1.00	137.90	159.37		137.90		0.79
Cyanidin	5.48 ± 0.53	144.49		144.49	159.40		1.76
Idaein	6.36 ± 0.87	143.06		143.06	159.64		0.19
Keracyanin	9.63 ± 1.81	144.67	157.04	144.67	165.47		-0.84
Delphinidin	8.33 ± 0.98	140.03		146.39	140.03	158.62	1.10
Callistephin	18.69 ± 1.52	184.06	151.50		184.06		0.78
Pelargonidin	17.95 ± 2.11	182.16			182.16		2.36
Catechol	6.61 ± 0.78	126.93		126.93	126.93		0.88
Pyrogallol	3.22 ± 0.25	119.87		137.26	119.87	125.70	0.21

Notes: Structural optimization of each flavonoid and its radical was determined by calculating the minimum energy conformation by the MM2 method. MOPAC2002 in Chem3D Ultra was used to determine the final minimum energy conformation of the flavonoids and was calculated by applying the semi-empirical Hamiltonian PM3 calculation to obtain the final heat of formation of each compound. The lowest ΔH_f of chrysin and galangin were obtained from their hydroxyl groups in the A ring. All other flavonoids were calculated from their B ring hydroxyl groups.

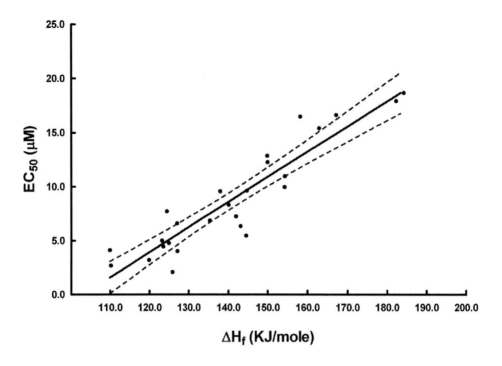

Figure 18. Correlation between EC_{50} (cytoprotective activity of flavonoids) and ΔH_f (the lowest heat of formation of the ArO-H bond of flavonoids from the A- or B-ring hydroxyl group) ($r^2 = 0.85$, n = 26).

As illustrated in Figure 18, a correlation was demonstrated between the calculated ΔH_f (the lowest differences in the enthalpy between each flavonoid's parent compound and its radical) values and the experimentally determined cytoprotective activity of flavonoids. There is a strong linear correlation between the lowest ΔH_f and the EC_{50} values and from the regression analysis a correlation with $r^2 = 0.85$ (n = 26) was obtained for the following equation 8:

$$EC_{50} = 0.23\ (\pm\ 0.02)\ \Delta H_f - 23.97\ (\pm\ 2.81)\ (n = 26) \tag{8}$$

These findings suggest that a relatively low O-H bond dissociation enthalpy (BDE, approximation by the lowest ΔH_f), which facilitates the H-abstraction reaction between flavonoids and reactive oxygen species (ROS) and other hydroxyl groups may well have contributed to this reaction in the consequent steps.

However, substitution of the hydroxyl group at the C_3 position of quercetin by the monosaccride glucose in isoquercetin and disaccharide rutinose in rutin decreases the lowest ΔH_f values, but this does not result in an increase in the cytoprotective activity.

This is probably due to the fact that glycosylation decreases the lipophilicity, and to the loss of the free hydroxyl group at the 3-position of the C-ring. An appropriate solubility, which improves the mobility of the antioxidant across cell membranes, is another important factor in explaining the cytoprotective effects of flavonoids.

Correlation between Partition Coefficient (Log P) and Cytoprotective Activity (EC50) of Flavonoids

As shown in Table 1, flavonoids with log P values that were high (log P > 3.0) or low (log P <1.0) had low cytoprotective activity indicating that the cytoprotective activity of flavonoids is associated with their affinity and distribution in lipid membranes. This is presumably because a) at high values of log P, the flavonoid is dispersed in a lipid phase and not located at the lipid-water interface and, b) at low value of log P, the flavonoid is located in an aqueous phase and has insufficient solubility in the lipid phase. This can be important in terms of paracellular transport of flavonoids and the ability to enter the cell to participate in intracellular protection from oxidative damage. It has long been recognized that for a chemical to be biologically active, it must first be transported from its site of administration to its site of action and then it must bind to or react with its receptor or target, i.e. biological activity is a function of partitioning and reactivity (Barratt, 1998). It should be noted that the effect of membrane partitioning is not necessarily a direct relationship with lipophilicity. Beyeler and coworkers (Beyeler et $al.$, 1988), for example, reported that the effects of cianidanols on rat hepatic monooxygenase increased with lipophilicity, reached a plateau, decreased and then leveled off for the most lipophilic compounds.

As illustrated in Figure 19, for the 26 compounds tested, no correlation could be found between EC_{50} and log P (equation 9).

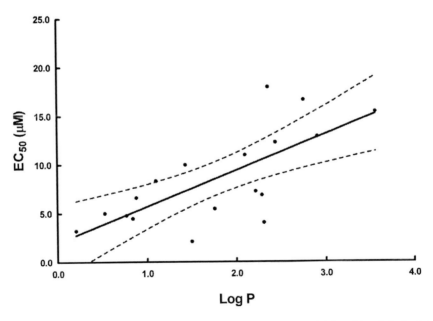

Figure 19. Correlation between EC_{50} (cytoprotective activity of flavonoids) and log P (calculated by CLogP program) of flavonoids ($r^2 - 0.15$, n = 26).

$$EC_{50} = 1.42\ (\pm 0.68)\ \log P + 7.23\ (\pm 1.20) \tag{9}$$

$$n = 26,\ r^2 = 0.15,\ p = 0.048$$

As mentioned above, glycosylation decreases the lipophilicity of flavonoid aglycones significantly and also decreased their cytoprotective activity depending on the nature of the sugar involved. Therefore, the balance of lipophilicity and lipophobicity allowing concentration at the interface is an important factor in the estimation of the antioxidant activity of flavonoids.

As demonstrated in Figure 20, there is a moderate linear correlation between the cytoprotective activities (EC_{50} values) and the partition coefficient (log P) values of flavonoid aglycones and from the regression analysis a correlation with $r^2 = 0.51$ (n = 18) was obtained for the following equation 10:

$$EC_{50} = 3.72 \ (\pm 0.92) \ \log P + 1.97 \ (\pm 1.82) \ (n = 18) \tag{10}$$

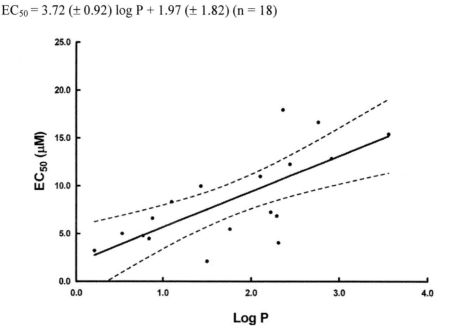

Figure 20. Correlation between EC_{50} (cytoprotective activity of flavonoids) and log P (calculated by CLogP program) of flavonoid aglycone (without sugar substitution) ($r^2 = 0.51$, n = 18).

Generally speaking, the cytoprotective activity of flavonoid aglycones decreases with increasing log P. However, this result also demonstrated that the cytoprotective activity of flavonoid aglycones is not solely dependant on their partition coefficient.

Quantitative Structure-Activity Relationship (QSAR) Model

The QSAR paradigm has been useful in elucidating the mechanisms of chemical-biological interactions in various biomolecules, particularly enzymes, membranes, organelles and cells (Hansch and Gao, 1997; Zhang et al., 2002; Zhang and Wang, 2002). The description of QSARs has been undertaken in order to find predictive models and/or mechanistic explanations for chemical as well as biological activities (Soffers et al., 2001). The underlying premise of SARs and QSARs is that the properties of a chemical are implicit in its molecular structure and the behavior of chemical compounds is dominated by their physicochemical properties (Hansch et al., 2000). If a QSAR model is deficient in modeling either partition or reactivity, only a partial correlation with the in vivo response is likely to be observed. The ΔHf, which represents the stability of the free radical formed after H-abstraction, Log P is an important parameter in chemical toxicology as it can indicate metabolic fate and biological transport properties.

The cytoprotection assay is a more biological measurement that accounts for some aspects of uptake, metabolism and location of flavonoids within cells. Here, a QSAR is modeled by starting from log P values and then incorporating ΔH_f.

With the introduction of ΔH_f (ease of H-abstraction) into equation 9, a two-parameter predictive model in the cytoprotection system could be derived based on the 26 compounds tested by step-wise regression as shown in equation 11.

$$EC_{50} = -0.45 \ (\pm 0.33) \log P + 0.25 \ (\pm 0.02) \ \Delta H_f - 25.75 \ (\pm 3.04) \qquad (11)$$

$$n = 26, \ r^2 = 0.86, \ p = 0.0000000001$$

The use of the new parameter (ΔH_f) increased the correlation coefficient r^2-value from 0.15 to 0.86. The significant increase in correlation coefficient upon the introduction of ΔH_f confirmed the importance of O-H bond strength (bond dissociation energy approximated by ΔH_f), which contributed most to the model. However, log P gave a negative contribution to the EC_{50} values in this model.

A QSAR model could also be derived by the introduction of ΔH_f to equation 9 for flavonoid aglycones through step-wise regression as shown in equation 12.

$$EC_{50} = 0.38 \ (\pm 0.65) \log P + 0.24 \ (\pm 0.03) \ \Delta H_f - 26.25 \ (\pm 4.04) \qquad (12)$$

$$n = 18, \ r^2 = 0.88, \ p = 0.000000072$$

Equation 12 indicated that log P gave a positive contribution to the EC_{50} values of flavonoid aglycones.

In conclusion, a QSAR model was derived from the cytoprotective activity and calculated theoretical parameters (enthalpy of hemolytic O-H bond cleavage ΔH_f and the partition coefficient).

It demonstrated that the H-abstraction was not the sole mechanism responsible for the cytoprotective activity of flavonoids. It seems that the relative contribution of lipophilicity (log P) is much smaller than that of ΔH_f.

These results demonstrated that it is feasible to estimate the cytoprotective activities of a flavonoid from the lipophilicity and the difference of heat of formations by using equation 11. The lipophilicity and heat of formation can be calculated purely by computer programs. Therefore, the QSAR model derived here could be useful in the selection of natural flavonoids with potential cytoprotective effects.

CONCLUSION

In summary, the cytoprotection assay provides information regarding cellular activity of antioxidants, which is important to our understanding of this area of antioxidant research. Traditionally, the antioxidant activity of phytochemicals has been measured using a range of chemically-defined laboratory-based assays. A cytoprotection assay that is a more biologically relevant method than the chemical antioxidant assays has been developed and can be adapted for use in other cell lines appropriate to tissues of interest. Using the cytoprotection assay, the effects of compounds on cells is determined, providing information regarding the cellular response to antioxidants, taking into account some aspects of uptake, metabolism, location of antioxidant compounds within cells and intracellular effects on signalling pathways and enzyme activity.

In the present study we showed that with a cell-based bioassay it is possible to identify natural-occurring flavonoids that are gastroprotective in a model of oxidant injury. By directly evaluating the effects of different classes of flavonoids using a lower dose and more chronic H_2O_2 exposure, we showed that removal of excess ROS or suppression of their generation by flavonoids may be effective in preventing oxidative cell death.

In this study, we carried out a theoretical investigation into the possible mechanisms governing the cytoprotective activity of 24 different subclasses of flavonoids by computational chemistry, and explored the correlation between experimentally determined cytoprotective activities and physicochemical properties.

It is reasonable to conclude that multiple mechanisms regulate the protective actions of flavonoid compounds although they contribute to the cytoprotective activity to different degrees. The cytoprotective activities of flavonoids were strongly correlated to their calculated enthalpy of hemolytic O-H bond cleavageΔH_f but weakly correlated to their lipophilicity. It is concluded that the relative contribution of lipophilicity (log P) is much smaller than that of ΔH_f to their cytoprotective capacity. However, the balance of lipophilicity and lipophobicity is still critical in determining their abilities to protect human cells from oxidative damage.

Judging from the improvement in the correlation coefficient in the stepwise multiple-linear regression, we can conclude that the more precise the mechanistic information included in the QSAR model, the better the coefficient of relation that is obtained.

It is reasonable to conclude that multiple mechanisms regulate the cytoprotection actions of flavonoids although they contribute to cytoprotective activity to different degrees. These results suggest the possibility of predicting the degree of contribution of different physicochemical factors among flavonoids by their *in vitro* actions against oxidative stress-induced cellular damage.

ACKNOWLEDGMENTS

This work was funded by the Foundation for Research Science and Technology Wellness Food Programme Contract C06X0405. We thank Dr. Tony McGhie and Dr. Jeffery Greenwood for critically reviewing the manuscript.

REFERENCES

Agullo, G; Gamet-Payrastre, L; Manenti, S; Viala, C; Rémésy, C; Chap, H; Payrastre, B. Relationship between flavonoid structure and inhibition of phosphatidylinositol 3-kinase: A comparison with tyrosine kinase and protein kinase C inhibition. *Biochemical Pharmacology,* (1997) 53, 1649-1657.

Andree, HA; Reutelingsperger, CP; Hauptmann, R; Hemker, HC; Hermens, WT; Willems, GM. Binding of vascular anticoagulant alpha (VAC alpha) to planar phospholipid bilayers. *Journal of Biological Chemistry,* (1990) 265, 4923-4928.

Antunes, F; Cadenas, E. Cellular titration of apoptosis with steady state concentrations of H_2O_2: submicromolar levels of H_2O_2 induce apoptosis through fenton chemistry independent of the cellular thiol state. *Free Radical Biology and Medicine,* (2001) 30, 1008-1018.

Arora, A; Nair, MG; Strasburg, GM. Structure-activity relationships for antioxidant activities of a series of flavonoids in a liposomal system. *Free Radical Biology and Medicine,* (1998) 24, 1355-1363.

Awad, HM; Boersma, MG; Boeren, S; van Bladeren, PJ; Vervoort, J; Rietjens, IM. Structure-activity study on the quinone/quinone methide chemistry of flavonoids. *Chemical Research in Toxicology,* (2001) 14, 398-408.

Awad, HM; Boersma, MG; Boeren, S; Van Bladeren, PJ; Vervoort, J; Rietjens, IM. Quenching of quercetin quinone/quinone methides by different thiolate scavengers: Stability and reversibility of conjugate formation. *Chemical Research in Toxicology,* (2003) 16, 822-831.

Bakalbassis, EG; Chatzopoulou, A; Melissas, VS; Tsimidou, M; Tsolaki, M; Vafiadis, A. Ab initio and density functional theory studies for the explanation of the antioxidant activity of certain phenolic acids. *Lipids,* (2001) 36, 181-190.

Barbouti, A; Doulias, PT; Nousis, L; Tenopoulou, M; Galaris, D. DNA damage and apoptosis in hydrogen peroxide-exposed Jurkat cells: bolus addition versus continuous generation of H_2O_2. *Free Radical Biology and Medicine,* (2002) 33, 691-702.

Barratt, MD. Integrating computer prediction systems with *in vitro* methods towards a better understanding of toxicology. *Toxicology Letters,* (1998) 102-103, 617-621.

Behl, C; Davis, JB; Lesley, R; Schubert, D. Hydrogen peroxide mediates amyloid ¯ protein toxicity. *Cell,* (1994) 77, 817-827.

Berkowitz, J; Ellison, GB; Gutman, D. Three methods to measure RH bond energies. *Journal of Physical Chemistry,* (1994) 98, 2744-2765.

Beyeler, S; Testa, B; Perrissoud, D. Flavonoids as inhibitors of rat liver monooxygenase activities. *Biochemical Pharmacology,* (1988) 37, 1971-1979.

Bodor, N; Gabanyi, Z; Wong, CK. A new method for the estimation of partition coefficient. *Journal of the American Chemical Society,* (1989) 111, 3783-3786.

Boersma, MG; Vervoort, J; Szymusiak, H; Lemanska, K; Tyrakowska, B; Cenas, N; Segura Aguilar, J; Rietjens, IM. Regioselectivity and reversibility of the glutathione conjugation of quercetin quinone methide. *Chemical Research in Toxicology,* (2000) 13, 185-191.

Bordwell, FG; Zhang, X-M. From equilibrium acidities to radical stabilization energies. *Account in Chemical Research,* (1993) 26, 510-517.

Borges dos Santos, RM; Simoes, JAM. Energetics of the O-H bond in phenol and substituted phenols: A critical evaluation of literature data. *Journal of Physical Chemistry Reference Data,* (1998) 27, 707-739.

Borkowski, T; Szymusiak, H; Gliszczynska-Swiglo, A; Tyrakowska, B. The effect of 3-O-[beta]-glucosylation on structural transformations of anthocyanidins. *Food Research International,* (2005) 38, 1031-1037.

Bors, W; Heller, W; Michel, C; Saran, M. Radical chemistry of flavonoid antioxidants. *Advances in Experimental Medicine and Biology,* (1990) 264, 165-170.

Brigati, G; Lucarini, M; Mugnaini, V; Pedulli, GF. Determination of the substituent effect on the O-H bond dissociation enthalpies of phenolic antioxidants by the EPR radical equilibration technique. *Journal of Organic Chemistry,* (2002) 67, 4828-4832.

Briviba, K; Devasagayam, TP; Sies, H; Steenken, S. Selective para hydroxylation of phenol and aniline by singlet molecular oxygen. *Chemical Research in Toxicology* (1993) 6, 548-553.

Brouillard, R. The in vivo expression of anthocyanin colour in plants. *Phytochemistry,* (1983) 22, 1311-1323.

Brouillard, R; Delaporte, B. Chemistry of anthocyanin pigments. 2. Kinetic and thermodynamic study of proton transfer, hydration, and tautomeric reactions of malvidin 3-glucoside. *Journal of the American Chemical Society,* (1977) 99, 8461-8468.

Brouillard, R; Dubois, J-E. Mechanism of the structural transformations of anthocyanins in acidic media. *Journal of the American Chemical Society,* (1977) 99, 1359-1364.

Buchwald, P; Bodor, N. Octanol-water partition of nonzwitterionic peptides: predictive power of a molecular size-based model. *Proteins,* (1998) 30, 86-99.

Cantoni, O; Cattabeni, F; Stocchi, V; Meyn, RE; Cerutti, P; Murray, D. Hydrogen peroxide insult in cultured mammalian cells: Relationships between DNA single-strand breakage, poly(ADP-ribose) metabolism and cell killing. *Biochimica et Biophysica Acta,* (1989) 1014, 1-7.

Car, R. Introduction to density-functional theory and *ab-Initio* molecular mynamics. *Quantitative Structure-Activity Relationships,* (2002) 21, 97-104.

Castle, L; Perkins, MJ. Inhibition kinetics of chain-breaking phenolic antioxidants in SDS micelles. Evidence that intermicellar diffusion rates may be rate-limiting for hydrophobic inhibitors such as -tocopherol. *Journal of the American Chemical Society,* (1986) 108, 6382-6384.

Chen, JW; Zhu, ZQ; Hu, TX; Zhu, DY. Structure-activity relationship of natural flavonoids in hydroxyl radical-scavenging effects. *Acta Pharmacologica Sinica,* (2002) 23, 667-672.

Chipman, DM; Liu, R; Zhou, X; Pulay, P. Structure and fundamental vibrations of phenoxyl radical. *Journal of Chemical Physics,* (1994) 100, 5023-5035.

Choi, JS; Young, CH; Sik, KS; Jung, JM; Won, KJ; Kyung, NJ; Ah, JH. The structure-activity relationship of flavonoids as scavengers of peroxynitrite. *Phytotherapy Research,* (2002) 16, 232-235.

Colic, M; Pavelic, K. Molecular mechanisms of anticancer activity of natural dietetic products. *Journal of Molecular Medicine,* (2000) 78, 333-336.

Convard, T; Dubost, JP; Le Solleu, H; Kummer, E. SmilogP: A program for a fast evaluation of theoretical log P from smiles code of a molecule. *Quantitative Structure-Activity Relationships,* (1994) 13, 34-37.

Cos, P; Ying, L; Calomme, M; Hu, JP; Cimanga, K; Van Poel, B; Pieters, L; Vlietinck, AJ; Berghe, DV. Structure-Activity Relationship and Classification of Flavonoids as Inhibitors of Xanthine Oxidase and Superoxide Scavengers. *Journal of Natural Products,* (1998) 61, 71-76.

Darzynkiewicz, Z; Juan, G; Li, X; Gorczyca, W; Murakami, T; Traganos, F. Cytometry in cell necrobiology: analysis of apoptosis and accidental cell death (necrosis). *Cytometry,* (1997) 27, 1-20.

Davies, KJ. The broad spectrum of responses to oxidants in proliferating cells: a new paradigm for oxidative stress. *IUBMB Life,* (1999) 48, 41-47.

Denisov, ET; Khydyakov, IV. Mechanisms of action and reactivities of the free radicals of inhibitors. *Chem Rev,* (1987) 87, 1313-1357.

Devaux, PF. Static and dynamic lipid asymmetry in cell membranes. *Biochemistry,* (1991) 30, 1163-1173.

Dewar, MJS; Thiel, W. Ground states of molecules. 38. The MNDO-method approximations and parameters. *Journal of the American Chemical Society,* (1977) 99, 4899-4907.

Dewar, MJS; Zoebisch, EG; Healy, EF; Stewart, JJP. AM1: A new general purpose quantum mechanical molecular model. *Journal of the American Chemical Society,* (1985) 107, 3902-3909.

Diplock, AT; Charleux, JL; Crozier-Willi, G; Kok, FJ; Rice-Evans, C; Roberfroid, M; Stahl, W; Vina-Ribes, J. Functional food science and defence against reactive oxidative species. *British Journal of Nutrition,* (1998) 80 Suppl 1, S77-112.

Dziezak, JD. Preservatives: Antioxidants, the ultimate answer to oxidation. *Food Technology,* (1986) 40 (9), 94-106.

Eadsforth, CV; Moser, P. Assessment of reverse-phase chromatographic methods for determining partition coefficients. *Chemosphere,* (1983) 12, 1459-1475.

Eastwood, MA. Interaction of dietary antioxidants in vivo: how fruit and vegetables prevent disease? *Quarterly Journal of Medicine,* (1999) 92, 527-530.

Fossen, T; Andersen, ØM Spectroscopic techniques applied to flavonoids, in Andersen, VM; Markham, KR editor. *Flavonoids : Chemistry, biochemistry, and applications.* Boca Raton, FL: Taylor and Francis Group; 2006; pp 37-142.

Fujita, T; Iwasa, J; Hansch, C. A new substituent constant, T, derived from partition coefficients. *Journal of the American Chemical Society,* (1964) 86, 5175-5180.

Gardner, AM; Xu, FH; Fady, C; Jacoby, FJ; Duffey, DC; Tu, Y; Lichtenstein, A. Apoptotic vs. nonapoptotic cytotoxicity induced by hydrogen peroxide. *Free Radical Biology and Medicine,* (1997) 22, 73-83.

Ghose, AK; Crippen, GM. Atomic physicochemical parameters for three-dimensional-structure-directed quantitative structure-activity relationships. 2. Modeling dispersive and hydrophobic interactions. *Journal of Chemical Information and Computer Sciences,* (1987) 27, 21-35.

Gille, JJP; Joenje, H. Cell culture models for oxidative stress: Superoxide and hydrogen peroxide versus normobaric hyperoxia. *Mutation Research/DNAging,* (1992) 275, 405-414.

Giugliano, D. Dietary antioxidants for cardiovascular prevention. *Nutrition, Metabolism and Cardiovascular Diseases,* (2000) 10, 38-44.

Gotoh, N; Noguchi, N; Tsuchiya, J; Morita, K; Sakai, H; Shimasaki, H; Niki, E. Inhibition of oxidation of low density lipoprotein by vitamin E and related compounds. *Free Radical Research,* (1996) 24, 123-134.

Halliwell, B. Reactive oxygen species and the central nervous system. *Journal of Neurochemistry,* (1992) 59, 1609-1623.

Hansch, C; Gao, H. Comparative QSAR: Radical reactions of benzene derivatives in chemistry and biology. *Chemical Review,* (1997) 97, 2995-3060.

Hansch, C; McKarns, SC; Smith, CJ; Doolittle, DJ. Comparative QSAR evidence for a free-radical mechanism of phenol-induced toxicity. *Chemico-Biological Interactions,* (2000) 127, 61-72.

Hansch, C; Quinlan, JE; Lawrence, GL. The linear free energy relationship between partition coefficients and the aqueous solubility of organic liquids. *Journal of Organic Chemistry,* (1968) 33, 347-350.

Harborne, JB; Williams, CA. Advances in flavonoid research since 1992. *Phytochemistry,* (2000) 55, 481-504.

Havsteen, BH. The biochemistry and medical significance of the flavonoids. *Pharmacology and Therapeutics,* (2002) 96, 67-202.

Hedrickson, HP; Kaufman, AD; Lunte, CE. Electrochemistry of catechol-containing flavonoids. *Journal of Pharmaceutical and Biomedical Analysis* (1994) 12, 325-334.

Herrmann, KM. Flavonoids and flavones in food plants: A review. *Journal of Food Technology,* (1976) 11, 443-448.

Herrmann, KM. Occurrence and content of hydrocinnamic and hydrobenzoic acid compounds in foods. *Critical Review of Food Science and Nutrition,* (1989) 28, 315-347.

Hopia, A; Heinonen, M. Antioxidant activity of flavonol aglycones and their glycosides in methyl linoleate. *Journal of the American Chemical Society,* (1999) 76, 139-144.

Hoshino, T. An approximate estimate of self-association constants and the self-stacking conformation of Malvin quinonoidal bases studied by 1H NMR. *Phytochemistry,* (1991) 30, 2049-2055.

Hoshino, T; Goto, T. Effects of pH and concentration on the self-association of malvin quinonoidal base -- electronic and circular dichroic studies. *Tetrahedron Letters,* (1990) 31, 1593-1596.

Hoshino, T; Matsumoto, U; Goto, T. Self-association of some anthocyanins in neutral aqueous solution. *Phytochemistry,* (1981) 20, 1971-1976.

Hyslop, PA; Zhang, Z; Pearson, DV; Phebus, LA. Measurement of striatal H_2O_2 by microdialysis following global forebrain ischemia and reperfusion in the rat: Correlation with the cytotoxic potential of H_2O_2 *in vitro. Brain Research,* (1995) 671, 181-186.

Japertas, P; Didziapetris, R; Petrauskas, A. Fragmental methods in the design of new compounds. Applications of the advanced algorithm builder. *Quantitative Structure-Activity Relationships,* (2002) 21, 23-37.

Kim, DK; Cho, ES; Um, HD. Caspase-dependent and -independent events in apoptosis induced by hydrogen peroxide. *Experimental Cell Research,* (2000) 257, 82-88.

Klopman, G; Zhu, H. Estimation of the aqueous solubility of organic molecules by the group contribution approach. *Journal of Chemical Information and Computer Sciences,* (2001) 41, 439-445.

Kroon, PA; Clifford, MN; Crozier, A; Day, AJ; Donovan, JL; Manach, C; Williamson, G. How should we assess the effects of exposure to dietary polyphenols*in vitro*? *American Journal of Clinical Nutrition,* (2004) 80, 15-21.

Leo, A; Hansch, C; Elkins, D. Partition coefficients and their uses. *Chemical Review,* (1971) 71, 525-616.

Leo, AJ. Hydrophobic parameter: Measurement and calculation. *Methods in Enzymology,* (1991) 202, 544-591.

Leo, AJ. Calculating log P_{oct} from structures. *Chemical Review,* (1993) 93, 1281-1306.

Lien, EJ; Ren, S; Bui, H-H; Wang, R. Quantitative structure-activity relationship analysis of phenolic antioxidants. *Free Radical Biology and Medicine,* (1999) 26, 285-294.

Lipinski, CA; Lombardo, F; Dominy, BW; Feeney, PJ. Experimental and computational approaches to estimate solubility and permeability in drug discovery and development settings. *Advanced Drug Delivery Reviews,* (1997) 23, 3-25.

Lucarini, M; Pedrielli, P; Pedulli, GF. Bond dissociation energies of O-H bonds in substituted from equilibration studies. *Journal of Organic Chemistry,* (1996) 61, 9259-9263.

Lucarini, M; Pedulli, F. Bond dissociation enthalpy of ˜tocopherol and other phenolic antioxidants. *Journal of Organic Chemistry,* (1994) 59, 5063-5070.

MacFarland, JW. On the parabolic relationship between drug potency and hydrophobicity. *Journal of Medicinal Chemistry,* (1970) 13, 1192-1196.

Mahoney, LR; DaRooge, MA. Kinetic behavior and thermochemical properties of phenoxy radicals. *Journal of the American Chemical Society,* (1975) 97, 4722-4731.

Mannhold, R; Petrauskas, A. Substructure versus whole-molecule approaches for calculating log P. *QSAR and Combinatorial Science* (2003) 22, 466-475.

Manthey, JA. Biological properties of flavonoids pertaining to inflammation. *Microcirculation,* (2000) 7, S29-34.

Masuda, J; Nakamura, K; Kimura, A; Takagi, T; Fujiwara, H. Introduction of solvent-accessible surface area in the calculation of the hydrophobicity parameter log P from an atomistic approach. *Journal of Pharmaceutical Sciences,* (1997) 86, 57-63.

McMurry, J Describing a reaction: Bond dissociation energies., in McMurry, J editor. *McMurry Organic Chemistry.* Belmont, California: Brooks/Cole Publishing Company; 1992; pp 156-159.

Metodiewa, D; Jaiswal, AK; Cenas, N; Dickancaite, E; Segura-Aguilar, J. Quercetin may act as a cytotoxic prooxidant after its metabolic activation to semiquinone and quinoidal product. *Free Radical Biology and Medicine,* (1999) 26, 107-116.

Meylan, WM; Howard, PH. Atom/fragment contribution method for estimating octanol-water partition coefficients. *Journal of Pharmaceutical Sciences,* (1995) 84, 83-92.

Middleton, E, Jr.; Kandaswami, C; Theoharides, TC. The Effects of Plant Flavonoids on Mammalian Cells:Implications for Inflammation, Heart Disease, and Cancer. *Pharmacological Reviews,* (2000) 52, 673-751.

Mora, A; Paya, M; Rios, JL; Alcaraz, MJ. Structure-activity relationships of polymethoxyflavones and other flavonoids as inhibitors of non-enzymic lipid peroxidation. *Biochemical Pharmacology,* (1990) 40, 793-797.

Moridani, MY; Galati, G; O'Brien, PJ. Comparative quantitative structure toxicity relationships for flavonoids evaluated in isolated rat hepatocytes and HeLa tumor cells. *Chemico-Biological Interactions,* (2002) 139, 251-264.

Mulder, P; Saastad, OW; Griller, D. O-H bond dissociation energies in *para*-substituted phenols. *Journal of the American Chemical Society,* (1988) 110, 4090-4092.

Murakami, S; Muramatsu, M; Tomisawa, K. Inhibition of gastric H+,K+-ATPase by flavonoids: A structure-activity study. *Journal of Enzyme Inhibition,* (1999) 14, 151-166.

Noguchi, N; Okimoto, Y; Tsuchiya, J; Cynshi, O; Kodama, T; Niki, E. Inhibition of oxidation of low-density lipoprotein by a novel antioxidant, BO-653, prepared by theoretical design. *Archives of Biochemistry and Biophysics,* (1997) 347, 141-147.

Osterberg, T; Norinder, U. Prediction of drug transport processes using simple parameters and PLS statistics: The use of ACD/logP and ACD/ChemSketch descriptors. *European Journal of Pharmaceutical Sciences,* (2001) 12, 327-337.

Packer, L; Rimbach, G; Virgili, F. Antioxidant activity and biologic properties of a procyanidin-rich extract from pine (Pinus maritima) bark, pycnogenol. *Free Radical Biology and Medicine,* (1999) 27, 704-724.

Pople, JA; Beveridge, DL; Doboshlc, PA. Molecular orbital theory of the electronic structure of organic compounds. 11. Spin densities in paramagnetic species. *Journal of the American Chemical Society,* (1968) 90, 4201-4209.

Porter, WL; Black, ED; Drolet, AM. Use of polyamide oxidative fluorescence test on lipid emulsions: Contrast in relative effectiveness of antioxidants in bulk versus dispersed systems. *Journal of Agricultural and Food Chemistry,* (1989) 37, 615-624.

Qin, Y; Wheeler, RA. Density-functional methods give accurate vibrational frequencies and spin densities for phenoxyl radical. *Journal of Chemical Physics,* (1995) 102, 1689-1698.

Riemersma, RA; Rice-Evans, CA; Tyrrell, RM; Clifford, MN; Lean, MEJ. Tea flavonoids and cardiovascular health. *Quarterly Journal of Medicine,* (2001) 94, 277-282.

Schwarz, K; Frankel, EN; German, JB. Partition behaviour of antioxidative phenolic compounds in heterophasic systems. *Lipid -Fett,* (1996) 98, 115-121.

Simonian, NA; Coyle, JT. Oxidative stress in neurodegenerative diseases. *Annual Review of Pharmacology and Toxicology,* (1996) 36, 83-106.

Soffers, AEMF; Boersma, MG; Vaes, WHJ; Vervoort, J; Tyrakowska, B; Hermens, JLM; Rietjens, IMCM. Computer-modeling-based QSARs for analyzing experimental data on biotransformation and toxicity. *Toxicol In Vitro,* (2001) 15, 539-551.

Stack, D Phenolic metabolism., in Dey, PM; Harborne, JB editor. *Plant Biochemistry.* London, UK.: Academic Press; 1997; pp 387-417.

Stewart, JPJ. Optimization of parameters for semiempirical methods. II. Applications. *Journal of Computational Chemistry,* (1989) 10, 221-264.

Sugihara, N; Kaneko, A; Furuno, K. Oxidation of flavonoids which promote DNA degradation induced by bleomycin-Fe complex. *Biological and Pharmaceutical Bulletin,* (2003) 26, 1108-1114.

Sun, YM; Zhang, HY; Chen, DZ; Liu, CB. Theoretical elucidation on the antioxidant mechanism of curcumin: A DFT study. *Organic Letters,* (2002) 4, 2909-2911.

Timberlake, CF; Bridle, P. Flavylium salts, anthocyanidins and anthocyanins. I. - Structural transformations in acid solutions. *Journal of the Science of Food and Agriculture,* (1967) 18, 473-478.

Tournaire, C; Croux, S; Maurette, MT; Beck, I; Hocquaux, M; Braun, AM; Oliveros, E. Antioxidant activity of flavonoids: efficiency of singlet oxygen (1 delta g) quenching. *Journal of Photochemistry and Photobiology B,* (1993) 19, 205-215.

van Acker, SA; de Groot, MJ; van den Berg, DJ; Tromp, MN; Donne-Op den Kelder, G; van der Vijgh, WJ; Bast, A. A quantum chemical explanation of the antioxidant activity of flavonoids. *Chemical Research in Toxicology,* (1996) 9, 1305-1312.

van Acker, SA; Koymans, LM; Bast, A. Molecular pharmacology of vitamin E: Structural aspects of antioxidant activity. *Free Radical Biology and Medicine,* (1993) 15, 311-328.

van Engeland, M; Nieland, LJ; Ramaekers, FC; Schutte, B; Reutelingsperger, CP. Annexin V-affinity assay: a review on an apoptosis detection system based on phosphatidylserine exposure. *Cytometry,* (1998) 31, 1-9.

Vaya, J; Mahmood, S; Goldblum, A; Aviram, M; Volkova, N; Shaalan, A; Musa, R; Tamir, S. Inhibition of LDL oxidation by flavonoids in relation to their structure and calculated enthalpy. *Phytochemistry,* (2003) 62, 89-99.

Vedernikova, I, Tollenaere, JP; Haemers, A. Quantum mechanical evaluation of the anodic oxidation of phenolic compounds. *Journal of Physical Organic Chemistry,* (1999) 12, 144-150.

Vermes, I; Haanen, C; Reutelingsperger, C. Flow cytometry of apoptotic cell death. *Journal of Immunological Methods,* (2000) 243, 167-190.

Ververidis, F; Trantas, E; Carl, D; Guenter, V; Georg, K; Panopoulos, N. Biotechnology of flavonoids and other phenylpropanoid-derived natural products. Part I: Chemical diversity, impacts on plant biology and human health. *Biotechnology Journal,* (2007) 2, 1214-1234.

Wang, R; Fu, Y; Lai, L. A new atom-additive method for calculating partition coefficients. *Journal of Chemical Information and Computer Sciences,* (1997) 37, 615-621.

Wayner, DD; Lusztyk, JE; Page, D; Ingold, KU; Mulder, P; Laarhoven, LJJ; Aldrichs, HS. Effects of solvation on the enthalpies of reaction of *tert*-butoxyl radicals with phenol and on the calculated O-H bond strength in phenol. *Journal of the American Chemical Society,* (1995) 117, 8738-8744.

Wayner, DDM; Parker, VD. Bond energies in solution from electrode potentials and thermochemical cycles. A simplified and general approach. *Account in Chemical Research,* (1993) 26, 287-294.

Wedworth, SM; Lynch, S. Dietary flavonoids in atherosclerosis prevention. *The Annals of Pharmacotherapy,* (1995) 29, 627-628.

Winkel-Shirley, B. Flavonoid Biosynthesis. A Colorful Model for Genetics, Biochemistry, Cell Biology, and Biotechnology. *Plant Physiology,* (2001) 126, 485-493.

Wright, JS; Carpenter, DJ; McKay, DJ; Ingold, KU. Theoretical calculation of substituent effects on the O-H bond strength of phenolic antioxidants related to vitamin E. *Journal of the American Chemical Society,* (1997) 119, 4245-4252.

Wright, JS; Johnson, ER; DiLabio, GA. Predicting the activity of phenolic antioxidants: Theoretical method, analysis of substituent effects, and application to major families of antioxidants. *Journal of the American Chemical Society,* (2001) 123, 1173-1183.

Yong, DC (2001) *Computational chemistry: A practical guide for applying techniques to real-world problems.* Wiley Interscience, New York.

Zhang, H-Y; You-Min, S; Xiu-Li, W. Substituent Effects on O-H Bond Dissociation Enthalpies and Ionization Potentials of Catechols: A DFT Study and Its Implications in the Rational Design of Phenolic Antioxidants and Elucidation of Structure-Activity Relationships for Flavonoid Antioxidants. *Chemistry - A European Journal,* (2003a) 9, 502-508.

Zhang, HY. Selection of theoretical parameter characterizing scavenging activity of antioxidants on free radicals. *Journal of the American Oil Chemists' Society* (1998) 75, 1705-1709.

Zhang, HY; Ge, N; Zhang, ZY. Theoretical elucidation of activity differences of five phenolic antioxidants. *Zhongguo Yao Li Xue Bao,* (1999) 20, 363-366.

Zhang, HY; Sun, YM; Wang, XL. Electronic effects on O-H proton dissociation energies of phenolic cation radicals: A DFT study. *Journal of Organic Chemistry,* (2002) 67, 2709-2712.

Zhang, HY; Wang, LF. Theoretical elucidation on structure-antioxidant activity relationships for indolinonic hydroxylamines. *Bioorganic and Medicinal Chemistry Letters,* (2002) 12, 225-227.

Zhang, HY; Wang, LF; Sun, YM. Why B-ring is the active center for genistein to scavenge peroxyl radical: A DFT study. *Bioorganic and Medicinal Chemistry Letters,* (2003b) 13, 909-911.

Zhang, J; Stanley, RA; Melton, LD. Lipid peroxidation inhibition capacity assay for antioxidants based on liposomal membranes. *Molecular Nutrition and Food Research,* (2006) 50, 714-724.

Ziegler, T. Approximate density functional theory as a practical tool in molecular energetics and dynamics. *Chemical Review,* (1991) 91, 651-667.

In: Biological Aspects of Human Health and Well-Being
Editor: Tsisana Shartava

ISBN: 978-1-61209-134-1
© 2011 Nova Science Publishers, Inc.

Chapter VI

SIALYLATION MECHANISM IN BACTERIA: FOCUSED ON CMP-*N*-ACETYLNEURAMINIC ACID SYNTHETASES AND SIALYLTRANSFERASES

Takeshi Yamamoto[*]

Glycotechnology Business Unit, Japan Tobacco Inc.
Higashibara 700, Iwata, Shizuoka438-0802, Japan

ABSTRACT

Sialic acidsare important components of carbohydrate chains and are linked to terminal positions of the carbohydrate moiety of glycoconjugates, including glycoproteins and glycolipids.Various studies have focused onclarifying the structure–function relationship of sialic aicds and have revealed that *N*-acetylneuraminic acid (Neu5Ac) is the major sialic acids component of glycoconjugates,and that the sialylated carbohydrate chains of glycoconjugates play significant roles in many biological processes, including immunological responses, viral infections, cell–cell recognition,and inflammation.

Sialylated glycoconjugates are formed by specific sialyltransferases in the cell. All sialyltransferases use cytidinemonophosphate *N*-acetylneuraminic acid(CMP-Neu5Ac) as the common donor substrate. Up to the present, sialyltransferases have been cloned from various sources, including mammalian organs, bacteria and virus. As to the sialyltransferases, all of the sialyltransferases have been classified into five families in the CAZy (carbohydrate-active enzymes) database (family29, 38, 42, 52 and 80), and all of the marine bacterial sialyltransferases are classified into the family 80.

Generally, the enzymes with a bacterial origin are more stable and productive in *Escherichia coli*protein expression systems than the mammalian-derived enzymes. In addition, the bacterial-derived sialyltransferases show broader acceptor substrate specificity than the mammalian enzymes. These advantages highlight the capacity of bacterial enzymes as efficient tools for the *in vitro* enzymatic synthesis of sialosides.

The recent increase in research focusing on sialyltransferases from a diverse range of bacteria has led to the identification of many bacterial sialyltransferases. Several bacterial

---

[*]Tel: +81-538-32-7389; Fax: +81-538-33-6046; E-mail: takeshi.yamamoto@jt.com

CMP-Neu5Ac synthetases have also recently been identified. This articlereviews the bacterial CMP-Neu5Ac synthetases and sialyltransferasesthat show promise as tools for the productionof sialosides.

INTRODUCTION

Sialic acids (Sia)sare a family of monosaccharides comprising over 50 naturally occurring derivatives of neuraminic acid, 5-amino-3,5-dideoxy-D-*glycero*-D-*galacto*-2-nonulosonic acid (Neu) [1,2]. Structurally, the Sia derivatives of Neu carry a variety of substitutions at the amino and/or hydroxyl groups. The amino acid group is often acetylated, glycolylated,or deaminated. The hydroxyl groups can be acetylated at O7, O8, or O9, singly or in combination [3,4], and can also be modified by acetate, lactate, phosphate or sulfate esters.

The three major members of the Sia group are *N*-acetylneuraminic acid (Neu5Ac), *N*-glycolylneuraminic acid (Neu5Gc), and 2-keto-3-deoxy-D-*glycero*-D-*galacto*-nonulosonic acid (KDN) [1,3]. Although, Sias are widely distributed in higher animals and some microorganisms, only Neu5Ac is ubiquitous, and Neu5Gc is not found in bacteria [3]. Usually, Sias exist in the carbohydrate moiety of glycoconjugates, such as glycoproteins and glycolipids, and are linked to the terminal positions of the carbohydrate chains of the glycoconjugates [5]. Sialylated carbohydrate chains play important roles in many biological processes, including immunological responses, viral infections, cell–cell recognition,and inflammation[6–9]. The structures of typical sialic acids are shown in Figure 1.

Therelationship between the structure and function of sialylated carbohydrate chains is demonstrated by influenza virus infections. It is well known that the influenza virus binds to cell receptorsof host cellsvia Neu5Ac-linked glycoproteins or glycolipids through viral hemagglutininrecognition of host cell Neu5Ac[10, 11].

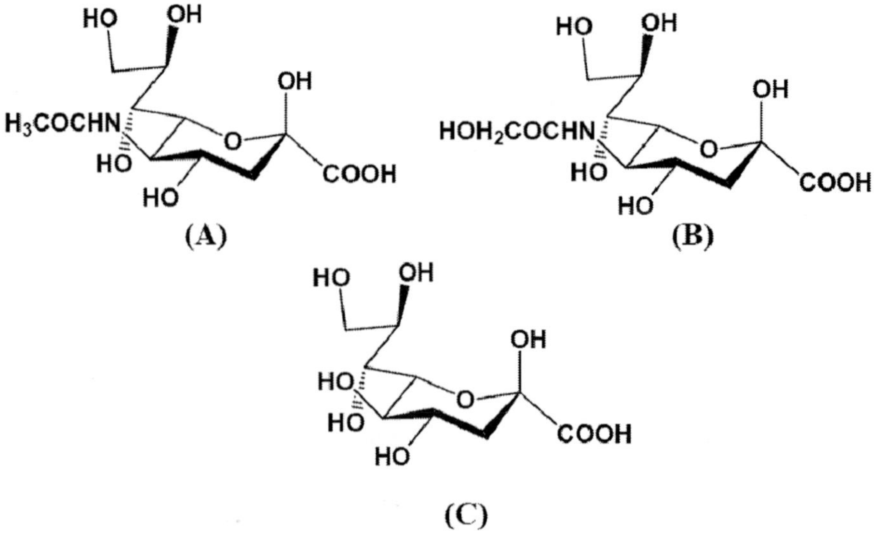

Figure 1. The structure of three major sialic acids.(A) *N*-acetylneuraminic acid (Neu5Ac); (B) *N*-glycolylneuraminic acid (Neu5Gc); (C) 2-keto-3-deoxy-D-glycero-D-galacto-noninic acid (KDN).

The influenza virus also recognizes the carbohydrate chain structure of the host cell [10, 11]. For example, the avian influenza virusesrecognize Neu5Acα2-3Galβ1-3/4GlcNAc structures, and the human influenza viruses recognize Neu5Acα2-6Galβ1-3/4GlcNAc structures [12, 13]. The host cell specificity of the influenza viruses is determined by the linkage of Neu5Ac to thegalactose residues and by the number of sialic acid residues and core structures [10].For this reason, the distribution of Neu5Ac on the host cell surface is an important determinant of host tropism. An understanding of the mechanism underlying the control of Neu5Ac expression in human cells and tissues is essential for understanding the process of influenza pathogenesis.

Another example is the Guillain-Barré syndrome (GBS). GBS is theprototypic postinfectious autoimmune disease [14]. Epidemiological studies clearly demonstrate that GBS patients develop the syndrome following infection by the Gram-negative bacterium*Campylobacter jejuni* [15]. GBS patients develop neuropathy, which is the most common cause of generalized paralysis [16]. The most common *C. jejuni* serotype associated with GBS is *C. jejuni* O:19 [17]. The serotypes of *C. jejuni* associated with GBS-derived neuropathies express ganglioside-like lipooligosaccharide (LOS) structureson the cell surface. Detailed investigations of LOS structures isolated from GBS patients reveal that the core oligosaccharides mimic gangliosides located in neural tissue. Terminal oligosaccharide moieties identical to those of the gangliosides GM1, GD3, and GT1a have been found in *C. jejuni* O:19 strains [18–21]. The development of GBS in patients following infection with *C. jejuni* is thought to be related to the molecular mimicry of gangliosides andthe cross-reactionof antibodies against LOS, leading to neuropathy. The molecular mimicry basis of GBS is consistent with the finding thatmost GBS patients develop autoantibodiesthat react with GM1 subsequent to *C. jejuni* enteritis [22, 23]. Molecular mimicry has been proposed as a pathogenic mechanism underlying autoimmune disease [24].It has also been demonstrated in the development of Fisher syndrome after infection with *Haemophilus influenzae* in the production of anti-GQ1b autoantibodies mediated by the GQ1b-mimicking LOS on the bacterial surface [25].

An abundant supply of sialylated oligosaccharides is essential for a detailedinvestigation of the biological function of sialylation. Chemical and enzymatic glycosylationhave the two major routes for the preparation of sialylatedoligosaccharides. Although chemical glycosylation has the advantage of high flexibility and wide applicability, the reaction processes are complicated, as the chemical reactionsoften require multiple protection and de-protection steps [26–28]. By comparison, enzymaticglycosylation using glycosyltransferases is a single-step process with high positional and anomerselectivity and high reaction yield. For example, with sialylation, the transfer of Neu5Ac by sialyltransferases to the appropriate substrate from CMP-Neu5Ac as the donor can be readily achieved in the final step under mild reaction conditions [29–31].

This articlereviews the pathways for biosynthesis of Neu5Ac and sialylatedglycoconjugatcs, and recent progress in the study of bacterial CMP-Neu5Ac synthetases and sialyltransferases.

Biosynthesis of Sialic Acids and Sialyl-Glycoconjugates

The bacterial biosynthesis of sialyl-glycoconjugates involves several steps, including the synthesis of the sugar nucleotide CMP-Neu5Ac from Neu5Ac, a typical sialic acid,by CMP-Neu5Ac synthetase, and the subsequent transfer of Neu5Ac from this donor substrateto appropriate acceptor substrates by sialyltransferase [1].

The biosynthesis of Neu5Ac in bacteria first involves the conversion ofN-acetylglucosamine 6-phosphate to N-acetylmannosamine 6-phosphate by N-acylglucosamine 6-phosphate 2-epimerase. N-acetylmannosamine 6-phosphate is then dephosphorylated to N-acetylmannosamine, which is then conjugated to phospho*enol*pyruvateto form Neu5Ac by Neu5Ac synthase. In an alternative pathway in bacteria, Neu5Ac is generated from N-acetylmannosamine and pyruvate byN-acetylneuraminate lyase [32]. Neu5Ac isconjugated to cytidine triphosphate (CTP) by CMP-Neu5Ac synthetase to formthe activated donor substrate, CMP-Neu5Ac. Neu5Ac is transferred from CMP-Neu5Ac to the appropriate acceptor substrates by sialyltransferase. A schematic diagram of the biosynthesis of CMP-Neu5Ac in bacteria is shown in Figure 2.

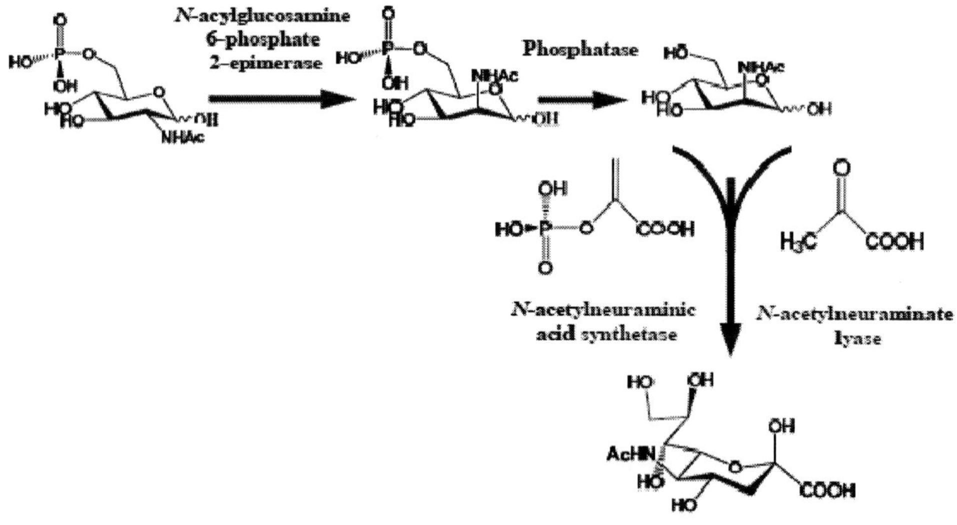

Figure 2. The pathway of Neu5Ac biosynthesis in bacteria. As described in the text, Neu5Ac is synthesized from N-acetylglucosamine 6-phosphate in 3 enzymatic reaction steps.

CMP-Neu5Ac Synthetase

CMP-Neu5Ac is a common donor substrate for all known bacterial and mammalian sialyltransferases. CMP-Neu5Ac is produced by CMP-Neu5Ac synthetase (EC 2.7.7.43;also known asN-acylneuraminate cytidyltransferase, CMP-sialate pyrophosphorylase, and acylneuraminate cytidyltransferase) from CTP and Neu5Ac as described in the previous section. CMP-Neu5Ac synthetase is considered an important enzyme in the field of glycobiology, as an abundant supply of CMP-Neu5Ac is essential for the synthesis of biologically important sialylated oligosaccharides by means of sialyltransferase [1]. For example, sialylated oligosaccharides are used in the development of clinical medicines and in the study of many biochemical processes [29].

CMP-Neu5Ac synthetases are obtained from several sources [33–35]. Those of bacterial and mammalian origin have many enzymatic and protein propertiesin common [36]. For example, all the known CMP-Neu5Acsynthetases display maximum enzymatic activity under basic conditions, at around pH 8.5, and require divalent cations for activity. The enzymatic reaction mechanisms for CMP-Neu5Acsynthetaseare well characterized [37].

In the presence of a divalent cation, CMP-Neu5Acsynthetase catalyzes the nucleophilic attack of the alpha-phosphate of cytidine triphosphate (CTP) by the anomeric oxygen of beta-Neu5Ac to produce pyrophosphate and CMP-Neu5Ac. Molecular cloning reveals the presence of five conserved regions, known as motifs I to V, among the known CMP-Neu5Acsynthetases [38].

Motifs I to III are also conserved in the CMP-3-deoxy-D-manno-octulosonate (KDO) synthetases (EC 2.7.7.38) [38]. Motif I is involved in nucleotide binding in the CMP-Neu5Ac synthetases and CMP-KDO synthetases [39]. KDO is an eight-carbon monosaccharidethat is an essential component of cell wall capsular polysaccharides in Gram-negative bacteria [38]. KDO is activated by CMP to generate CMP-KDO [40], which is the donor substrateof KDO transferase [41, 42]. KDO and Neu5Ac are unique,as the activated forms are nucleoside monophosphate diesters, rather than the more common nucleoside diphosphate diesters.

The CMP-Neu5Acsynthetases and CMP-KDO synthetases are highly specific for their respective sugar substrates: CMP-KDO synthetases do not recognize sialic acid as a substrate, and CMP-Neu5Acsynthetases do not recognize KDO as a substrate [43].

While the CMP-Neu5Ac synthetases have similar enzymatic features, several obvious differences have been demonstratedbetween mammalian- and bacterial-derived enzymes, including differences in substrate specificity and tertiary structure [44].

Origin of Bacterial CMP-Neu5Ac Synthetase

Genes encodingCMP-Neu5Ac synthetase have been cloned from several types of bacteria, including*Escherichia coli* [45], *Neisseria meningitides* [46], *Haemophilusducreyi* [47], *Streptococcus agalactiae* [48],and *Clostridium thermocellum*[48]. The CMP-Neu5Ac synthetase cloned from *C. thermocellum*, an anaerobic, Gram-positive, nonpathogenic, thermophilic bacterium, is thermostable. It is highly active at 50 °C and retains full activity following incubation for 24 h at 37 and 50 °C [44]. CMP-Neu5Ac synthetasederived from *C. thermocellum* is suggested to be a powerful tool for the preparation of CMP-Neu5Ac.

The crystal structure of bacterial CMP-Neu5Ac synthetase, obtained from *N. meningitidis*, has been determined in the presence and absence of the substrate analog cytidine diphosphate CDP [49]. The crystal structure of a mammalian equivalent, murine CMP-Neu5Ac synthetase, has also been resolved [50].

For reference, the crystal structure of CMP-KDO synthetase, obtained from *E.coli*, has been determined in the presence of substrates and substrate analogs [39].

Features of Bacterial CMP-Neu5Ac Synthetases

CMP-Neu5Ac synthetase has been purified to apparent homogeneity from the cytoplasmic fraction of *E. coli* O18:K1 [34], and the complete nucleotide sequence has been reported [44]. Several CMP-Neu5Ac synthetase genes have subsequently been cloned from various strains of *E. coli*. The characteristics of *E. coli* synthetase have been revealed by use ofthe native enzyme and several recombinant enzymes. *E. coli* synthetase requires divalent cationsfor enzyme activity [34]. Of the divalent cations tested, Mg^{2+} and Mn^{2+} are essential for enzyme activity,but Ca^{2+}, Cd^{2+}, and Zn^{2+} haveno effect. Maximum *E. coli* synthetase activity is observed with 20–40 mMMg^{2+} at a pHof 9.0–10.0 [34]. Although the crystal structure of *E. coli* CMP-Neu5Ac synthetasehas not yet been determined, site-directed mutagenesis and chemical modification studies have revealed the functional role of specific amino acids. For example, Arg12 may be located within the activesite of *E. coli* synthetase and play an important role in substrate binding [51]. Furthermore, the mutation of Cys129 to Ser enhanced the sensitivity of the enzyme to heat and chemical denaturation [52]. Recent evidence reveals that *E. coli* CMP-Neu5Ac synthetase is bifunctional, with both CMP-Neu5Ac synthetase and platelet-activating factor acetylhydrolase activity [53].Systems for the high expression of native and recombinant *E. coli*CMP-Neu5Ac synthetases have been established to generate functional levels of this enzyme. For example, over 25,000 units of wild-type CMP-Neu5Ac synthetase from *E. coli* K-235 has been produced by using optimized culture medium [54].

N. meningitidis CMP-Neu5Ac synthetasehas been cloned and expressedin *E. coli* [55, 56]. Recombinant *N. meningitidis* CMP-Neu5Ac synthetase has been expressed at very high levels of over 70,000 units of enzyme per liter of culture medium, and has been purified with a high yield [57]. Purified *N. meningitidis* CMP-Neu5Ac synthetase has been used in the gram-scale synthesis of CMP-Neu5Ac [57].

N. meningitidis and *E. coli* CMP-Neu5Ac synthetase have specific enzymatic propertiesin common. For example, both require divalent cationsfor activity and have an optimal pHof 9.0. Severaldifferences have also been demonstrated. Firstly, the substrate specificity of *N. meningitidis* CMP-Neu5Ac synthetase is broaderthan that of *E. coli* synthetase. *E. coli*CMP-Neu5Ac synthetase does not recognize *N*-glycolylneuraminic acid as a substrate, whereas *N. meningitidis* CMP-Neu5Ac synthetase uses*N*-glycolylneuraminic acid and *N*-acetylneuraminic acid as substrates [58]. Secondly, *N. meningitidis* CMP-Neu5Ac synthetase synthesizes several sugar nucleotides, including CMP-Neu5Ac and derivatives with a substitution at C5, C8, and C9 of Neu5Ac [56]. By comparison, *E. coli* synthetase accepts only a few C9-substituted derivatives of Neu5Ac as substrate [59]. CMP-Neu5Ac synthetase from *N. meningitidis* has been crystallized as a dimer, and structural information has revealed the amino acid residuescomprising the active site, the mononucleotide-binding pocket, the substrate-binding site, and the dimerization domain of this class of enzyme [49].

SIALYLTRANSFERASES

The known bacterial sialyltransferases do not contain the sialyl motifs found in mammalian sialyltransferases, despite the specificity of the bacterial sialyltransferasesfor

CMP-Neu5Ac, a donor substrate recognized by mammalian sialyltransferases [60].Two short motifs, referred to as theD/E-D/E-G and HP motifs, have been recently identified in the bacterial sialyltransferasesand are shown to be functionally important for enzyme catalysis and donor substrate binding [61]. These two motifs are structurally distinct from the motifs found inmammalian sialyltransferases.

Like the mammalian sialyltransferases, viral $\alpha2,3$-sialyltransferase, obtained from myxoma-virus–infected RK13 cells, contains the motifs L and S in its catalytic domain and is closely related to mammalian $\alpha2,3$-sialyltransferase (ST3Gal IV) on the basis of amino acidsequencesimilarity [62]. The substrate specificity of the viral and mammalian sialyltransferases is very different, despite the common presence of the sialylmotifs. Viral $\alpha2,3$-sialyltransferase transfers Neu5Ac to the fucosylated substrates Lewis[X] and Lewis[a] [62].

Origin and Classification of the Bacterial Sialyltransferases

Genes encoding sialyltransferases have been cloned from various types of bacteria, including *Neisseria gonorrhoeae* [63], *N. meningitidis*[64], *C. jejuni* [65], *E. coli*[66], *Photobacterium damselae*[67],*Photobacterium phosphoreum* [68], *Photobacterium leiognathi* [69], *Photobacterium* sp. [70], *Vibrio* sp. [71], *Pasteurella multocida*[72], *H. influenzae* [73], and *Streptococcus agalactiae* [74].All bacterial sialyltransferases are classified into one of four families in the CAZy (carbohydrate-active enzymes) database [75]: (1) glycosyltransferase (GT) family38 (polysialyltransferase from *E. coli* and *N. meningitidis*); (2) GT family 42 (lipooligosaccharide $\alpha2,3$-sialyltransferase and $\alpha2,3$-/$\alpha2,8$-sialyltransferase from *C. jejuni* and *H. influenzae*); (3) GT family 52 ($\alpha2,3$-sialyltransferase from *H. influenzae*, *N. gonorrhoeae*, and *N. meningitidis*); and (4) GT family 80 ($\alpha2,6$-sialyltransferase and $\alpha2,3$-/$\alpha2,6$-sialyltransferase from *Ph.damselae* and *Ph. multocida*).

Tertiary Structure of Bacterial Sialyltransferases

Despite the relatively few reports onbacterial sialyltransferases compared with those on mammalian sialyltransferases, the crystal structures of five bacterial sialyltransferases have been described. The first was that of the bifunctional enzyme $\alpha2,3$-/$\alpha2,8$-sialyltransferase (CstII) from *C. jejuni* OH4384, complexed with a substrate analog [76]. The second was that of the multifunctional enzyme $\alpha2,3$-sialyltransferase (Δ24PmST1) from *P. multocida* strain P-1059, in the presence and absence of CMP. This enzyme shows $\alpha2,3$-sialyltransferase activity, $\alpha2,6$-sialyltransferase activity, sialidase activity, and trans-sialidase activity. In addition to a description of the crystal structure of this enzyme incomplexes with both acceptor and donor substrate analogs, the substrate binding sites and catalytic mechanism have also been reported [77, 78]. The third structure described was that of monofunctional *α2,3*-sialyltransferase (CstI) from*C. jejuni* in apo- and substrate-analog–bound forms [79]. The fourth was that of monofunctional *α2,6*-sialyltransferase (Δ16pspST6) from*Photobacterium* sp. as a CMP donor product and lactose-bound complex[80].The fifth was that of $\alpha2,6$-sialyltransferase from *P. multocida*in a complex with CMP-3FNeuAc and lactose [81].

The overall structures of multifunctional α2,3-sialyltransferase (Δ24PmST1) and monofunctional *α*2,6-sialyltransferase (Δ16pspST6) are similar [80]. These enzymes are both classified in GT family 80 in the CAZy database. There is similarity in the overall architecture of bifunctional α2,3-/α2,8-sialyltransferase(CstII) and monofunctional *α*2,3-sialyltransferase (CstI),which are both classified in GT family 42. However, there is no structural similarity between the GT family 42 and GT family 80.

Reaction of GT Family 80 Sialyltransferases

In general, the mechanism of reaction of the inverting glycosyltransferases resemblesthat of the inverting glycosylhydrolases, as both possess an acidic amino acid that activates the acceptor hydroxyl group by deprotonation [82]. As mentioned previously, sialyltransferases use CMP-Neu5Ac (cytidine 5'-monophospho-β-D-*N*-acetylneuraminic acid) as a donor substrate and transfer Neu5Ac to an acceptor substrate through α-linkage. Therefore, sialyltransferases are classified into inverting glycosyltransferases. Superimposition of the substrate binding sites of Δ16pspST6 and Δ24PmST1 reveals the catalytic role of specific amino acid residues in the sialyltransferases [80]. Firstly, Asp232of Δ16pspST6, which corresponds to Asp141 of Δ24PmST1 in the primary amino acid sequence, may play a key role as the catalytic base in the transfer reaction [80]. Secondly, His405of Δ16pspST6, which corresponds to His311 of Δ24PmST1, may act as a catalytic residue by protonating phosphate oxygen as a catalytic acid [80].

In the crystal structure, CMP is located in a deep cleft between domains 2 and 3of Δ16pspST6 andΔ24PmST1, and nine amino acid residues (Gly361, Lys403, His405, Pro406, Ser430, Phe431, Ser449, Ser450m, and Leu451bb) interact with it. Moreover, all except Ser430 and Leu451 are conserved among the sialyltransferases in GT family 80 [83].

Lactose is recognized by five amino acid residues (Trp270, Asp141, His112, Asn85, and Met144) in Δ24PmST1. These residues make contact with only the galactose moiety of lactose. Three of the five inΔ24PmST1—Trp270, Asp141, and His112—are conserved in Δ16pspST6 as Trp365, Asp232, and His204, respectively, and occupystructurally identical positions [80].The main difference in the galactose binding site between Δ24PmST1 and Δ16pspST6 is the presence of Asn85 in Δ24PmST1 and His123 in Δ16pspST6. His123 in Δ16pspST6 lies within the hydrogen-bonding distance of O3 and O5 of the galactose binding site and is thought to determine the orientation of lactose [80]. Pro34, the corresponding residue in Δ24PmST1, cannot form a hydrogen bond with the galactose binding site. His123 of Δ16pspST6 is conserved among other α2,6-sialyltransferases belonging to GT family 80 [83]. However, in these α2,3-sialyltransferases, including those cloned from *P. phosphoreum*, *Photobacterium* sp., and *Vibrio* sp., the amino acids corresponding to the His123 of Δ16pspST6 are substituted for arginine. For this reason, the conserved amino acid residues of the sialyltransferases belonging to GT family 80 are thought to be involved in the linkage of sialic acid [80, 83].

Sialyltransferases Belong to GT Family 80

As described in the previous section, bacterial sialyltransferases are classified into four groups as GT families 38, 42, 52, and 80 in the CAZy database. Many bacterial sialyltransferases in GT family 80 show unique enzymatic characters. The primary structure of sialyltransferases belonging to the GT family 80 and conserved motifs are shown in Figure 3.

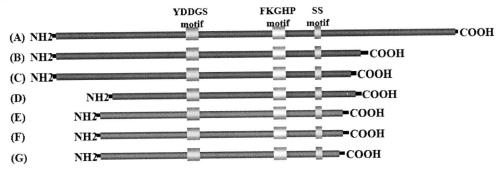

A   *P. damselae*β-galactoside α2,6-sialyltransferase (accession no. BAA25316),

B   *Photobacterium* sp. JT-ISH-224 β-galactoside α2,6-sialyltransferase (accession no.AB293985),

C   *P. leiognathi* JT-SHIZ-145 β-galactoside α2,6-sialyltransferase (accession no. AB306315),

D   *P. multocida* multi-functional sialyltransferase (accession No. AAY89061),

E   *P. phosphoreum* JT-ISH-467 α/β-galactoside α2,3-sialyltransferase (accession no. BAF63530),

F   *Photobacterium* sp. JT-ISH-224 α/β-galactoside α2,3-sialyltransferase (accession no. AB293984),

G   *Vibrio* sp. JT-FAJ-16 α/β-galactoside α2,3-sialyltransferase (accession no. AB308042).

Figure 3. Primary structure of sialyltransferases belonging to the GT family 80. As described in the text, there are three conserved motifs in these enzymes.

α2,6-SialyltransferaseObtained from *P. damselae*

The monofunctional α2,6-sialyltransferase produced by*P. damselae* JT0160 has unique acceptor substrate specificity compared with the equivalent mammalian enzymes.The *P. damselae* α2,6-sialyltransferase recognizes lactose and *N*-acetyllactosaminide as acceptor substrates with almost equal K_m values. By comparison, rat liver*α2,6*-sialyltransferase has a K_m value approximately 33 times higher for lactose than for *N*-acetyllactosaminide[84].These results indicate that *P. damselae* α2,6-sialyltransferase does not recognize the 2-acetamido group in the *N*-acetylglucosaminyl residue.Although fucosylated acceptors are not the native substrates of mammalian sialyltransferases, the *P. damselae* α2,6-sialyltransferase transfers Neu5Ac to the galactose residue of carbohydrate chains at position 6 in 2′-fucosyllactose and 3′-sialyllactose[85]. Furthermore, the *P. damselae* α2,6-sialyltransferasealso transfers Neu5Ac to both asialo-*N*-linked and asialo-*O*-linked glycoproteins[86].

P. damselae α2,6-sialyltransferase recognizes both CMP-Neu5Ac and CMP-KDNas donor substrates[87].It also recognizes many CMP-sialic acid derivativeswiththe non-natural modification of an azido or acetylene group at positions C5, C7, C8, and/or C9[88]. This feature highlights this enzyme as a valuable tool for the *in vitro* enzymatic synthesis of sialosides.

α2,6-SialyltransferaseObtained from P. leiognathi

The maximal enzyme activity ofmost sialyltransferases of mammalian and bacterial origin is achieved under acidic conditions, at around pH 6.0. By comparison, the maximal activity of monofunctional α2,6-sialyltransferase produced by *P.leiognathi*is achievedat pH 8 [69]. CMP-Neu5Ac (the common donor substrate) and the sialylglycoconjugates (the sialyltransferasereaction products) are both more stable under basic conditions than under acidic conditions. The high functionality of *P. leiognathi*-derivedα2,6-sialyltransferaseunder basic conditionsprovides a unique advantage of this enzymeforthe efficient production of sialosides.

An increasein the concentration of sodium chloride also increases the activity of*P. leiognathi*α2,6-sialyltransferase. Approximately 500 mM NaClin the reaction mixture is required formaximum activity of this enzyme. The conditions in the periplasm are thought to be similar to those of the environment in which the bacterium grows. The optimal conditions for the α2,6-sialyltransferase from *P. leiognathi* are very similar tothose of average seawater (pH 8.0, 500 mM NaCl). The enzyme of *P. leiognathi* JT-SHIZ-145,a Gram-negative marine bacterium, has a candidate signal peptide in the NH_2-terminal region of its deduced sequence. Thus, this enzyme is thought to be translocated across the cytoplasmic membrane to the periplasm [69].

α/β-Galactoside α2,3-sialyltransferase from Vibrio sp.

Theα/β-galactoside α2,3-sialyltransferase from *Vibrio* sp. has optimal enzyme activity at pH 5.5 and 20 °C.One of the most striking featuresof this enzyme is its similar specificity for both methyl-α-D-galactopyranoside and methyl-β-D-galactopyranoside [71]. Similar specificity has been reported for α 2,3-sialyltransferase from *Neisseria meningitidis*immunotype L1. Since the α2,3-sialyltransferase transfers Neu5Ac to both the α- and β-anomers of sugars, it can be described as an α/β-galactoside α2,3-sialyltransferase.

Among the monosaccharides and disaccharides, the preferred acceptor substrate of α/β-galactoside α2,3-sialyltransferase from *Vibrio* sp.is methyl-α-D-galactopyranoside.There is no significant difference in enzyme activity for lactose, *N*-acetyllactosamine, and methyl-β-D-galactopyranosyl-β-1,3-*N*-acetylglucosaminide. This finding suggests a lack of sensitivity to the second sugar at the non-reducing terminus and to the linkage between the terminal two sugars. Moreover, α/β-galactoside α2,3-sialyltransferase from *Vibrio* sp. shows twice the relative activity for methyl-α/β-D-galactopyranosides than for methyl-α/β-D-glucopyranosides. Together, these findings reveal that the affinity of this enzyme for the acceptor substrate is affected by the orientation of the hydroxyl group at the C4 position of the sugar. It has also been shown that the relative activity of α/β-galactoside α2,3-sialyltransferase from *Vibrio* sp. for methyl-α-D-galactopyranoside is twice that for *N*-acetylgalactosamine and methyl-β-D-mannopyranoside. So, the enzyme may be affected by the acetoamido group at the C2 position and by the orientation of the hydroxyl group at thatposition.This enzyme shows a wide range of acceptor substrate specificities andtransfersNeu5Ac to both *N*-linked and *O*-linked asialo-glycoprotein [71].

Multi-Functional α2,3 Sialyltransferase from P. multocida

This enzyme is a putative sialyltransferase encoded by the *Pm0188* gene from *P. multocida* genomic strain Pm70 and has been identified by a BLAST search using the amino acid sequence of a *P. damselae*α2,6-sialyltransferase as a probe[72]. Recombinant *P. multocida* sialyltransferase shows four functions: (1) α2,3-sialyltransferase activity that efficiently transfers a Neu5Ac from CMP-Neu5Ac to galactosides at a wide pH range (pH 6.0–10.0), with optimal activity at pH 7.5–9.0; (2) α2,6-sialyltransferase activity that efficiently transfers a Neu5Ac from CMP Neu5Ac to galactosides at pH 4.5–7.0; (3) *trans*-sialidase activity that transfers the Neu5Ac residue from α2,3-linked sialyl galactosides to another galactoside, with optimal activity at pH5.5–6.5; and (4) neuraminidase activity that specifically cleaves α2,3-sialyl linkages, with optimal activity at pH 5.0–5.5 [72].

P. multocida sialyltransferase shows flexible donor substrate specificity when functioning as a sialyltransferase, and accepts CMP-Neu5Ac (the common donor substrate of both mammalian and bacterial sialyltransferases), CMP-KDN, and CMP-Neu5Gc. By comparison, CMP-Neu5Gc and CMP-KDN are poor donor substrates for *N. meningitidis*α2,3-sialyltransferase, and *N. gonorrhoeae*α2,3-sialyltransferase is not able to recognize CMP-KDN as a donor substrate.

CONCLUSION

Known systemsfor the glycosylationof proteinshave historically been restricted to those in eukaryotes. Recent studies, however, have revealed that glycoproteins are expressed also in prokaryotes [89, 90], although the bacterial *N*-linked sugar chains differ structurally from their eukaryotic counterparts [91, 92]. Many bacterial enzymesinvolved in glycan synthesis have been reported, and in general, they are more stable and productive than theirmammalian equivalents. The efficient production of glycans and the modification of glycoconjugates through the use ofbacterial enzymes have become more widely applicable.Several methods for the large-scale preparation of oligosaccharides with bacterial enzymes have been reported [93, 94]. For example, the large-scale synthesis of important oligosaccharides, including the carbohydrate moieties of gangliosides GM1, GM2, and H-antigen, have been synthesized by using metabolically engineered *E. coli* strains that overexpress the heterologous bacterial glycosyltransferase gene [95–98]. Developments should allow more types of importantoligosaccharides to be prepared in large quantities with bacterial enzymes.

As described in this article, glycosyltransferases and related enzymes having broad acceptor specificity from prokaryote are now available and the number of these enzymes is increasing. Thus, easy enzymatic production of sialylated glycans and modification of glycoconjugate by addition of sialic acid at low cost have been becoming more and more widely applicable. For example, as mentioned before, sialyllactose or sialyllactosamine are known to act as receptors for various types of influenza viruses. Recently, several glycopolymers carrying sialic acids have been developed as inhibitors against influenza virus infection and the effect of these glycopolymers have been reported using human influenza viruses, including clinically isolated strains [99, 100]. In the near future, production of sialyloligosaccharides with bacterial enzymes will become indispensable for the industrial application of sialyloligosaccharides.

REFERENCES

[1] Angata, T., and Varki, A. (2002) Chemical diversity in the sialic acids and related α-keto acids: an evolutionary perspective. *Chem. Rev.102*, 439–469.

[2] Vimr, E. R., Kalivoda, K. A., Denzo, E. L. and Steenbergen, S. M. (2004) Diversity of microbial sialic acid metabolism. *Microbiol. Mol. Biol. R., 68*, 132-153.

[3] Schauer, R. (2004) Sialic acid: fascinating sugars in higher animals and man. *Zoology, 107*, 49-64.

[4] Corfield, A.P. and Schauer, R.Occurrence of sialic acids.In: Schauer, R., editor.*Sialic Acids: Chemistry, metabolism and function.*New York: Springer-Verlag Vienna; 1982; 5–50.

[5] Schauer, R. (1982) Chemistry, metabolism, and biological functionsof sialic acids in *Advances in Carbohydrate Chemistry andBiochemistry,* Vol. 40, pp. 131-234, Academic Press, New York.

[6] Paulson, J.C. (1989) Glycoproteins: What are the sugar chainsfor? *Trends. Biochem. Sci.,14*, 272-276.

[7] Hakomori, S. (1990) Bifunctional role of glycosphingolipids. *J.Biol.Chem.,265*, 18713-18716.

[8] Lasky, L.A. (1992) Selectins: Interpreters of cell-specific carbohydrateinformation during inflammation. *Science,258*, 964-969.

[9] Feizi, T. (1985) Demonstration by monoclonal antibodies thatcarbohydrate structures of glycoproteins and glycolipids are onco-developmental antigens. *Nature,314*, 53-57.

[10] Suzuki, Y. (2005) Sialobiology of influenza: molecular mechanism of host range variation of influenza virus. *Biol. Pharm. Bull.,28*, 399-408.

[11] Weis, W., Brown, J. H., Cusack, S., Paulson, J. C., Skehel, J. J. and Wiley, D. C. (1988) Structure of influenza virus haemagglutinin complexed with its receptor, sialic acid. *Nature, 333*, 426-431.

[12] Connor, R. J., Kawaoka, Y., Webster, R. G. and Paulson, J. C. (1994) Receptor specificity in human, avian, and equine H2 and H3 influenza virus isolates. *Virology, 205*, 17–23.

[13] Matrosovich, M., Tuzikov, A., Bovin, N., Gambaryan, A., Klimov, A., Castrucci, M. R., Donatelli, I. and Kawaoka, Y. (2000) Early alterations of the receptor-binding properties of H1, H2, and H3 avian influenza virus hemagglutinins after their introduction into mammals. *J. Virol.,74*, 8502–8512.

[14] Yuki, N. (2001) Infectious origins of, and molecular mimicry in, Guillain-Barré and Fisher syndromes. *Lancet Infect. Dis.,1*, 29-37.

[15] Yuki, N., Suzuki, K., Koga, M., Nishimoto, Y., Odaka, M., Hirata, K., Taguchi, K., Miyatake, T., Furukawa, K., Kobata, T. and Yamada, M. (2004) Carbohydrate mimicry between human ganglioside GM1 and *Campylobacter jejuni*lipo-oligosaccharidecauses Guillain-Barré Syndrome.*Proc. Natl. Acad. Sci. USA, 101,* 11404-11409.

[16] Ropper, A. H. (1992) The Guillain-Barré-syndrome.*N. Engl. J. Med.,326*, 1130-1136.

[17] Kuroki, S., Saida, T., Nukina, M., Haruta, T., Yoshida, M., Kobayashi, Y. and Nakanishi, H. (1993) *Campylobacter jejuni* strains from patients with Guillain-Barré syndrome belong mostly to penner serogroup 19 and contain beta-*N*-acetylglucosamine residues.*Ann. Neurol.,33*, 243-247.

[18] Aspinall, G. O., McDonald, A. G., Raju, T. S., Pang, H., Mills, S. D., Kurjanezyk, L. A. and Penner, J. L. (1992) Serological diversity and chemical structures of *Campylobacter jejuni* low-molecular-weight lipopolysaccharides. *J. Bacteriol.*, *174*, 1324-1332.

[19] Aspinall, G. O., McDonald, A. G., Pang, H., Kurjanezyk, L. A. and Penner, J. L. (1994) Lipopolysaccharides of *Campylobacter jejuni* serotype O:19: structures of core oligosaccharide regions from the serostrain and two bacterial isolates from patients with the Guillain-Barré syndrome.*Biochemistry*, *33*, 241-249.

[20] Aspinall, G. O., McDonald, A. G. and Pang, H. (1994) Lipopolysaccharides of *Campylobacter jejuni* serotype O:19: structures of O antigen chains from the serostrain and two bacterial isolates from patients with the Guillain-Barré syndrome.*Biochemistry*, *33*, 250-255.

[21] Gilbert, M., Brisson, J. R., Karwaski, M. F., Michniewicz, J., Cunningham, A. M., Wu, Y., Young, N. M. and Wakarchuk, W. W. (2000) Biosynthesis of ganglioside mimics in *Campylobacter jejuni* OH4384. *J. Biol. Chem.,275*, 3896-3906.

[22] Ho, T. W., Willison, H. J., Nachamkin, I., Li, C. Y., Veitch, J., Ung, H., Wang, G. R., Liu, R. C., Cornblath, D. R., Asbury, A. K., Griffin, J. W. and McKhann, G. M. (1999) AntiGD1a antibody is associated with axonal but not demyelinating forms of Guillain-Barré syndrome. *Ann. Neurol.,45*, 168-173.

[23] Ogawara, K., Kuwabara, S., Mori, M., Hattori, T., Koga, M. and Yuki, N. (2000) Axonal Guillain-Barré syndrome: relation to anti-ganglioside antibodies and *Campylobacter jejuni* infection in Japan. *Ann. Neurol.*, *48*, 624-631.

[24] Yuki, N. (2005) Carbohydrate mimicry: a new paradigm of autoimmune diseases. *Curr. Opin. Immunol.*, *17*, 577-582.

[25] Koga, M., Gilbert, M., Li, J., Koike, S., Takahashi, M., Furukawa, K., Hirata, K. and Yuki, N. (2005) Antecedent infections in Fisher syndrome: a common pathogenesis of molecular mimicry. *Neurology*, *64*, 1605-1611.

[26] Kanie, O. and Hindsgaul, O. (1992) Synthesis of oligosaccharides, glycolipidsand glycopeptides. *Curr. Opin. Struct. Biol.*, *2*,674-681.

[27] Ito, Y., Nunomura, S., Shibayama, S. and Ogawa, T. (1992) Studies directed toward the synthesis of polysialogangliosides: The regio- and stereocontrolled synthesis of rationally designed fragments of the tetrasialoganglioside GQlb. *J. Org. Chem.*, *57*, 1821–1831.

[28] Wang, Z., Zhang, X.-F., Ito, Y., Nakahara, Y. and Ogawa, T. (1996) A new strategy for stereoselective synthesis of sialic acid-containing glycopeptide fragment.*Bioorg. Med. Chem.*, *4*, 1901–1908.

[29] Izumi, M., and Wong, C. -H. (2001) Microbial sialyltransferases for carbohydratesynthesis.*Trends Glycosci. Glycotechnol.,13*, 345–360.

[30] Koeller, K. M. and Wong, C.–H. (2001) Enzymes for chemical synthesis.*Nature,409*, 232–240.

[31] Sears, P. and Wong, C.-H. (2001) Toward automated synthesis of oligosaccharides and glycoproteins.*Science*, *291*, 2344–2350.

[32] Vann, W. F., Tavarez, J. J., Crowley, J., Vimr, E. and Silver, R. P. (1997) Purification and characterization of *Escherichia coli* K1 *neuB* gene product *N*-acetylneuraminic acid synthetase. *Glycobiology*, *7*, 697-701.

[33] Rodriguez-Aparicio, L. B., Luengo, J. M., Gonzalez-Clemente, C. and Reglero, A. (1992) Purification and characterization of nuclear cytidine 5'-monophosphate *N*-acetylneuraminic acid synthetase from rat liver. *J. Biol. Chem.,267*, 9257-9263.

[34] Vann, W. F., Silver, R. P., Abeijon, C., Chang, K., Aaronson, W., Sutton, A., Finn, C. W., Lindner, W. and Kotsatos, M. (1987) Purification, properties, and genetic location of *Escherichia coli* cytidine 5'-monophosphate *N*-acetylneuraminic acid synthetase. *J. Biol. Chem.,262*, 17556-17562.

[35] Munster, A. K., Eckhardt, M., Potvin, B., Muhlenhoff, M., Stanley, P. and Gerardy-Schahn, R. (1998) Mammalian cytidine 5'-monophosphate *N*-acetylneuraminic acidsynthetase: A nuclear protein with evolutionarily conserved structural motifs.*Proc. Natl. Acad. Sci. USA, 95*, 9140-9145.

[36] Kean, E. L. (1991) Sialic acid activation. *Glycobiology, 1*, 441-447.

[37] Bravo, I. G., Barralo, S., Ferrero, M. A., Rodriguez-Aparicio, L. B., Martinez-Blanco, H. and Regleo, A. (2001) Kinetic properties of the acylneuraminate cytidyltransferase from *Pasteurella haemolytica* A2. *Biochem. J.,358*, 585-598.

[38] Munster-Kuhnel, A. K., Tiralongo, J., Krapp, S., Weinhold, B., Ritz-Sedlacek, V., Jacob, U. and Gerardy-Schahn, R.(2004) Structure and function of vertebrate CMP-sialic acid synthetases.*Glycobiology, 14*, 43R-51R.

[39] Jelakovic, S. and Schlz, G. E. (2001) The structure of CMP: 2-keto-3-deoxy-manno-octonic acid synthetase and of its complexes with substrates and substrate analogs. *J. Mol. Biol., 312*, 143-155.

[40] Raetz, C. R. H. and Whitfield, C. (2002) Lipopolysaccharide endotoxins. *Annu. Rev. Biochem., 71*, 635-700.

[41] Clementz, T. and Raetz, C. R. H. (1991) A gene coding for 3-deoxy-manno-octonic acid transferase in *Escherichia coli*. *J. Biol. Chem.,266*, 9687-9696.

[42] White, K. A., Kaltashov, I. A., Cotter, R. J. and Raetz, R. H. (1997) A mono functional 3-deoxy-D-manno-octulosonic acid (Kdo) transferase and a Kdo kinase in extracts of *Haemophilus influenzae*. *J. Biol. Chem., 272*, 16555-16563.

[43] Ambrose, M. G., Freese, S. J., Reinhold, M. S., Warner, T. G. and Vann, W. F. (1992) 13C NMR investigation of the anomeric specificity of CMP-*N*-acetylneuraminic acid synthetase from *Escherichia coli*. *Biochemistry, 31*, 775-780.

[44] Mizunar, R. M. and Pohl, N. (2007) Cloning and characterization of a heat-stable CMP-*N*-acetylneuraminic acid synthetase from *Clostridium thermocellum*. *Appl. Microbiol. Biotechnol., 76*, 827-834.

[45] Zapata, G., Vann, W. F., Aaronson, W., Lewis, M. S. and Moos, M. (1989) Sequence of the cloned *Escherichia coli* K1 CMP-N-acetylneuraminic acid synthetase gene. *J. Biol. Chem., 264*, 14769-14774.

[46] Edwards, U. and Frosch, M. (1992) Sequence and functional analysis of the cloned *Neisseria meningitidis* CMP-NeuAc synthetase. *FEMS Microbiol. Lett., 96*, 161-166.

[47] Tullius, M. V., Munson, R. S. Jr., Wang, J. and Gibson, W. (1996) Purification, cloning, and expression of a cytidine 5'-monophosphate N-acetylneuraminic acid synthetase from *Haemophilus ducreyi.J. Biol. Chem., 271*, 15373-15380.

[48] Yu, H., Ryan, W., Yu, H. and Chen, X. (2006) Characterization of a bifunctional cytidine 5'-monophosphate N-acetylneuraminic acid synthetase cloned from *Streptococcus agalactiae. Biotechnology letters, 28*, 107-113.

[49] Mosimann, S. C., Gilbert, M., Dombroswki, D., To, R., Wakarchuk, W. W. and Strynadka, N. C. J. (2001) Structure of a sialic acid-activating synthetase, CMP-acylneuraminate synthetase in the presence and absence of CDP. *J. Biol. Chem., 276*, 8190-8196.

[50] Krapp, S., Munster-Kuhnel, A. K., Kaiser, J. T., Huber, R., Tiralongo, J., Gerardy-Schahn, R. and Jacob, U. (2003) The crystal structure of murine CMP-5-N-acetylneuraminic acid synthetase. *J. Mol. Biol., 334*, 625-637.

[51] Stoughton, D. M., Zapata, G., Picone, R. and Vann, W. F. (1999) Identification of Arg-12 in the active site of *Escherichia coli* K1 CMP-sialic acid synthetase. *Biochem. J., 343*, 397-402.

[52] Zapata, G., Roller, P. P., Crowley, J. and Vann, W. F. (1993) The role of cysteine residues 129 and 329 in *Escherichia coli* K1 CMP-NeuAc synthetase. *Biochem. J., 295*, 485-491.

[53] Liu, G., Jin, C. and Jin, C. (2004) CMP-N-acetylneuraminic acid synthetase from *Escherichia coli* K1 is a bifunctional enzyme. *J. Biol. Chem.,279*, 17738-17749.

[54] Kittelmann, M., Klein, T., Kragl, U., Wandrey, C. and Ghisalba, C. (1995) CMP-N-acetyl neuraminic-acid synthetase from *Escherichia coli*; fermentative production and application for the preparative synthesis of CMP-neuraminic acid. *Appl. Microbiol. Biotechnol., 44*, 59-67.

[55] Gilbert, M., Watson, D. C. and Wakarchuk, W. W. (1997) Purification and characterization of the recombinant CMP-sialic acid synthetase from *Neisseria meningitidis. Biotechnol. Lett., 19*, 417-420.

[56] Yu, H., Yu, H., Kaepel, R. and Chen, X. (2004) Chemoenzymatic synthesis of CMP-sialic acid derivatives by a one-pot two-enzyme system: comparison of substrate flexibility of three microbial CMP-sialic acid synthetases. *Bioorg. Med. Chem., 12*, 6427-6435.

[57] Karwaski, M. F., Wakarchuk, W. W. and Gilbert, M. (2002) High-level expression of recombinant Neisseria CMP-sialic acid synthetase in *Escherichia coli. Protein expression and Purification, 25*, 237-240.

[58] Gilbert, M., Bayer, R., Cunningham, A. M., DeFrees, S., Gao, Y., Watson, D. C., Young, N. M. and Wakarchuk, W. W. (1998) The synthesis of sialylated oligosaccharides using CMP-Neu5Ac synthetase/sialyltransferase fusion. *Nat. Biotechnol., 16*, 769-772.

[59] Shames, S. L., Simon, S. L., Christopher, C. W., Schmid, W., Whitesides, G. M. and Yang, L. L. (1991) CMP-N-acetylneuraminic acid synthetase of *Escherichia coli*: high level expression, purification and use in the enzymatic synthesis of CMP-N-acetylneuraminic acid and CMP-neuraminic acid derivatives. *Glycobiology, 1*, 187-191.

[60] Datta, A. K., Sinha, A. and Paulson, J. C. (1998) Mutation of the sialyltransferase S-sialylmotif alters the kinetics of the donor and acceptor substrates.*J. Biol. Chem.,273*, 9608-9618.

[61] Freiberger, F., Claus,H., Gunzel, A., Oltmann-Norden,I., Vionnet,J., Muhlenhoff,M., Vogel, U., Vann, W.F., Gerardy-Schahn,R. and Stummeyer, K.(2007) Biochemical characterization of a *Neisseria meningitidis*polysialyltransferase reveals novel functional motifs in bacterialsialyltransferases.*Mol. Microbiol., 65*, 1258–1275.

[62] Sujino, K., Jackson, R. J., Chan, N. W. C., Tsuji, S. and Palcic, M. M. (2000) A novel viral a2,3-sialyltransferase (v-ST3Gal I): transfer of sialic acid to fucosylated acceptors. *Glycobiology, 10*, 313-320.

[63] Gilbert, M.,Watson, D.C., Cunningham,A.M., Jennings,M.P., Young, N.M., Wakarchuk, W.W. (1996)Cloning of the lipooligosaccharide α-2,3-sialyltransferase from the bacterial pathogens*Neisseria meningitidis*and *Neisseria gonorrhoeae.J. Biol. Chem., 271*, 28271–28276.

[64] Edwards, U., Muller, A., Hammerschmidt, S., Gerardy-Schahn,R. and Frosch, M. (1994) Molecular analysis of the biosynthesis pathway of the (α-2,8)polysialic acid capsule by *Neisseria meningitidis*serogroup B. *Mol.Microbiol., 14*, 141–149.

[65] Gilbert, M., Brisson, J.R., Karwaski, M.F., Michniewicz,J., Cunningham, A.M., Wu, Y., Young, N.M. and Wakarchuk, W.W. (2000) Biosynthesisof ganglioside mimics in*Campylobacter jejuni*OH4384: identificationof the glycosyltransferase genes, enzymatic synthesis of modelcompounds, and characterization of nanomole amounts by 600-MHz ^{1}H and ^{13}C NMR analysis.*J. Biol. Chem., 275*, 3896–3906.

[66] Shen, G.J., Datta, A.K., Izumi, M., Koeller, K.M. and Wong, C.-H. (1999) Expression ofα2,8/2,9-Polysialyltransferase from *Escherichia coli* K92. characterization of the enzyme and its reaction products.*J. Biol. Chem.,274*, 35139–35146.

[67] Yamamoto, T., Nakashizuka, M. and Terada, I.(1998) Cloning and expression of a marine bacterial beta-galactoside alpha 2,6-sialyltransferase gene from *Photobacterium damsela*JT0160.*J. Biochem. (Tokyo), 123*, 94–100.

[68] Tsukamoto, H., Takakura, Y. and Yamamoto, T.(2007) Purification, cloningand expression of an α-/β-galactoside α2,3-sialyltransferase from aluminous marine bacterium,*Photobacterium phosphoreum.J. Biol.Chem., 282*, 29794–29802.

[69] Yamamoto, T., Hamada, Y., Ichikawa, M., Kajiwara, H., Mine, T., Tsukamoto, H. and Takakura, Y.(2007) A β-galactoside α2,6-sialyltransferaseproduced by a marine bacterium, *Photobacterium leiognathi*JTSHIZ-145, is active at pH 8.*Glycobiology, 17*, 1167–1174.

[70] Okino, N., Kakuta, Y., Kajiwara, H.,Ichikawa, M., Takakura, Y., Ito, M. and Yamamoto, T.(2007) Purification, crystallization and preliminarycrystallographic characterization of the alpha 2,6-sialyltransferasefrom Vibrionaceae*Photobacterium*sp. JT-ISH-224. *Acta Crystallogr.Sect. F. Struct. Biol. Cryst. Commun., 1*, 662–664.

[71] Takakura, Y., Tsukamoto, H. and Yamamoto, T.(2007) Molecular cloning,expression and properties of an α/β-galactoside α2,3-sialyltransferasefrom *Vibrio*sp. JT-FAJ-16. *J. Biochem. (Tokyo), 142*, 403–412.

[72] Yu, H., Chokhawala, H., Karpel, R., Yu, H., Wu, B., Zhang, J., Zhang, Y., Jia, Q. and Chen, X. (2005) A multifunctional *Pasteurella multocida*sialyltransferase: a powerful tool for the synthesis of sialosidelibraries.*J. Am. Chem. Soc., 127*, 17618–17619.

[73] Hood, D.W., Cox, A.D., Gilbert, M., Makepeace, K., Walsh, S., Deadman, M. E., Cody, A., Martin, A., Mansson, M., Schweda, E.K., Brisson, J.R., Richards, J.C.,

Moxon, E.R. and Wakarchuk, W.W. (2001) Identificationof a lipopolysaccharide α-2,3-sialyltransferase from *Haemophilus influenzae. Mol. Microbiol., 39*, 341–350.

[74] Watanabe, M., Miyake, K., Yamamoto, S., Kataoka, Y., Koizumi, S., Endo, T., Ozaki, A. and Iijima, S. (2002) Identification of sialyltransferases of *Streptococcus agalactiae.J. Biosci. Bioeng., 93*, 610-613.

[75] Coutinho, P.M., Deleury, E., Davies, G.J. and Henrissat,B. (2003) An evolvinghierarchical family classification for glycosyltransferases.*J. Mol. Biol., 328*, 307–317.

[76] Chiu, C.P., Watts,A.G., Lairson,L.L., Gilbert, M., Lim, D., Wakarchuk, W.W., Withers, S.G. and Strynadka,N.C. (2004) Structural analysis of thesialyltransferase CstII from *Campylobacter jejuni*in complex with asubstrate analog.*Nat. Struct. Mol. Biol., 11*, 163–170.

[77] Ni, L., Sun, M., Yu, H., Chokhawala, H., Chen, X. and Fisher, A.J. (2006) Cytidine 5'-monophosphate (CMP)-induced structural changes in amultifunctional sialyltransferase from *Pasteurella multocida.Biochemistry, 45*, 2139–2148.

[78] Ni, L., Chokhawala, H., Cao, H., Henning, R., Ng,L., Huang, S., Yu, H., Chen, X. and Fisher, A.J. (2007) Crystal structures of *Pasteurella multocida*sialyltransferase complexes with acceptor and donor analogues revealsubstrate binding sites and catalytic mechanism.*Biochemistry, 46*, 6288–6298.

[79] Chiu, C.P., Lairson, L.L., Gilbert, M., Wakarchuk, W.W., Withers, S.G. and Strynadka, N.C. (2007) Structural analysis of the alpha-2,3-sialyltransferase Cst-I from *Campylobacter jejuni*in apo and substrate-analogue bound forms.*Biochemistry, 46*, 7196–7204.

[80] Kakuta, Y., Okino, N., Kajiwara, H., Ichikawa, M., Takakura, Y., Ito, M. and Yamamoto, T.(2008) Crystal structure of *Vibrionaceae Photobacterium*sp. JT-ISH-224α2,6-sialyltransferase in a ternary complex with donorproduct CMP and acceptor substrate lactose: catalytic mechanismand substrate recognition.*Glycobiology, 18*, 66-73.

[81] Kim, D. U., Yoo, J. H., Lee, Y. J., Kim, K. S. and Cho, H. S. (2007) Structural analysis of sialyltransferase PM0188 from *Pasteurellamultocida* complexed with donor analogue and acceptor sugar. *BMB Report, 41*, 48-54.

[82] Lairson, L.L. and Withers, S.G. (2004) Mechanistic analogies amongst carbohydratemodifying enzymes.*Chem. Commun.*, 2243–2248.

[83] Yamamoto, T., Ichikawa, M. and Takakura, Y. (2008) Conserved amino acid sequences in the bacterial sialyltransferases belonging to Glycosyltransferase family 80.*Biochem Biophys Res Commun., 365*, 340-343.

[84] Yamamoto, T., Nakashizuka, M., Kodama, H., Kajihara, Y. and Terada, I. (1996)Purification and characterization of a marine bacterial beta-galactoside alpha 2,6-sialyltransferase from *Photobacterium damsela* JT0160.*J. Biochem.(Tokyo), 120*, 104–110.

[85] Kajihara, Y., Yamamoto, T., Nagae, H., Nakashizuka, M., Sakakibara, T. and Terada, I. (1996) A novel α-2,6-sialyltransferase: Transfer of sialic acid to fucosyl and sialyl trisaccharides.*J. Org. Chem.,61*, 8632–8635.

[86] Yamamoto, T., Nagae, H., Kajihara, Y. and Terada, I.(1998) Mass production of bacterial alpha2,6-sialyltransferase and enzymatic syntheses of sialyloligosaccharides.*Biosci. Biotechnol. Biochem.,62*, 210–214.

[87] Kajihara, Y., Akai, S., Nakagawa, T., Sato, R., Ebata, T., Kodama, H. and Sato, K. (1999) Enzymatic synthesis of Kdn oligosaccharides by a bacterial α (2,6)-sialyltransferase. *Carbohydr. Res.,315*, 137–141.

[88] Yu, H., Huang, S., Chokhawala, H., Sun, M., Zheng, H. and Chen, X. (2006) Highly efficient chemoenzymatic synthesis of naturally occurring and non-natural alpha-2,6-linked sialosides: a P. damsela alpha-2,6-sialyltransferase with extremely flexible donor-substrate specificity.*Angew Chem Int Ed Engl., 45*, 3938-3944.

[89] Erickson, P. R. and Herzberg, M. C. (1993) Evidence for the covalent linkage of carbohydrate polymers to a glycoprotein from*Streptococcus sanguis. J. Biol. Chem.,268*, 23780–23783.

[90] Young, N. M., Brisson, J. R., Kelly, J., Watson, D. C., Tessier, L., Lanthier, P. H., Jarrell, H. C., Cadotte, N., St Michael, F., Aberg, E. and Szymanski, C. M. (2002) Structure of the *N*-linked glycanpresent on multiple glycoproteins in the gram-negative bacterium, *Campylobacter jejuni. J. Biol. Chem.,277*, 42530–42539.

[91] Wacker, M., Linton, D., Hitchen, P. G., Nita-Lazar, M., Haslam, S. T., North, S. J., Panico, M., Morris, H. R., Dell, A., Wren, B. W. and Aebi, M. (2002) N-linked glycosylation in *Campylobacter jejuni* and its functional transfer into *E. coli. Science, 298*, 1790–1793.

[92] Power, P. M., Roddam, L. F., Dieckelmann, M., Srikhanta, Y. N., Tan, Y. C., Berrington, A. W. and Jennings, M. P. (2000) Genetic characterization of pilin glycosylation in *Neisseria meningitidis.Microbiology, 146*, 967–979.

[93] Koizumi, S. Endo, T. Tabata, K. Nagano, H. Ohnishi, J. and Ozaki, A.(2000) Large-scale production of GDP-fucose and Lewis X by bacterial coupling.*J. Ind.Microbiol. Biotechnol., 25*, 213-217.

[94] Koizumi, S. Endo, T. Tabata, K. and Ozaki, A.(1998) Large-scale production of UDP-galactose and globotriose by coupling metabolically engineered bacteria.*Nat. Biotechnol., 16*, 847-850.

[95] Dumon, C., Bosso, C., Utille, J. P., Heyraud, A. and Samain, E. (2006) Production of Lewis X tetrasaccharides by metabolically engineered *Escherichia coli.Chembiochemistry, 7*, 59-65.

[96] Antoine, T., Priem, B., Heyraud, A., Greffe, L., Gilbert, M., Wakarchuk, W. W., Lam, J. S. and Samain, E. (2003) Large-scale in vivo synthesis of the carbohydrate moieties of gangliosides GM1 and GM2 by metabolically engineered *Escherichia coli.Chembiochemistry, 4*, 406-412.

[97] Drouillard, S., Driguez, H. and Samain, E. (2006) Large-scale synthesis of H-antigen oligosaccharides by expressing *Helicobacter pylori* alpha1,2-fucosyltransferase in metabolically engineered *Escherichia coli* cells.*Angew. Chem. Int. Ed. Engl., 45*, 1778-1780.

[98] Fierfort, N. and Samain, E. (2008) Genetic engineering of Escherichia coli for the economical production of sialylated oligosaccharides.*J. Biotechnol., 134*, 261-265.

[99] Hidari, K. I. P. J., Murata, T., Yoshida, K., Takahashi, Y., Minamijima, Y., Miwa, Y., Adachi, S., Ogata, M., Usui, T., Suzuki, Y. and Suzuki, T. (2008) Chemoenzymatic

synthesis, characterization, and application of glycopolymers carrying lactosamine repeats as entry inhibitors against influenza virus infection. *Glycobiology, 18,* 779-788.

[100] Totani, K., Kubota, T., Kuroda, T., Murata, T., Hidari, K. I. –P. J., Suzuki, T., Suzuki, Y., Kobayashi, K., Ashida, H., Yamamoto, K. and Usui, T. (2003) Chemoenzymatic synthesis and application of glycopolymers containing multivalent sialyloligosaccharides with a poly(L-glutamic acid) backbone for inhibition of infection by influenza viruses. *Glycobiology, 13,* 315-326.

In: Biological Aspects of Human Health and Well-Being   ISBN: 978-1-61209-134-1
Editor: Tsisana Shartava   © 2011 Nova Science Publishers, Inc.

Chapter VII

THE HEPATOCELLULAR DYSFUNCTION CRITERIA: HEPATOCYTE CARBOHYDRATE METABOLIZING ENZYMES AND KUPFFER CELL LYSOSOMAL ENZYMES IN 2'NITROIMIDAZOLE EFFECT ON AMOEBIC LIVER ABSCESS (ELECTRON MICROSCOPIC – ENZYME APPROACH)

Rakesh Sharma[1]

Gastroenterology Department,
All India Institute of medical Sciences, New Delhi-110029 India

ABSTRACT

Aim: to understand the 2'-nitroimidazole cytotoxicity and liver cell interaction, we proposed a "Hapatocellular Dysfunction Criteria". Based on it, forty eight patients with amoebic liver abscess on 2'-nitroimidazole therapy were studied for their carbohydrate metabolizing enzymes in serum and hepatocellular enzymes in liver biopsy tissues. *Materials and Methods:* Proven ten cases were studied for hepatocellular cytomorphology by electron microscopy. The clinical status of amoebiasis was assessed by enzyme linked immunosorbent assay (ELISA) antibody titers and stool examination. *Results and Discussion:* Out of forty eight, forty five patients showed elevated carbohydrate metabolizing enzyme levels in serum. The enzymes hexokinase (in 80% samples), aldolase (in 50% samples), phosphofructokinase (in 60% samples), malate dehydrogenase (in 75% samples), isocitrate dehydrogenase (ICDH) (in 80% patients) were elevated while succinate dehydrogenase and lactate dehydrogenase (LDH) levels remained unaltered. Lysosomal enzymes β-glucuronidase, alkaline phosphatase, acidphosphatase, showed enhanced levels in the serum samples. In proven ten amoebic

[1] The manuscript is part of Ph.D "Effect of imidazole on liver cells during development of amoebic liver abscess" submitted by author. Present address: Center of Nanobiotechnology, Florida State University, Tallahassee, USA. Email: rksz2004@yahoo.com

liver abscess biopsies, the hepatocytes and Kupffer cell preparations showed altered enzyme levels. Hepatocytes showed lowered hexokinase (in 80%), LDH (75%), and higher content of aldolase (in 60%), pyruvate kinase (in 70%), malate dehydrogenase (in 66%), ICDH (in 85%), citrate dehydrogenase (in 70%), phosphogluconate dehydrogenase (66%). Kupffer cells showed higher enzyme levels of β-glucuroronidase (in 80%), leucine aminopeptidase (in 70%), acid phosphatase (in 80%) and aryl sulphatase (in 88%). In these 10 repeat biopsy samples from patients on 2'-nitronidazole clinical recovery, the electron microscopy cytomorphology observations showed swollen bizarre mitochondria, proliferative endoplasmic reticulum, and anisonucleosis. 2'-Nitroimidazole showed reverse effect in favor of liver cell regeneration by recovering hepatic damage. *Conclusion:* The proposed "Hepatocellular Dysfunction Criteria" showed different clinical enzyme activities in these patients as they could distinguish nonspecific amoebic hepatitis from amoebic liver abscess.

Keywords: liver, enzymes, electron microscopy, carbohydrate metabolizing enzymes, heaptocellular dysfunction criteria.

INTRODUCTION

Hepatic damage due to Entamoeba histolytica has been considered to occur in stages viz. initial loss in liver cell metabolic integrity, apoptosis, proliferation, and necrosis leading to debris formation [1, 2]. The first initial stage of liver cell damage has been less understood due to its low sensitivity to clinical investigations and available diagnosis at the very late necrotic stage [3, 4]. Early studies reported the validity of diffused damage of parenchymal and nonparenchymal liver cells as initial stage of hepatic damage using electron microscopy and biochemical markers in serum with emphasis of pathophysiology [5]. The present study proposed a 'hepatocellular dysfunction criteria' assuming that initially liver cells loose metabolic integrity and undergoes apoptosis followed by detectable necrosis and complete cell death to end as abscess formation as a result of entameba histolytica infection in liver. The major players are the energy metabolizing and lysosomal enzymes as initial liver damage. The liver cell enzyme profile was used to define the hepatic amoebiasis development (a hepatic damage entity by Entamoeba histolytica infection) by "Hepatocellular Dysfunction Criteria". The criterion was also used to distinguish nonspecific amoebic hepatitis and hepatic liver abscess. The novelty of hepatocellular dysfunction criteria was the detailed cytomorphic-biochemical information with altered enzyme profiles during initial cytomorphic changes in hepatic cells by electron microscopy to confirm the role of cell organelle in disease development. The criterion can be used to evaluate other liver diseases. However, the possibility of various hepatocellular mechanisms was reported as initial loss in metabolic integrity in experimental animal model and intracellular hepatocyte and Kupffer cell metabolite alterations were evidenced by altered enzymes of carbohydrate, lipid and protein metabolism mainly energy metabolism [6-8]. Moreover, glucose and calcium hemostasis has been the primary targets of hepatic damage in amoebic induced metabolic integrity loss leading to regulatory failure in glycolytic, TCA cycle, gluconeogenesis and Ca^{++} mediated cAMP related biodegradation of molecules [9-11].

Hepatocellular Dysfunction Criteria: The criterion was a step by step sequence of hepatocellular damage: 1. initial loss of metabolic integrity; 2. programmed regulatory failure of cell metabolism; 3. liver tissue inflammation and immunity loss; 4. necrosis and active death; and 5. pus formation. Sequence: The purpose of metabolic integrity in hepatocytes is keep intact by maintaining balance between glucose formation and glucose breakdown to maintain energy flow of NADH and ATP molecules. The dysfunction of liver cells leads to enhanced glucose breakdown and more demand of ATP and NADH.

With course of time, high energy demand at the cost of cell metabolic resources leads to metabolic integrity or energy loss and step by step programmed cell death. In the present study, the hepatic dysfunction criterion focus is on 3 major components: 1. major event of high glycolytic rate is immediate result of high glucose turnover in amoebic recruited hepatocytes such as glycolysis followed by secondary metabolic cycles viz. tricarboxylic acid, glycogenolysis, gluconeogenesis, pentose phosphate pathways; 2. changes in peripheral biomarkers such as immunity, phagocytosis, altered enzymes; 3. after severe energy loss changes in liver cellular morphology and pathology development. The study highlights the sensitivity of different enzymes to distinguish hepatic dysfunction and response of nitroimidazole. We proposed a simple scheme of "Hepatocellular Dysfunction Criteria" as shown in Table 1.

Glucokinase, aldolase, pyruvate kinase and lactate dehydrogenase serve as glycolytic regulatory enzymes. TCA cycle being as source of hexose synthesis and gluconeogenesis as secondary source of glucose synthesis, it serves mainly as available central control tool of glucose homeostasis in degenerated or recovered cell [13-16]. The citrate synthase, malate dehydrogenase, isocitrate dehydrogenase and succinate dehydrogenase enzymes serve as TCA cycle regulatory enzymes. The phosphogluconate dehydrogenase enzyme regulates the phosphgluconate pathway. These regulatory enzymes in hepatocytes likely define the nature of hepatocyte damage in ALA patients and heaptocyte response after nitroimidazole therapy[17].

Other component of hepatocyte dysfunction is phagocytosis evaluation by lysosomal enzymes. The lysosomal enzymes β-glucuronidase, leucine aminopeptidase, acid phosphatase, alkaline phosphatase and aryl sulphatase in Kupffer cells, were analyzed in amoebiasis. For validation, the enzymes were estimated in serum and biopsy liver cell after sonication in their sonipreps. Various studies reported the hepatic damage in hepatic diseases induced in animal models based upon carbohydrate metabolizing enzymes indicating their importance in clinical diagnosis [18-20].

In present study, emphasis has been concentrated only upon 2'-nitroimidazole induced carbohydrate metabolizing enzymes and description of disease by 'hepatocellular dysfunction criteria' of disease. Based on these findings nitroimidazole was evaluated for its drug value and liver cells' recovery by nitroimidazole therapy. Earlier studies reported only nitroimidazole evaluation by nitroimidazole circulating plasma concentration, half life, amoebicidal action [21]. The paucity of information on nitroimidazole-liver tissue interaction can be predicted partly by data of initial sequential biochemical changes in liver cells which further lead to pathogenesis [22]. The novelty of representative serum enzyme measurement was to establish 'hepatocellular dysfunction criteria' as clinical profile of early evaluation of initial stages of disease and sequential diagnosis of hepatic damage during disease progression before hepatic abscess develops. The major finding was that carbohydrate metabolizing enzymes showed metabolic integrity loss in hepatocyte cells and phagocytosis of Kupffer cell lysosomes. The serum enzyme profile serves as quick assessment.

Table 1. A step by step scheme of "hepatocellular dysfunction criteria" to evaluate liver damage Hepatocellular Dysfunction Criteria

Morphological changes	Clinical changes	Liver biochemical changes
1. Physical examination:	abdominal pain	--
--	fever	IHA/ELISA +ve
intestinal damage	hepatomegaly	liver function tests(elevated)
hepatic infiltration		
Loss of cellular metabolic integrity		
2. Electron microscopy:	hepatomegaly with	altered hepatocyte enzymes:
Mitochondria(M)	diffused injury	ATP/ADP & NADPH/NADP
endoplasmic reticulum(ER)		gluconeogenesis
peroxisome (P)		glycogenolytic
lysosome (L)		lysosomal enzymes
nuclear changes (N)		oxygen flux related
Hepatocellular enlargement (Apoptosis)		
3. Cellular organelle damage:	poor drug response	slow metabolic disorder:
mitochondria(M)		oxidative phosphorylation
microsomes (MI)		drug metabolizing enzymes
lysosome (L)		initial phagocytosis
nuclear (N)		protein synthesis
cytosol (C)		glucose/protein/respiratory
		burst
Hepatocellular proliferation(inflammation)		
4. Liver pathology changes:	raised diaphragm	
mitochondrial damage	amebic liver scan +ve	stimulation of Kupffer cells
exfoliative ER		activation of Kupffer cells
anisonucleosis		loss of metabolic control
autophagy& lysosomal irritation		increased water accumulation
cytosolic granulation		increased molecule imbalance
fatty liver appearance with		increased lipid synthesis
membrane damage		
Hepatocellular degeneration and necrosis (pus formation)		
5. Hepatocytology:		
Cell proliferation	advancing abscess	surgical aspirates(altered
Cell debris	cavity fills on ultrasound/	proteins, lipids, enzymes)
No cells	hot area on liver scan	
	No drug response	
Nitroimidazole single dose therapy schedule		
6. Liver cell recovery	negative liver scan/ultrasound	normal liver function test
OR	if unchanged or poor recovery	abnormal ELISA, enzymes
	Surgical intervention	

METHODS AND MATERIALS

Patients: Forty eight subjects with liver abscess or amoebic hepatitis subjects were screened from medicine clinic of hospital (see the Figure 1). The samples were collected for stool, ELISA, serum, liver biopsy (only in 10 cases) collection. Only proved 10 cases of amoebic liver abscess were examined for biochemical estimations in isolated liver cells from biopsy specimens. All patients were treated with nitroimidazole therapy 2x5 g one-time dose. At the end of dose, repeat biopsy and serum collection was done. The ELISA was done by method of Prakash et al.[23]. Liver biopsy was taken by Manghini needle [24].

Biochemical assays: In biopsy sonipreps and serum samples the hepatocellular enzymes were estimated as described elsewhere [25]. All substrates of enzymes were obtained from Sigma Chemical Company, St Louis, USA. The estimations were done on UV spectrophotometer Cecil Inc. England. Electron microscopy: The biopsy specimens at stored in glutaraldehyde at -20 C. For fixing samples were fixed in dental wax and emersed in pool of 0.1 M phosphate buffer containing 0.1 M Ca++. The samples were cut 1 mm cube by Gillette blade. The cubes were fixed for 2 hours in 0.2 M phosphate buffer in 4 changes at 30 minutes interval in capped vials. The vials were washed for 2 hours in 0.2 M phosphate buffer 4 time at 30 minutes interval in capped vials. The vials were post fixed in buffered 1% osmium tetroxide for 1-2 hours at 4C, dehydrated in cold 10% ethanol for about 5 minutes followed by dehydrations in cold 50%, 70%, 80%, 95% ethanol for about 5 minutes each and tissues were kept at room temperature. Further tissues were dehydrated at 100% ethanol for 15 minutes. Electron Microscope PA model was used for screening hepatic cell damage by epoxy resin blocks and osmium tetroxide staining [26-28].

RESULTS

All the 48 patients had clinical findings viz. right abdominal pain, intercostal tenderness, high liver function tests. Stool Entamoeba histolytica +ve, ELISA +ve (antibody titers 1: 400 in 45 cases). Out of 48 only 10 showed enlarged liver and abscess formation, confirmed by ultrasound and liver scan. Results are compared with normal healthy persons as shown in figure 1 and Table 1.

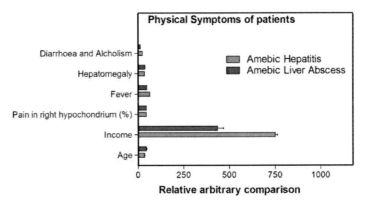

Figure 1. Patient demography data of patients as shown in Figure as bars.

Table 1. Patient demography data of patients as shown mean±sd in table for symptoms and sensitivity in percentage in brackets

Patients	amebic hepatitis	amebic liver abscess
Clinical features		
Mean age + sd (years)	35 ±15	45 ± 20
Average income (monthly)	1750 ± 100	1435 ± 225
Pain right hypochondrium (%)	46/72 (75%)	46/48 (95%)
Fever (> 98 °F)	65/72 (92%)	48/48 (100%)
Hepatomegaly (+3cm)	36/72 (50%)	39/48 (80%)
Diarrhea and alcoholism(>pint/day)	2/72 (35%)	13/48 (23%)

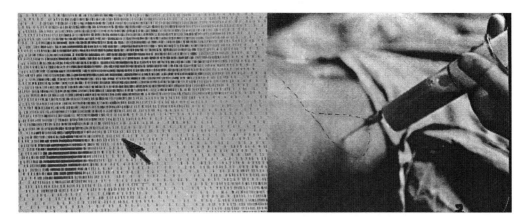

Figure 2. The amebic liver scan (on left panel) is shown for assessing the position of hepatic abscess location and for biopsy is shown (on right panel).

Table 2. The liver enzyme specific activities of carbohydrate metabolizing enzymes in serum of patients with amoebic liver abscess and 2'-nitroimidazole response

Enzyme	Control	Amoebic	2'nitroimidazole
		Liver Abscess	response (n=48)
Glucokinase	15.5 ± 1.2	14.3 ± 1.0 (80)	15.0 ± 1.1
Phosphofructokinase	9.65 ± 1.1	$13.9\pm2.1*(60)$	6.7 ± 0.4 (80)
Aldolase	6.9 ± 0.5	$11.9\pm1.4*(50)$	7.4 ± 1.5 (80)
Lactate Dehydrogenase	256.5 ± 13.4	$626.6\pm16.5*(62)$	$289.5\pm23.4(80)$
Pyruvate Kinase	30.4 ± 4.2	$76.4\pm12.2*(60)$	$36.6 \pm 9.2(100)$
Malate Dehydrogenase	92.4 ± 12.4	$236.5 \pm 29.6*(75)$	$90.4 \pm 24.6(80)$
Isocitrate Dehydrogenase	1.85 ± 1.02	$5.6\pm2.2*(80)$	$2.6 \pm 1.4(60)$
Citrate Synthase	3.2 ± 0.2	$4.1\pm0.1*(80)$	$3.3 \pm 0.4(70)$
Phosphogluconate Dehydrogenase	49.6 ± 11.5	$69.7\pm13*(68)$	$47.8 \pm 9.6(80)$
Succinic Dehydrogenase	44.8 ± 10	$79.2\pm9.5**(10)$	43.9 ± 9.8 (60)
βGlucuronidase	5.2 ± 1.2	$11.2\pm2.0*(90)$	$6.1 \pm 0.3(50)$
Acid Phosphatase	96.4 ± 9.2	$172.4\pm11.8*(90)$	$108.3\pm10.1(55)$
Leucine Aminopeptidase	25.6 ± 10.2	$36.8\pm6.2*(60)$	$26.2 \pm 5.1(80)$
Aryl Sulphatase	10.5 ± 1.2	$21.6\pm1.9*(66)$	$18.2 \pm1.1(80)$

Brackets represent the percentage of patients showing altered enzyme levels and indicate the % sensitivity of enzyme to distinguish hepatic dysfunction and nitroimidazole response. *P < 0.05 and **P < 0.01 indicate comparison of enzyme levels in control vs liver abscess.

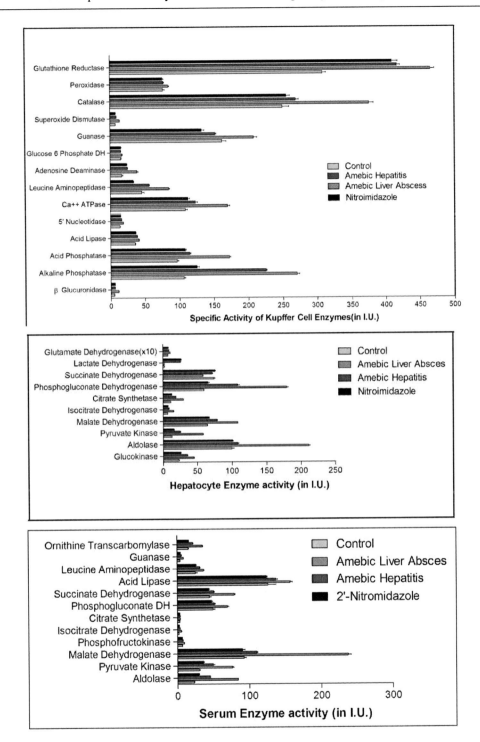

Figure 3. The histogram bars show the effect of nitroimidazole on biomarker enzymes and comparison of different enzymes in hepatocytes and Kupffer cells in amebic hepatitis vs amoebic abscess subjects relative to control patients(panel on top and middle) and serum biomarker enzymes (panel at bottom).

Table 3. Specific activities of carbohydrate metabolizing enzymes in hepatocytes from patients with amoebic liver abscess on 2'nitroimidazole therapy

Enzyme	Control	Amoebic Liver Abscess (n=48)	2'nitroimidazole response (n=48)
Glucokinase	22.4 ± 2.1	44.9 ±4.0 (70)	25.0 ±1.1
Aldolase	99.5 ± 11.1	211.2±13.8 *(60)	96.7±2.4(80)
Lactate Dehydrogenase	2.5± 0.4	2.2 ±0.5*(75)	2.8 ±0.4(80)
Pyruvate Kinase	12.4 ± 0.6	58.3 ± 3.8*(70)	16.6 ±.2(100)
Malate Dehydrogenase	64 ±1.8	108.5 ±2.1*(66)	85.4 ±4.6(80)
Isocitrate Dehydrogenase	6.5 ± 0.8	15.4 ±1.1*(85)	7.6 ±1.4(80)
Citrate Synthase	11.2 ± 0.02	29.5 ±0.8*(70)	13.3 ±0.4(70)
Phosphogluconate Dehydrogenase	59.6 ± 1.5	179.2 ±5.8*(66)	58.8 ±9.6(80)
Succinic Dehydrogenase	74.2 ± 6.2	58.5 ±2.1**(80)	72.5 ±5.8(70)

Brackets represent the percentage of patients showing altered enzyme levels and indicate the % sensitivity of enzyme to distinguish hepatic dysfunction and nitroimidazole response. *P < 0.05 and **P < 0.01 indicate comparison of enzyme levels in control vs liver abscess.

Table 4. Specific activities of lysosomal enzymes in Kupffer cells from patients with amoebic liver abscess

Enzyme	Control N = 15	Amoebic Liver Abscess n = 48	2'nitroimidazole response(n=48)
βGlucuronidase	5.2 ± 1.2	11.2 ±2.0*(80)	5.6 ±0.3(50)
Acid Phosphatase	96.0±9.2	172.4±11.8*(80)	98.3±10.1(60)
Leucine Aminopeptidase	45.6±13.4	84.4±11.1*(70)	26.2 ±5.1(80)
Aryl Sulphatase	45.5±14.1	65.5±5.0* (88)	48.5 ± 3.5(88)

In serum and liver cells of these patients, exhibited more or less specific characteristic changes in enzymes. Glucokinase enzyme levels were lowered insignificantly (P > 0.05; 80%) in serum, in heaptocytes increased significantly (P < 0.05; 70%). In nitroimidazole treated patient samples the levels were more or less normal in both serum and biopsy (100%). Phosphofructokinase levels were elevated in serum and biopsy (P < 0.05; 60%). In nitroimidazole treated samples the phosphofructokinase levels were nonspecific. Aldolase levels in serum and hepatocytes were increased significantly (P < 0.05; 50%). In nitroimidazole treated patients the levels were normal (80%). Lactate dehydrogenase levels were decreased in hepatocytes (P > 0.05; 75%) but elevated in serum (P < 0.05; 62%). In nitroimidazole treated patient samples the lactate dehydrogenase levels were reversed to normal. Pyruvate kinase levels increased significantly (P < 0.05; 70%) in hepatocytes and in serum remained increased significantly (p < 0.05; 70%). In nitroimidazole treated patient samples the levels were normal (100%). Malate dehydrogenase levels were increased significantly in hepatocytes (P < 0.001; 66%) and in serum also remained high significantly (P < 0.05; 75%). In nitroimidazole treated patient samples, the levels were normal (80 %). Isocitrate dehydrogenase levels were increased significantly in hepatocytes (P < 0.001; 85%) and in serum also remained high significantly (P < 0.05; 75%). In nitroimidazole treated samples, the isocitrate dehydrogenase enzyme levels were normal (80 %) in serum. Citrate synthase levels were increased significantly in hepatocytes (P < 0.001; 66%) and in serum also remained high significantly (P < 0.05; 75%). In nitroimidazole treated samples, the citrate synthase levels were normal. Phosphogluconate dehydrogenase levels were increased significantly in hepatocytes (P < 0.05; 66%) and in serum also remained high significantly (P < 0.05; 68%). In nitroimidazole treated samples, the phosphogluconate dehydrogenase levels were normal (80 %) in both. Succinate dehydrogenase levels were decreased significantly in hepatocytes (P < 0.01; 80%) and in serum also remained high significantly (P < 0.05; 70%). In nitroimidazole treated samples, the succinate dehydrogenase levels were normal (60 %) in both. The enzymes are shown in Figure 3 and Tables 2-4.

β-glucuronidase levels were increased significantly in isolated hepatocytes (P < 0.05; 80%) and in serum also remained high significantly (P < 0.05; 90%). In nitroimidazole treated patient samples, the β-glucuronidase levels were normal (50 %) in Kupffer cells and also normal in serum (60%). Acid phosphatase levels were increased significantly in Kupffer cells (P < 0.05; 80%) and in serum also remained high significantly (P < 0.05; 90%). In nitroimidazole treated patient samples, the acid phosphatase levels were normal in both. Leucine aminopeptidase levels were increased significantly in Kupffer cells (P < 0.05; 70%) and in serum also remained high significantly (P < 0.05; 60%).

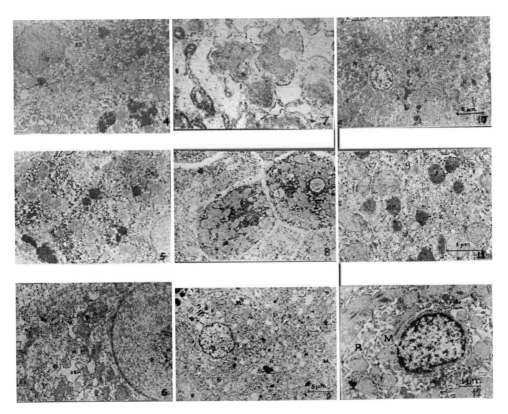

Figure 4. Ultrastructural changes are shown in hepatocyte organelles: exfoliation of endoplasmic reticulum(top on left); anisonucleosis (mid and bottom on left); intercellular junction gaps (top and mid panels in center); nuclear inclusions (bottom on center); swollen and bizarre mitochondria(top on right); inclusions in peroxisome (mid on right); mitochondrial atrophy with lipid vesicles (bottom on right).

In nitroimidazole treated patient samples, the levels were normal in serum (80 %) and remained high in Kupffer cells (60%). Aryl Sulfatase levels were increased significantly in Kupffer cells and serum (P < 0.05; 88% and 66%). In nitroimidazole treated patient samples, the levels were normal in both Kupffer cells and serum (80%). The enzymes are shown in Figure 2 and Table 3.

The ultrastructure of liver cells showed characteristic changes. Hepatocyte mitochondria became swollen, bizarre with dense matrix and showed destroyed cristae, endoplasmic reticulum dilated vesicles, giant nuclei with diffuse proliferation of endoplasmic reticulum and clear anisonucleosis features. Kupffer cell hyperplasia was observed with swollen lysosomal contents as shown in Figure 4.

In above, altered glycolytic enzymes in cytosol, TCA cycle enzymes in mitochondria, lysosomes and increased synthesis of enzymes by endoplasmic reticulum showed correlation with clear liver cell degeneration of microbodies. After nitroimidazole treatment, observations of both electron microscopy and biochemical parameters suggested the reversed hepatocellular changes towards normal recovery.

DISCUSSION

Nonsuppurative damage may be amoebic hepatitis or developing into localized cavitatory abscess formation. Due to nonspecific entity, early stages of both cannot be rules out even resembling other liver diseases. Thus only hepatocytes and Kupffer cells are as sole targets that exhibit their intracellular biochemical changes which could be analyzed in serum as clinically significant. Initially, metabolic integrity loss leads to oversecretion of liver cell enzymes including lysosomal enzymes. Soon after, the ultrastructural changes in liver cells by electron microscopically are suggestive of acute organelle degeneration.

It is evident that ultrastructural parenchymal damage was diffused in cells distant from abscess, associated with regenerative change, consistently observed in all the cases which upon nitroimidazole therapy showed reversed changes. Only few electron microscopy studies are available describing hyperplasia as unequivocal diffuse parenchymal injury resembling with nonsuppurative hepatomegaly exhibiting diffused sinusoidal and portal infiltration [26, 27, 29]. Still such changes may be misleading as these can be expected to occur in any inflammatory disease of liver associated with suppuration. So, in present study, biochemical evidence in liver biopsy samples along with the nitroimidazole reversed effect on clinical recovery are evidenced as possible biomarker tools in support of said hepatic abscess damage occurred. Degenerative changes of hepatocytes were consistently observed suggestive of necrosis. Degenerative heaptocytes changes have been reported in many diseases viz. viral hepatitis, fatty liver, protein calorie malnutrition as sequential series of steps-initial metabolic loss, apoptosis, proliferation and necrosis [30]. At maximum, these conditions were excluded among the present group of patients. Nonsuppurative and suppurative types of hepatioc amoebiasis were described with no difference between two entities.

Evidence of hepatic regenerative reaction was characterized by unusual degree of anisonucleosis and the presence of giant nucleus in hepatocytes [30]. Endoplasmic reticulum showed a diffuse and intense proliferative activity in these liver cells with normal appearance of mitochondria. However, intramitochondrial inclusion bodies were absent but they have been reported as very prominent features of nonsuppurative group [26]. In hepatic amoebiasis, the abscess was caused by Entamoeba histolytica but regarding ultrastructural changes, liver injury may not be claimed developed by E. histolytica parasite. The cause of the ultrastructural changes shown in this study can be supported by initial biochemical alterations reflecting loss of metabolic integrity probably induced by toxins from amoeba or bacteria in gut. Since ultrastructural changes in liver were completely reversible after specific nitroimidazole therapy for 7-9 days, it is quite reasonable that pathogenesis of diffuse heaptocytic damage was due to Entamoeba histolytica.

The initial loss of hepatocellular metabolic integrity leads to hepatic injury. Glucose metabolic dysfunction and nitric oxide formation with Ca^{++} homeostasis have been cited as main initial determinants of disease [31]. Present study addresses the question of glucose metabolizing pathways viz. glycolysis, TCA cycle, gluconeogenesis. In hepatic cells, metabolic alterations of these cycles in disease are best correlated with respective enzymes and ultrastructural changes in cells. The following description is broad explanation of different enzymes secreted from hepatocytes as a result of disease development. Hexokinase, a rate limiting enzyme in cytosol for glucose turnover was estimated due to high glucose conversion into glucose 6P showed high enzyme secretion from heptocytes. Aldolase, rate controlling enzyme which splits fructose-1, 6 Diphosphate into two 3- phosphoglyceric acid and dehydroxyacetone phosphate, showed high enzyme secretion from heptocytes.

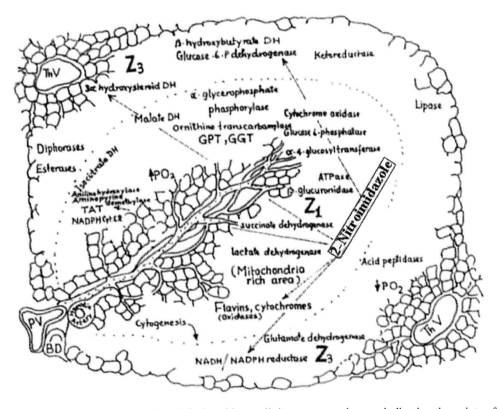

Figure 5. The sketch of nitroimidazole induced intracellular enzymes changes indicating the points of metabolism and metabolic control during recovery after nitroimidazole treatment . The enzymes distribution in organelle and enzyme location in different hepatic sites explains the liver damage and enzymatic basis of hepatocellular dysfunction criteria.

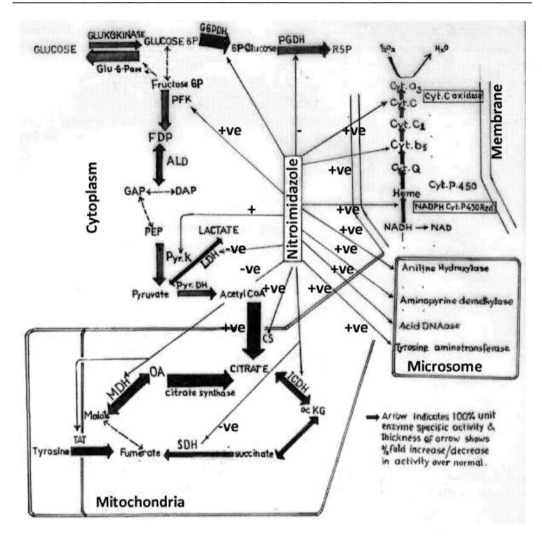

Figure 6. The sketch of relative biomarker enzyme changes in liver during development of amebic liver abscess and nitroimidazole effect. The +ve sign denotes the relative increase in enzyme activity and –ve sign denotes the decrease in enzyme activity in liver cells in different organelles. The figure sketch also represents a sequence of metabolic steps of hepatocellular dysfunction criteria.

Phosphofructokinase, rate controlling enzyme which phosphorylates fructose 6P into fructose 1, 6 DiP, enhanced as its more demand in hepatocellular glucose turnover which probably induced higher secretion of phosphofructokinase. Pyruvate kinase, transferring phosphoryl group from phosphoenolpyruvate to ADP with pyruvate formation, showed high enzyme secretion from heptocytes. It may be attributed due to high concentration of glycolytic intermediates as terminal step was blocked or due to 2, 3 Di-Phosphoglycerate regulated abnormal oxygen dissociation. Malate dehydrogenase and isocitrate dehydrogenase enzymes, catalyzing malate oxidation into oxaloacetate and oxidative decarboxylation of isocitrate into α-ketoglutarate respectively both require NAD^+. Both enzymes were elevated in damaged cells due to reducing equivalent low ratio of $NADH/NAD^+$ pushing in forward direction. Succinate dehydrogenase, oxidizing succinate to fumerate using FAD, showed decreased enzyme levels due to low iron sulphur proteins resulting with lowered electron transport system in inner mitochondrial membrane i.e. supply of electrons to molecular oxygen by electron transfer from $FADH_2$ to Fe^{+++} (SDH) in amoebic recruited liver cells. Citrate synthase, synthesizing citrate from oxaloacetate and acetyl CoA by aldol condensation followed by hydrolysis, showed elevated levels due to rapid turnover of oxaloacetate and acetyl CoA molecules during cytolysis. Moreover, high TCA cycle activity in hepatocyte during liver damage conditions was described earlier [1]. In serum, the enzymes exhibit their significance but aldolase, pyruvate kinase and LDH, MDH, ICDH observed as distinguishing the diffused injury or abscess formation [32]. Phosphogluconate dehydrogenase elevated levels may be attributed due to ribulose 5P formation and transaldolase and transketolase control, for phosphogluconate pathway.

Kupffer cell degeneration consistently observed suggestive of hepatic necrosis has been reviewed widely [30]. In amoebic hepatitis, liver lysosomal enzymes have been reported elevated due to their secretion around amoebae or stimulation for any cellular defense apart from respiratory burst and chemotaxis by these cells [33]. The electron microscopic observations indicated that kupffer cells accumulate around and exhibited hyperplasia condition showing degenerated nucleus, enlarged lysosomal vesicles. The enlarged lysosomal vesicles were further correlated by higher lysosomal enzyme levels in serum. In liver biopsies these enzymes were significantly enhanced as shown in Table 2-3. Acid phosphatase, leucine aminopeptidase enzymes catalyzing phosphorylation and amino-peptidization, seem to be secreted more and suggestive of active protein degradation. β-glucuronidase and aryl sulphatase high enzyme secretion was suggestive of continuous breakdown of aryl substituted and β-glucuronidation reactions during cytolysis [10]. No data is available on liver cell enzymes estimated in liver cells from liver biopsy of amoebic patients. However, serum choline esterase, alkaline phosphatase, glucose-6P-dehydrogenase, ornithine carbamyl transferase, cyclooxygenase2, lactate dehydrogenase enzymes have been reported significant in patients with addition of new members of enzymes [34-41].

There appear two main reasons of enzyme level recovery by nitroimidazole. First, these enzyme changes were recovered by nitroimidazole therapy as drug kills the causative amoeba and kupffer cell macrophagal activity diminishes the hepatic enzyme secretion and results in decreased enzyme levels towards normal. Second, regenerating hepatocytes may also regulate the normal cell recovery process and initializes the signaling Kupffer cells to keep stored enough lysosomal enzymes.

CHALLENGES, LIMITATIONS AND FUTURISTIC APPROACHES

There are two major challenges. First challenge was to get enough biopsy to estimate several enzymes. Second challenge was to choose significant enzymes as representative of hepatocellular criteria. The main limitation was that the 'hepatocellular dysfunction criteria' was used in small number of patients. The repeat biopsy samples for electron microscopy observations needs more thoroughly controlled experiments. Other limitation was that the enzyme estimations in serum and biopsy samples may not be perfect representative samples and it needs to establish the measurable and actual enzyme activities in cells. In future, chip technology or silica or polymer coated enzyme estimation techniques may be more reliable in small sample collections with high degree of accurate enzyme estimation.

CONCLUSION

The diffused suppurative or nonsuppurative amoebic liver damage are two separate entities. The hepatocellular dysfunction criterion distinguishes the amoebic entities by enzyme levels in serum, biopsy samples and electron microscopy. Initially, glucose metabolic regulation participates and leads to metabolic integrity loss. It may be developed into hepatomegaly or extrahepatic involvement. 2'-nitroimidazole is drug of choice in hepatic amoebiasis and its action may be evaluated rapidly by enzyme biochemical estimations without time consuming drug monitoring and therapeutic assay techniques.

ACKNOWLEDGMENT

The author acknowledges the assistance provided by Professor Rakesh K Tandon to provide assistant research officer job during this work with grant assistance from ICMR and Dr V.S.Singh for valuable guidance and grant from ICMR.

REFERENCES

[1]   Rigothier, MC, Khun H, Tavares P, Cardona A, Huerre M, Guillén N. Fate of *Entamoeba histolytica* during Establishment of Amoebic Liver Abscess Analyzed by Quantitative Radioimaging and Histology. *Infection Immunity*, 2002: 3208–3215.

[2]   Boettner DR, Huston CD, LinfordAS, Buss SN, Houpt E, ShermanNE, Petri WA Jr.Entamoeba histolytica phagocytosis of human erythrocytes involves PATMK, a member of the transmembrane kinase family. *PLoS Pathog.* 2008;4(1):e8.

[3]   Kerr JFR, Cooksley WGE, Serle J, Halliday JW, Halliday WJ, Holder L, Roberts I, Burnett W, Powell LW. *Lancet* 2:827-828.

[4]   Stanley SL. Amoebiasis. *The Lancet*, 2003;361: 1025-1034.

[5]   Stanley SL. Pathophysiology of amoebiasis *TRENDS in Parasitology* 2001;17(6):280-285.

[6]   Blazquez S, Rigothier MC, Huerre M, Guillen N. Initiation of inflammation and cell death during liver abscess formation by Entamoeba histolytica depends on activity of the galactose/N-acetyl-D-galactosamine lectin. *International Journal for Parasitology* 2007;37: 425–433.

[7]   Birmelin M, Decker K. *Eur. J. Biochem.* 1983; 131:539-543.

[8]   Miyai K. In *Toxic Injury of the Liver Part A*: Ed. E Farber and MM Fisher, M. Decker, New York, pp 59-154.

[9]   Virk KJ, Mahajan RC, Dilawari JB, Ganguly NK Mechanism of tissue damage through free oxygen radicals during hepatic amoebiasis in guinea pigs. *GastroenterolJpn.* 1990 Apr ;25 (2):265-9.

[10] K J Virk, R C Mahajan, J B Dilawari, N K Ganguly. Role of beta-glucuronidase, a lysosomal enzyme, in the pathogenesis of intestinal amoebiasis: an experimental study.*Trans R Soc. Trop. Med. Hyg.* 1988 ;82 (3):422-5.

[11] Virk KJ, Ganguly NK, Mahajan RC, Bhushnurmath SR, Dilawari JB Generation of reactive oxygen species by Kupffer cells and blood monocytes during intestinal amebiasis in guinea pigs.*Gastroenterol Jpn.* 1988 Dec ;23 (6):688-94.

[12] Shibayama M, Campos-Rodrıguez R, Ramırez-Rosales A, Flores-Romo L, Cantellano ME, Martınez-Palomo A, Tsutsumi V. *Entamoeba histolytica:* Liver Invasion and Abscess Production by Intraperitoneal Inoculation of Trophozoites in Hamsters, *MesocricetusAuratus.EXPERIMENTAL PARASITOLOGY* 1998; 88, 20–27.

[13] Bhatnagar R, Schade U, Reitschel ET, Decker K. *Eur. J. Biochem.* 1982;125:125-130.

[14] Bhatnagar R, Schirmer R, Ernst M, Decker K. *Eur. J. Biochem.* 1981;119:171-175.

[15] Cahill GF, Owen GF. *Carbohydrate metabolism and its disorders.* Vol 1.Ed. Dickens F, Whelan WJ and Randli PJ. Academic Press, Landon. Pp 497.

[16] Owen OE, Patel MS, Block BSB, Kruclean TH, Raichle FA, Mazzoli MA. Ed. Hanson RW and Mehlman MA. *Gluconeogenesis.* Wiley New York. pp 533.

[17] Scholte HR, Busch HF, Luyt-Houwen IE, Vaandrager Verduin MH, Przyrembel H, Arts WF, Defects in Oxidative phosphorylation, biochemical investigations in skeletal muscles and expression of the lesion in other cells. *J. Inherited Metab. Dis.* 1987, 10(1):81-97.

[18] Tandon BN, Tandon HD, Puri BK. An electron microscopic study of liver in hepatomegaly presumably caused by amebiasis. *Exp. Mol. Pathol.* 1974; 22:118-132.

[19] Virk KJ, Ganguly NK, Dilawari JB, Mahajan RC Role of lysosomal enzymes in tissue damage during hepatic amoebiasis. An experimental study.*Liver.* 1989 Dec ;9 (6):338-45.

[20] Virk KJ, Ganguly NK, Prasad RN, Dilawari JB, Mahajan RC Kupffer cell functions during intestinal amoebiasis in guinea pigs.*J. Gastroenterol. Hepatol.* ;5 (5):518-24.

[21] Wood BA, Monro AM. Metronidazole: Biotransformation. *Brit. J. Vene Dis.* 1975, 51, 51-53.

[22] Hodgkiss RJ. Use of 2-nitroimidazoles as bioreductive markers for tumour hypoxia*Anti-Cancer Drug Design* (1998), 13 687-702.

[23] Bos HJ, Van Den Eijk AA, Steeremberry PA. Application of ELISA in serodiagnosis of amoebiasis. *Trans Roy Soc. Trop. Med. Hyg.* 1976;6:440.

[24] Mehghini Giorgie. *New Series.* 1959;Vol 4(9):682-692.

[25] Worthington Enzyme Manual. Worthington Biochemical Corporation, 730 Vassar Avenue, Lakewood, New Jersey 08701, Editor: Worthington, V. 1993:1-1947.

[26] Tandon BN, Tandon HD, Puri BK. An electron microscopic study of liver in hepatomegaly presumably caused by amoebiasis. *Exp. Mol. Pathol.* 1975; 22:118-132.

[27] Takeuchi A, Phillips BP. Electron microscopis studies of experimental amoeba histolytica infection in guinea pigs(1) Penetration of intestinal epithelium by trophozoites. *Am. J. Trop. Med. Hyg.* 1975,24(34): .

[28] Tondon BN, Tandon HD, Ravi VV, Gandhi PC. Diffused liver injury in amoebic liver abscess of liver. An electron microscopic study. *Exp. Mol. Pathol.* 1974, 21.

[29] Bradfield JWB, Souhami RL. Hepatocyte damage secondary to Kupffer cell phaogocytosis: In *The reticuloendothelial system and pathogenesis of liver diseases.* Ed: Leihr H, Grun M. Elsevier, Amsterdam, 1980: 165-171.

[30] Michalopoulos GK, DeFrances MC. Liver regeneration. *Science.* 1997, 276:60-66.

[31] Ramirez-Emiliano J, Flores-Villavicencio LL, Segovia J, Arias-Negrete S. Nitric oxide participation during amoebic liver abscess development. *Medicina* (B.Aires) 2007; 67(2): 167-176.

[32] Das P, Debnath A, Munoz ML. Molecular mechanisms of pathogenesis in amoebiasis. *Ind. J. Gastroenterol.* 1999, 18(4):161-166.

[33] Katzenstein D, Rickerson V, Braude A. New concepts of amoebic liver abscess derived from hepatic imaging, serodiagnosis and hepatic enzymes in 67 consecutive cases in San Diego. *Medicine* (Baltimore) 1982, 61(4): 237-246.

[34] Santhanagopalan T, Jayasekaran S, Joseph HA. Serum enzymes in amoebic liver disease.*Indian J. Pathol. Bacteriol.* 1968 Jan;11(1):61-4.

[35] Sosa A, Bernal A, Reyes A, Rosado A. Lysosomal enzyme participation in the pathogenesis of amebic liver abscess. *Gac. Med. Mex.* 1971 Jun;101(6):759-70.

[36] Mahajan RC, Singh K, Ganguly NK, Chitkara NL. Significance of serum lactic dehydrogenase enzyme in hepatic amoebiasis. *Indian J. Med. Res.* 1975 Jul;63(7):1006-9.

[37] Mehrotra MP, Singh MM, Pursnani ML, Sharma RD. Serum ornithine carbamyl transferase activity in hepatic amoebiasis. *J. Indian Med. Assoc.* 1979 May 16;72(10):228-31.

[38] Narang AP, Datta DV, Mathur VS. Antipyrine clearance, aminopyrine N-demethylase, and bilirubin UDP-glucuronyl transferase activity in patients with amoebic liver abscess. *Biopharm.Drug Dispos.* 1982 Jan-Mar;3(1):39-45.

[39] Datta DV, Narang AP, Arya P, Mahajan RC. Glucose 6-phosphate dehydrogenase activity in amoebic liver abscess - a preliminary communication. *J. Assoc. Physicians* India. 1982 Aug;30(8):527-8.

[40] Ventura-Juárez J, Jarillo-Luna RA, Fuentes-Aguilar E, Pineda-Vázquez A, Muñoz-Fernández L, Madrid-Reyes JI, Campos-Rodríguez R. Human amoebic hepatic abscess: in situ interactions between trophozoites, macrophages, neutrophils and T cells. *Parasite Immunol.* 2003 Oct;25(10):503-11.

[41] Sánchez-Ramírez BE, Ramírez-Gil M, Ramos-Martínez E, Rohana PT. Entamoeba histolytica induces cyclooxygenase-2 expression in macrophages during amebic liver abscess formation. *Arch Med. Res.* 2000 Jul-Aug;31(4 Suppl):S122-3.

In: Biological Aspects of Human Health and Well-Being   ISBN: 978-1-61209-134-1
Editor: Tsisana Shartava   © 2011 Nova Science Publishers, Inc.

Chapter VIII

THE EFFECT OF NITROIMIDAZOLE ON GLUCOKINASE ENZYME REGULATORY PROPERTIES: GLUCOKINASE AS BIOSENSOR[#]

Rakesh Sharma[2] and Vijay S. Singh[2]*
[1]Gastroenterology Department,
All India Institute of Medical Sciences, New Delhi, India 110029
[2]Biochemistry Department,
L.L.R.M.MedicalCollege, Meerut, Uttar Pradesh, India

ABSTRACT

Aim: To evaluate the cytotoxicity of nitroimidazole in isolated human hepatocytes in cultures by using glucokinase enzyme activity as hepatocyte biomarker and evidence of hormonal dependent glucokinase regulatory behavior. Hypothesis: Hepatocyte hormone dependent glucokinase may be a biomarker of cytotoxicity evaluation of nitroimidazole. Methods and Materials: The selected liver biopsies from 10 patients in ongoing research were processed for isolation and fractionation of hepatocytes. Three groups of liver biopsy heaptocytes were: untreated control (group I); liver biopsy from liver abscess infected (group II); and liver biopsy from nitroimidazole treated liver(group III). Results: The glucokinase enzyme activities showed inhibited enzyme activity by actinomycin D, enhanced activity by insulin with triamicilone, poor glucokinase enzyme activity enhancement by progesterone. Discussion: The glucokinase enzyme synthesis is mainly hormonal dependent and regulated at gene level. Its gene expression control by insulin is

---

[#] The manuscript is part of Ph.D dissertation "The effect of imidazoles in hepatic cells during development of liver abscess" submitted at LLRMMedicalCollege, Meerut.

[2] The hepatocytes were cultured from biopsies of patients (n=10) who did not have any liver infection, disease and liver damage as control cells. Other matched group of hepatocytes were cultured from biopsies of hepatic abscess patients (n=10). Another group of hepatocytes were cultured from biopsies of livers from volunteers who received nitroimidazole 2 x 3 gm one day dose three times. The procedure of liver biopsy was by use of Manghini needle aspiration as described elsewhere [36]http://precedings.nature.com/documents/2738/version/1.

[*] E-mail: rksz2004@yahoo.com

significant in beta cells but may be possible in glycogen synthesis in human hepatocytes. The nitroimidazole effect was negative in comparison with actinomycinD. The effect of nitroimidazole is less likely to influence the gene expression of glucokinase unlike the insulin. Conclusion: The nitroimidazole directly affects the nonhormonal human glucokinase enzyme activity and can be used as biomarker. The nitroimidazole effect on hormonal dependent glucokinase synthesis is significant in defining the liver regeneration.

Keywords: nitroimidazole, cytotoxicity, liver, hepatocytes, glucokinase, progesterone, insulin.

INTRODUCTION

The Glucokinase enzyme is first regulatory enzyme in glycolysis for glucose converted to glucose 6 Phosphate during breakdown of glucose in cells to make pyruvate. The glucokinase was first reviewed by Weinhouse 1976 [1]. Glucokinase plays significant role in energy balance and maintains metabolic integrity under hormonal control. The hormonal regulatory behavior glucokinase enzyme was first reported in hepatocytes by Schudt 1979 to explain role of glucocorticoids in hepatic infections and hepatic dysfunction [2,3]. Later insulin, glucagon hormones, SH reagents were reported to play significant role in glucokinase regulation [4,5,6]. We believe that hepatic dysfunction begins with low ATP and NADPH in the cell mainly due to altered glucokinase hormonal regulation. Over 20 years the hormonal regulation of glucokinase was evidenced as main factor of liver metabolism and its hepatocellular control was evidenced in diabetes and hepatic diseases [7,8,9,10,11,12,13,14,15]. Later in hepatocytes, glucokinase emerged as glucose sensor and metabolic signal of hepatocytes [16]. The hormonal dependence of glucokinase gene expression and its regulation was reported in beta cell and hepatocyte glucokinase [17,18,19,20,21,22,23]. In recent past the emphasis was on gene and enzyme protein synthesis under the influence of insulin hormone [24,25].

Recently, nitroimidazole derivatives emerged as potential hypoxia monitoring chemicals and nitroimidazoles demonstrate significant effect on hepatocellular enzymes, cellular metabolic integrity and energy balance [26,27]. It is strong belief that nitroimidazole compounds cause cytotoxity by induceing oxygen insuffiency in liver cells. The nitroimidazole induced oxygen deficiency kills the oxygen starved hepatic tumor cells while normal liver cells undergo regeneration and survive [28]. In last three decades several efforts continued in the direction of nitroimidazole and derivatives to study the hepatocytototoxicity in hypoxic tumors and hepatic infections by exposing liver cells with nitroimidazole derivatives [29]. In present study, our focus was to characterize glucokinase enzyme regulatory behavior in presence of additives and possible cytotoxicity effect of nitroimidazole using glucokinase as biosensor. The hepatocyte glucokinase behavior was compared in three groups in presence of actinomycin D, insulin with triamicilone. The three groups were: control untreated hepatocytes with hepatic abscess liver hepatocytes and nitroimidazole treated liver heaptocytes.

MATERIALS AND METHODS

The hepatocytes were isolated from human liver biopsies using collagenase enzyme digestion and their monolayer cultures were maintained for 48 hours as described elsewhere [30]. Additives added in hepatocyte cell cultures: For experiments on the effect of additives on hepatocytes, these hepatocytes in culture were added with different additives viz. 3 ng per ml actinomycin D, 20 nM insulin with 20 nM triamacinolone, 0.1 M progesterone at various concentrations in flacon flask. The hepatocytes were directly exposed to different additives in their culture media aseptically in sterile conditions. The Actinomycin D(3 nG/ml), insulin (20 mM)with triamicilone(20 nM) were chosen optimal concentrations of additives. At different intervals, the additives added to hepatocytes in culture to observe the glucokinase behavior as altered response in presence of additives.

After the exposure of additives, the hepatocytes were harvested from falcon flasks and hepatocytes were homogenized and fractionated for isolation of cytosol from cell organelles as described elsewhere [31]. The cytosolic fractions were isolated and used to estimate the glucokinase enzyme activity. The assay of glucokinase was done by following methods.

Enzyme assay: Glucokinase enzyme specific activity was measured by spectrophotometer in a reaction mixture 2.0 ml containing composition of 0.1 M tris buffer pH 7.4, 0.2 mM NAD, 5.0 mM ATP, 5.0 mM $MgCl_2$ and 0.1 unit of glucose-6-PDH and 0.5 ml(100 mM) glucose at final temperature of 30˚C. One enzyme unit was calculated by substracting the glucokinase activity at 0.5 mM glucose from glucokinase activity at 100 mM glucose activity. The glucokinase specific activity was measured on spectrophotometer at 340 nm. The enzyme activity of glucokinase $=\Delta OD_{340}$ per minute per mg enzyme protein in per ml cytosol x 10^3 orone specific activity unit of glucokinase was defined as enzyme catalyzing phosphorylation of 1 µmole of glucose per minute at 30°C [32].

The specific activity of enzyme in presence of additives was screened at different time intervals as indication of regulatory behavior of glucokinase.

Statistical methods: The comparison of glucokinase enzyme activity in different groups of heaptocytes was made by linear regression analysis using mean $\pm$ sd, student test and P value. The sensitivity, specificity, accuracy and precision were calculated by true and false observation. The reproducibility was calculated using experiments three times.

RESULTS

The hepatocyte yield from livers was 2.5 x10^8 per mg liver. The cytosol fraction showed the removal of all cell organelle from supernatant and resultant supernatant as cytosol was free of any remnant by microscopic observation by trypan blue exclusion.

The actinomycin D showed non-significant inhibitory change in glucokinase enzyme activity as shown in Figure 1 at the top histogram bars. The nitroimidazole showed enhanced glucokinase activity and reversed the inhibitory effect of actinomycin D. The glucokinase activity in hepatic abscess hepatocyte group II did not show any significant effect of nitroimidazole. Overall actinomycin D alone did not show any change in glucokinase enzyme activity in hepatic abscess hepatocyte group II or nitroimidazole treated hepatocyte group III.

The addition of triamcinolone with insulin in combination showed significant increase in glucokinase activity in all hepatocyte groups. It pointed out to the enhancement in hormone dependent glucokinase enzyme synthesis in hepatocytes and enzyme activity showed highest enzyme synthesis rate in group II hepatic abscess hepatocytes while nitroimidazole treated hepatocytes group III showed lesser increase in hormone dependent glucokinase in comparison with group II hepatocytes as shown in Figure 1. The increase in enzyme activity or enhanced glucokinase enzyme synthesis in hepatocytes showed that glucokinase enzyme behavior was hormone dependent. Different concentration of insulin and progesterone hormones further indicated the regulatory behavior of glucokinase enzyme.

The addition of progesterone hormone 0.1 μgm/ml in hepatocyte cultures in group II showed maximum enhanced glucokinase enzyme activity and nitroimidazole treated hepatocytes in group III showed lesser enhanced glucokinase activity but more than control hepatocyte enzyme activity as shown in histogram bars.

The control glucokinase enzyme activity in untreated hepatocytes did not show any difference in their glucokinase enzyme activity in presence of nitroimidazole additive. It was clear that hepatic abscess liver biopsy gave yield of hepatocytes with similar glucokinase enzyme activities as glucokinase enzyme activities in untreated and nitroimidazole treated hepatocytes. So, addition of actinomycinD and insulin with triamicilone enhanced the glucokinase enzyme synthesis in hepatocytes as shown in Figure 1 at the bottom histograms bars.

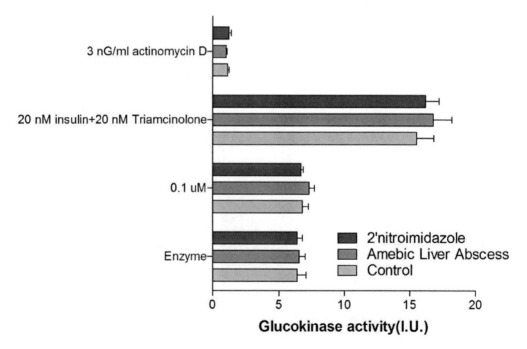

Figure 1. The nitroimidazole effect on glucokinase enzyme activity in guinea pig normal and infected hepatocytes is shown in presence of different enzyme regulatory additives actinomycin D, insulin-triamcilone. Notice the enhanced effect of insulin-triamicilone on glucokinase enzyme (P value <0.0001; r^2 0.9872) in nitroimidazole treated vs liver abscess biopsy heaptocytes (Group II vs group III).

DISCUSSION

Glucokinase enzyme plays major role in human liver metabolism. Glucokinase enzyme regulation in isolated human hepatocyte cultures was earlier evidenced by enhanced glucokinase enzyme synthesis in the presence of insulin, glucasamine, SH compounds progesterone [4,5,6,33]. Later glucokinase gene and gene expression gave clue of two types of enzyme synthesis in hepatocytes as hormonal and non-hormonal regulation over enzyme synthesis [7-15, 17-15]. Direct interaction of liver cells with amoebic trophozoits in presence of additives was reported only one time suggested similar glucokinase hormone dependent behavior [34]. However, the present study extends the same view of glucokinase enzyme regulatory behavior by insulin and progesterone hormones and demonstrates the cytotoxicity effect of nitroimidazole using glucokinase as biosensor enzyme. Different additives such as insulin-triamicinolone and actinomycin D exhibited the glucokinase specificity and increased glucokinase enzyme synthesis in hepatocytes. The enhanced enzyme activity may be attributed as a consequence of enhanced rate of over all enzyme protein synthesis. In previous report, enhanced protein synthesis as consequence of glucokinase regulation by insulin and corticosteroid hormones was reported in animal hepatocytes and now it is an established phenomenon initially proposed by Schudt et al.1979[2,3]. The insulin dependent enzyme activity enhancement of glucokinase was previously reported apparantly due to de novo enzyme synthesis [17,18,19,20]. The present study also showed similar increased glucokinase synthesis in hepatocyte cells in culture in response to insulin addition and progesterone addition. Therefore, insulin hormone controls probably the cellular glucokinase activities by specific interference with over all protein synthesis. The similar observation of hepatocytes glucokinase enzyme was noticed in case of progesterone hormone. It was clear in earlier experiments that there exist two factors of enzyme synthesis in the hepatocytes. First factor was possibly the enhanced hormone specific glucokinase enzyme synthesis in presence of hormones or hormone specific receptor mediated trigger of glucokinase specific protein synthesis. The second factor could be non-hormone dependent low protein synthesis in hepatocytes. The view of hormone dependent glucokinase enzyme synthesis is also supported by our observations of hepatocyte enzyme activities in presence of insulin, triamicilone. In the absence of progesterone hormone, glucokinase enzyme activity of hepatocytes declined spontaneously. It appeared that the glucokinase enzyme synthesis was arrested after withdrawal of insulin and triamicinolone. It may be possible that only glucokinase enzyme degradation proceeded in the absence of hormones and only enzyme synthesis was depending upon over all protein synthesis at lower rate. Glucokinase enzyme degradation in vivo in hepatocytes was reported much faster by withdrawal of hormones in earlier reports [8,9,10].

In present experiment, insulin hormone stimulated the glucokinase enzyme synthesis of hepatocytes in presence of triamicilone. The triamicilone was considered as sole enhancer of hormone action across the cell membrane surface receptor. The present study also supports the enhanced hormone dependent glucokinase enzyme synthesis in presence of triamicilone. Other important observation was that hepatocytes from hepatic abscess liver biopsies showed higher rate of glucokinase synthesis. This observation is on the same lines as reported by Ventura-Juarez et al. 2003 using direct interaction of hepatocytes with trophozoits [34]. In our study it was clear that glucokinase enzyme synthesis was enhanced while nitroimidazole tends to recover the glucokinase enzyme synthesis by lowering the insulin dependent glucokinase enzyme synthesis possibly over all protein synthesis. It needs further investigation if isolated hepatocytes have time dependent and/or concentration dependent influence of hormone receptors during glucokinase enzyme synthesis. However, earlier study reported that the presence of amoebic trophozoits in hepatocytes changes the glucose availability from culture media to hepatocytes possibly due to active phagocytosis and amoebic virulent nature [35]. In our earlier report, the trophozoits were nonvirulent. So, it was likely that the trophozoits remain competitive physically with hepatocytes for their glucose supply without influencing their virulence [36].

In recent past, the progesterone was cited valuable hormones in pancreatic function and regulator of glucokinase activity in beta cells to answer the effect of steroids on regulatory behavior of glucokinase in cAMP and phosphodiesterase cascade [37,38]. The progesterone hormone addition to hepatocytes exhibited the enhanced glucokinase enzyme synthesis as induced effect on de novo enzyme synthesis. The induced glucokinase enzyme synthesis effect may be attributed due to transcription of proteins as an accelerated function of progesterone receptor or insulin receptor or ion transport system [38]. Still it remains to investigate if ion transport system controls the receptor function. Similar induced enzyme effect may be presumed for insulin-triamcinolone additives with hepatocytes.

On contrary, actinomycin D inhibited glucokinase synthesis in all three hepatocyte cells in different experimental groups. Moreover, glucokinase enzyme synthesis regulation in multi-hormonal pretreated hepatocytes has been reported by regulated rate of glucokinase enzyme synthesis with available sufficient mRNA synthesis in hepatocytes [13,18]. In principle, degradation of glucokinase enzyme was appeared to get reduced in absence of insulin-triamcinolone in hepatocytes. The glucokinase enzyme synthesis was induced by de novo non-hormonal dependent over all protein synthesis after removal of these hormones in cultures.

The effect of nitroimidazole on enhanced non-hormone dependent glucokinase enzyme synthesis in presence of actinomycin D may be attributed by the fact that nitroimidazole dwarfs the inhibitory effect of actinomycin D on glucokinase enzyme synthesis as shown in Figure 1. While the enhanced hormone dependent glucokinase enzyme synthesis in presence of insulin and reduced non-hormone dependent gucokinase enzyme synthesis in hepatocytes in presence of triamicilone may be attributed by the combined effect of these additives. The nitroimidazole acts specifically on non-hormone glucokinase enzyme synthesis and it may be observed from Figure 1 in insulin-triamicilone exposed hepatocytes. The combined effect of insulin and actinomycin D on glucokinase enzyme activity showed lower enzyme activity lower than hepatic abscess exposed hepatocytes as shown in Figure 1. These observations put evidence that nitroimidazole action is cytotoxic in one way that nitroimidazole squeezes out glucose depleted reservoirs further induced by pre-hepatic abscess exposed hepatocytes. Other evidence is clear that hormone dependent glucokinase enzyme synthesis also reduced by nitroimidazole. However, it remains to investigate as what makes nitroimidazole to enhance the non-hormonal glucokinase enzyme activity otherwise mediated by progesterone hormonal influence.

Challenges, limitations, futuristic approach: The main challenge was limited amount of human biopsy available from patients on protocol. The liver biopsy tissue from volunteers and hepatic abscess subjects may differ in the experimental conditions of biopsy collection. Another challenge was hepatocyte isolation to get higher hepatocyte yield. The cell fractionation further poses challenge as achieve sufficient glucokinase enzyme activity due to loss of activity during centrifugation and handling cytosol supernatant in culture flasks. The limitation of the data was due to less subjects used for biopsy and repetition of experiments was risk due to ethical issues. Other major limit was that glucokinase enzyme activity may not be true representative of glucokinase enzyme hormone and non-hormone dependent enzyme synthesis. The glucokinase regulatory behavior studies are significant from several reasons. The glucokinase enzyme synthesis represents the beginning of utilization of glucose reservoir to supply energy ATP and redox potential NADPH in hepatocytes. The hormone regulatory glucokinase behavior may explain the reason of hypoxia, endocrine control of liver tumors and infections. It may be significant in evaluation of hormonal replacement treatment and hepatic cancer. In future, possible advanced tracer techniques and intracellular glucose metabolic imaging may answer better the localization of glucokinase role in liver and possibly in other organs.

REFERENCES

[1]   Weinhouse S. Regulation of glucokinase in liver. *Curr. Top Cell Regul.* 1976;11:1-50.

[2]   Schudt C. Hormonal regulation of glucokinase in primary cultures of rat hepatocytes. *Eur. J. Biochem.* 1979 Jul;98(1):77-82.

[3]   Schudt C. Regulation of glycogen synthesis in rat-hepatocyte cultures by glucose, insulin and glucocorticoids. *Eur. J. Biochem.* 1979 Jun;97(1):155-60.

[4]   Baltrusch S, Lenzen S, Okar DA, Lange AJ, Tiedge M. Characterization of Glucokinase-binding Protein Epitopes by a Phage-displayed Peptide Library. Identification of 6-phosphofructokinase/Fructose 2,6-Bisphosphatase as a novel interaction partner. *J. Biol. Chem.* 2001 Nov 23;276(47):43915-23.

[5]   Postic C, Shiota M, Magnuson MA. Cell-specific roles of glucokinase in glucose homeostasis. *Recent Prog. Horm Res.* 2001;56:195-217.

[6]   Gasa R, Fabregat ME, Gomis R. The role of glucose and its metabolism in the regulation of glucokinase expression in isolated human pancreatic islets. Biochem *Biophys. Res. Commun.* 2000 Feb 16;268(2):491-5.

[7]   Burke CV, Buettger CW, Davis EA, McClane SJ, Matschinsky FM, Raper SE. Cell-biological assessment of human glucokinase mutants causing maturity-onset diabetes of the young type 2 (MODY-2) or glucokinase-linked hyperinsulinaemia (GK-HI). *Biochem. J.* 1999 Sep 1;342 (Pt 2):345-52.

[8]   Nordlie RC, Foster JD, Lange AJ. Regulation of glucose production by the liver. *Annu.Rev. Nutr.* 1999;19:379-406.

[9]   Hers HG. Regulation of glucokinase and its role in liver metabolism. Trends Biochem Sci. 1992 Feb;17(2):59.

[10]   Tiedge M, Steffeck H, Elsner M, Lenzen S. Metabolic regulation, activity state, and intracellular binding of glucokinase in insulin-secreting cells. *Diabetes.* 1999 Mar;48(3):514-23.

[11]   Berman HK, Newgard CB. Fundamental metabolic differences between hepatocytes and islet beta-cells revealed by glucokinase overexpression. *Biochemistry.* 1998 Mar 31;37(13):4543-52.

[12] Clark SA, Quaade C, Constandy H, Hansen P, Halban P, Ferber S, Newgard CB, Normington K. Novel insulinoma cell lines produced by iterative engineering of GLUT2, glucokinase, and human insulin expression. *Diabetes.* 1997 Jun;46(6):958-67.

[13] Tanizawa Y, Koranyi LI, Welling CM, Permutt MA. Human liver glucokinase gene: cloning and sequence determination of two alternatively spliced cDNAs. *Proc. Natl.Acad. Sci. USA.* 1991 Aug 15;88(16):7294-7.

[14] Iynedjian PB, Marie S, Wang H, Gjinovci A, Nazaryan K. Liver-specific enhancer of the glucokinase gene. *J. Biol. Chem.* 1996 Nov 15;271(46):29113-20.

[15] Watada H, Kajimoto Y, Umayahara Y, Matsuoka T, Kaneto H, Fujitani Y, Kamada T, Kawamori R, Yamasaki Y. The human glucokinase gene beta-cell-type promoter: an essential role of insulin promoter factor 1/PDX-1 in its activation in HIT-T15 cells. *Diabetes.* 1996 Nov;45(11):1478-88.

[16] Matschinsky FM. Glucokinase as glucose sensor and metabolic signal generator in pancreatic beta-cells and hepatocytes. *Diabetes.* 1990 Jun;39(6):647-52..

[17] Iynedjian PB, Marie S, Gjinovci A, Genin B, Deng SP, Buhler L, Morel P, Mentha G. Glucokinase and cytosolic phosphoenolpyruvate carboxykinase (GTP) in the human liver. Regulation of gene expression in cultured hepatocytes. *J. Clin. Invest.* 1995 May;95(5):1966-73.

[18] Christ B, Nath A, Heinrich PC, Jungermann K. Inhibition by recombinant human interleukin-6 of the glucagon-dependent induction of phosphoenolpyruvate carboxykinase and of the insulin-dependent induction of glucokinase gene expression in cultured rat hepatocytes: regulation of gene transcription and messenger RNA degradation. *Hepatology.* 1994 Dec;20(6):1577-83.

[19] Lenzen S, Tiedge M. Regulation of pancreatic B-cell glucokinase and GLUT2 glucose transporter gene expression. *Biochem. Soc. Trans.* 1994 Feb;22(1):1-6.

[20] Leibiger B, Walther R, Leibiger IB. The role of the proximal CTAAT-box of the rat glucokinase upstream promoter in transcriptional control in insulin-producing cells. *Biol. Chem. Hoppe Seyler.* 1994 Feb;375(2):93-8.

[21] Takeda J, Gidh-Jain M, Xu LZ, Froguel P, Velho G, Vaxillaire M, Cohen D, Shimada F, Makino H, Nishi S, et al. Structure/function studies of human beta-cell glucokinase. Enzymatic properties of a sequence polymorphism, mutations associated with diabetes, and other site-directed mutants. *J. Biol. Chem.* 1993 Jul 15;268(20):15200-4.

[22] Iynedjian PB. Mammalian glucokinase and its gene. *Biochem. J.* 1993 Jul 1;293 (Pt 1):1-13.

[23] Gidh-Jain M, Takeda J, Xu LZ, Lange AJ, Vionnet N, Stoffel M, Froguel P, Velho G, Sun F, Cohen D, et al. Glucokinase mutations associated with non-insulin-dependent (type 2) diabetes mellitus have decreased enzymatic activity: implications for structure/function relationships. *Proc. Natl. Acad. Sci. USA.* 1993 Mar 1;90(5):1932-6.

[24] Printz RL, Magnuson MA, Granner DK. Mammalian glucokinase. *Annu. Rev. Nutr.* 1993;13:463-96.

[25] Magnuson MA. Glucokinase gene structure. Functional implications of molecular genetic studies. *Diabetes.* 1990 May;39(5):523-7.

[26] Brezden CB, Horn L, McClelland RA, Rauth AM. Oxidative stress and 1-methyl-2-nitroimidazole cytotoxicity.*Biochem. Pharmacol.* 1998 Aug 1; 56(3): 335-44.

[27] Ersoz G, Karasu Z, Akarca US, Gunsar F, Yuce G, Batur Y. Nitroimidazole-induced chronic hepatitis.*Eur. J. Gastroenterol. Hepatol.* 2001 Aug;13(8):963-6.

[28] Thiim M, Friedman LS. Hepatotoxicity of antibiotics and antifungals.*Clin. Liver Dis.* 2003 May;7(2):381-99.

[29] Moller P, Wallin H, Vogel U, Autrup H, Risom L, Hald MT, Daneshvar B, Dragsted LO, Poulsen HE, Loft S Mutagenicity of 2-amino-3-methylimidazo[4,5-f]quinoline in colon and liver of Big Blue rats: role of DNA adducts, strand breaks, DNA repair and oxidative stress. *Carcinogenesis.* 2002 Aug;23(8):1379-85.

[30] Hendriks HF, Brouwer A, Knook DL. Isolation, purification, and characterization of liver cell types. *Methods Enzymol.* 1990;190:49-58.

[31] Ladola M, Newsam RJ, Sumner IG. A rapid method for fractionation of isolated hepatocytes. *Annals of Biochem.* 1981;117:45-52.

[32] Belfiore F, Iannello S. A formula for quantifying the effects of substrate cycles (futile cycles) on metabolic regulation. Its application to glucose futile cycle in liver as studied by glucose-6-phosphatase/glucokinase determinations. *Acta Diabetol Lat.* 1990 Jan-Mar;27(1):71-80.

[33] Tiedge M, Krug U, Lenzen S. Modulation of human glucokinase intrinsic activity by SH reagents mirrors post-translational regulation of enzyme activity. *Biochim. BiophysActa.* 1997 Feb 8;1337(2):175-90.

[34] Ventura-Juárez J, Jarillo-Luna RA, Fuentes-Aguilar E, Pineda-Vázquez A, Muñoz-Fernández L, Madrid-Reyes JI, Campos-Rodríguez R. Human amoebic hepatic abscess: in situ interactions between trophozoites, macrophages, neutrophils and T cells. *Parasite Immunol.* 2003 Oct;25(10):503-11.

[35] Virk KJ, Ganguly NK, Dilawari JB, Mahajan RC Role of lysosomal enzymes in tissue damage during hepatic amoebiasis. An experimental study.*Liver.* 1989 Dec ;9 (6):338-45.

[36] Sharma R. The Hepatocellular Hypoxia Criteria: 2'Nitroimidazole Effect on Hepatocyte Carbohydrate Metabolizing Enzymes and Kupffer Cell Lysosomal Enzymes: Hypoxia Screening. Nature Precedings. 2008; Dec. http://precedings. nature.com/documents/2738/version/1.

[37] Shao J, Qiao L, Friedman JE. Prolactin, progesterone, and dexamethasone coordinately and adversely regulate glucokinase and cAMP/PDE cascades in MIN6 beta-cells. *Am.J. Physiol. Endocrinol. Metab.* 2004 Feb;286(2):E304-10.

[38] Magnaterra R, Porzio O, Piemonte F, Bertoli A, Sesti G, Lauro D, Marlier LN, Federici G, Borboni P. The effects of pregnancy steroids on adaptation of beta cells to pregnancy involve the pancreatic glucose sensor glucokinase. *J. Endocrinol.* 1997 Nov;155(2):247-53.

In: Biological Aspects of Human Health and Well-Being   ISBN: 978-1-61209-134-1
Editor: Tsisana Shartava   © 2011 Nova Science Publishers, Inc.

Chapter IX

POST-TRANSCRIPTIONAL EFFECTS OF ESTROGENS ON GENE EXPRESSION: MESSENGER RNA STABILITY AND TRANSLATION REGULATED BY MICRORNAS AND OTHER FACTORS

Nancy H. Ing[*]

TexasA&MUniversity,
Departments of Animal Science and
Veterinary Integrative Biosciences,
College Station, TX

ABSTRACT

Estrogens exert powerful effects on physiology by regulating gene expression. Their effects on the transcriptional activities of genes are well described in the literature. However, estrogens are also the hormones that are best known for post-transcriptional gene regulation. With the combination of transcriptional and post-transcriptional regulation, gene expression can be rapidly and powerfully controlled to maximize the utility of genomic information throughout the long lives of vertebrate animals. For some cell responses, up to 50% of the genes with altered expression are the result of changes in the stabilities of the messenger RNAs (mRNAs). For many genes including the estrogen receptor alpha (ER) gene, post-transcriptional regulation is the primary mode of alteration of expression. This indicates that post-transcriptional gene regulation is critical to estrogen actions because the ER protein determines the estrogen-responsiveness of animal tissues to a large extent. Estrogens have been shown to regulate the expression of certain genes by greatly altering the stabilities of mRNAs, including stabilizing ER mRNA. This effect may be ancient as it appears to be conserved from mammals to fish and frogs. Some studies have identified unique proteins that are induced by estrogens to bind and protect specific mRNAs from degradation. Recently, hundreds of microRNAs have been discovered and are estimated to actively regulate about one third of protein-

---

[*] Nancy H. Ing, D.V.M, Ph.D, Associate Professor. ning@cvm.tamu.edu

encoding mRNAs. MicroRNAs associate with proteins in complexes on mRNAs, where they usually destabilize the mRNA or block its translation. Estrogens regulate the expression of microRNA genes in responsive tissues during normal physiology and disease processes. Other cell signals alter the expression of certain microRNAs that affect ER gene expression. Elucidation of the molecular mechanisms responsible for these post-transcriptional effects is certain to reveal novel molecular targets for therapeutic control of estrogen actions.

INTRODUCTION

Estrogens are a family of hormones that potently regulate reproductive, cardiovascular, bone, brain and other physiology in vertebrates by altering gene expression [Tsai et al., 1998]. Estrogens regulate gene expression on several levels. The mechanisms by which estrogen regulates the rate of transcription of genes (messenger RNA (mRNA) synthesis) are well described. However, concentrations of mRNAs are also dependent upon their rates of degradation. Posttranscriptional regulation of gene expression at the level of mRNA stability is rapidly being recognized as a powerful and widespread phenomenon [Tsai et al., 1998; Watson et al., 2007]. For some cell responses, up to 50% of the changes in concentrations of mRNAs are the result of changes in the stabilities of the mRNAs [Cheadle et al., 2005]. There are also effects on the rates of translation of mRNAs with little or no change in mRNA concentration. For the sake of space, this chapter does not address post-translational effects of estrogens on gene expression. Estrogens are the hormones best known for post-transcriptional regulation of gene expression. For many genes, including that of estrogen receptor alpha (ER), post-transcriptional regulation is the primary mode of regulation of gene expression [Ing, 2005a]. In the following review, we will use the post-transcriptional regulation of the expression of the ER gene as an example because (1) the ER protein transduces the majority of estrogen effects, (2) the ER gene is predominantly regulated post-transcriptionally, and (3) several different post-transcriptional mechanisms regulate the expression of the ER gene. The post-transcriptional regulation of the ER gene is likely to be common to a set of mRNAs (an "mRNA regulon") that are coordinately expressed but remain to be identified [Keene, 2007].

Post-transcriptional regulation of mRNA stabilities is similar to transcriptional regulation of genes, but with cytoplasmic trans-acting factors (including proteins and microRNAs) acting on cis-elements within the mRNAs. Most of the information controlling translation efficiency and mRNA stability is carried within the 5' and 3' untranslated regions (UTRs) of the mRNA, respectively. Typically, 5'UTRs are short (100 to 200 bases long) and are involved with ribosome loading and the initiation of translation. In contrast, the 3' UTRs of mRNAs can be quite extensive (500 to > 5,000 bases) and in many cases the 3' UTR sequences compose the majority of the mRNA. For example, the 6351 base long ER mRNA of the sheep carries 4354 bases of 3'UTR [Mitchell and Ing, 2003]. 3'UTRs regulate mRNA stability and are usually encoded in only one exon so they are not subject to alternative splicing [Nagy and Maquat, 1998]. Although mRNAs are often graphically represented as straight lines, mRNAs adopt complex secondary and tertiary structures by the intramolecular pairing of bases. The 3'UTRs of mRNAs are unique in that they are not subjected to passage of ribosomes, so their secondary and tertiary structures are likely to be more stable than those in other mRNA regions. While overall conservation of 3' UTR sequences of mRNA homologs between species is lower than in coding sequences, there are relatively large (> 500 base long) regions within 3' UTRs that are highly conserved across species [Mitchell and Ing, 2003]. This conservation of sequences in non-coding RNA regions implies that the sequences have important function(s) and the primary function of 3' UTR sequences is the regulation of mRNA stability.

The transacting factors (protein and microRNAs) that bind mRNA cis-elements determine the function and fate of the mRNA. ER mRNA is similar to a lot of other mRNAs encoding hormone receptors in that it is inherently unstable because its very long 3'UTR that carries destabilizing mRNA cis-elements such as the A+U Rich Element (ARE) [Chen and Shyu, 1995; Mitchell and Ing, 2003]. Sheep and human ER mRNAs carry 10 and 14 putative AREs, respectively, and all are in the 3'UTR [Kenealy et al., 2000; Mitchell and Ing, 2003]. Transfer of the 3' UTRs of ER mRNA to other mRNAs destabilizes those mRNAs. AREs destabilize mRNAs by binding destabilizing factors such as A + U-rich binding factor 1 p37 (AUF1p37) and tristetraprolin which direct the mRNA to exosomes, where mRNAs are degraded [Zhao et al., 2000; Parker and Song, 2004; Mukherjee et al., 2002]. There are also examples of mRNA stabilizing proteins such as HuR which, in response to cell signals, competitively bind AREs and prevent destabilization [Sengupta et al., 2003; Lasa et al., 2000; Wilson et al., 2001; Chen et al., 2002]. In every cell, the lifespan of an mRNA depends upon the balance of these influences. The discovery of microRNAs as a major class of post-transcriptional regulators of gene expression in animal tissues has revolutionized the field of gene expression. Whether microRNAs interact with destabilizing proteins like AUF1p37 in the same or different ribonucleoprotein complexes remains to be determined. Estrogens regulate mRNA stabilities, mRNA binding proteins, and microRNAs, as discussed below. Knowledge gained in this rapidly emerging field of post-transcriptional gene regulation will greatly improve our understanding of estrogen actions in normal and pathological tissues.

Estrogen influence and ER regulation are important in numerous aspects of normal physiology as well as in the development of hormone-dependent diseases. For example, the incidences of breast and uterine cancers and endometriosis in women are related to their exposure to estrogens [McKean-Cowdin et al., 2001; Gurates and Bulun, 2003].The incidence of breast cancer (carcinoma) has doubled worldwide over the last 20 years and it will occur in 1 out of 9 women in the USA [Yan et al., 2008]. The ability of breast tumors to respond to estrogens is dependent upon the expression of ER. Breast cancer cases with the worst prognosis and advanced disease typically lack ER protein and do not respond to treatment with Selective Estrogen Receptor Modulators (SERMs) drugs like tamoxifen and raloxifene [Jordan and O'Malley, 2007]. Uterine leiomyomas (also called uterine fibroids) are non-malignant myometrial cancers that are remarkably common in women of reproductive ages, with autopsy studies indicating their presence in more than 75% of those women [Cramer and Patel, 1990]. In about 25% of women of reproductive ages, leiomyomas cause pain, bleeding and infertility. Leiomyomas are the most common reason for hysterectomies. Leiomyomas have increased concentrations of estrogen-responsive genes, including the ER gene, compared to neighboring normal myometrium. Recent studies have identified alterations in expression microRNA genes that correlate with these estrogen-dependent diseases and might be useful for prognoses [Wang et al., 2007b; Yan et al., 2008]. Studies are beginning to unravel the post-transcriptional processes by which estrogens act, which probably reflect alterations in the compositions of ribonucleoprotein complexes on mRNAs. Elucidation of the molecular mechanisms responsible for estrogen-dependent physiology and disease processes is likely to provide novel targets and approaches for therapies.

ESTROGENS STABILIZE AND DESTABILIZE SPECIFIC mRNAS

Expression of the ER gene in the uterus is tightly regulated by steroid hormones from the ovary. In the luteal phases of estrous and menstrual cycles, progesterone down-regulates ER gene expression [Miller et al., 1979]. The subsequent preovulatory surge of estrogen up-regulates concentrations of ER mRNA and protein to restore estrogen responsiveness to the uterus, which is critical to its support of developing embryos [Ing, 1999; Moore et al., 1983]. Estrogen up-regulation of ER gene expression occurs in several tissues of vertebrate species ranging from fish to mammals [Ing, 1999; Friend et al., 1997; Rodriguez-Pinon et al., 2005]. In our animal model, the ovariectomized ewe, one physiological dose of estradiol up-regulates ER mRNA abundance in the uterus by 5-fold in 24 h [Ing and Ott, 1999; Mitchell and Ing, 2003]. ER was required for the up-regulation because SERM drugs blocked the effect [Robertson et al., 2001; Farnell and Ing, 2003c; Farnell and Ing, 2003b; Farnell and Ing, 2003a]. Nuclear runoff experiments detected no increase in the rate of transcription of the ER gene. However, both pulse-chase and transcription inhibitor experiments demonstrated that estradiol treatment enhanced ER mRNA stability.

The mechanism of estrogen stabilization of ER mRNA was investigated to identify the cis-elements (mRNA sequences) and transacting factors (binding proteins) involved. Using a cell-free assay for mRNA stability, the ER mRNA sequences responsible for estradiol-enhanced stability were localized within the 3' UTR. The mRNA stability assay employed cytosolic extracts from uteri of control and estradiol-treated ewes. The assay reproduced the mRNA specificity and magnitude of the stabilization of ER mRNA in the uterus of the ewe. Extensive studies identified two 82 base long Minimal Estrogen Modulated Stability Sequences (MEMSS) within the 4354 base long 3'UTR of ER mRNA [Mitchell and Ing, 2003]. Both MEMSS contained a 10 base long U-rich element (URE) that was predicted to be positioned on the ends of stem-loop structures [Mitchell and Ing, 2003]. Also, both MEMSS conferred E2-enhanced stability when transferred to heterologous RNAs. The stabilizing factor(s) that bind MEMSS appears to be proteinaceous because proteinase K or 70°C heat treatment eliminated the enhanced stability of ER mRNA in uterine extracts from estradiol-treated ewes [Mitchell and Ing, 2003]. UV-crosslinking was used to detect four MEMSS-binding proteins that were induced by estradiol treatment. The predominant, estradiol-induced binding protein immunoprecipitated with antiserum to AUF1. The size of the binding protein identified it as AUF1p45. AUF1 mRNA is alternatively spliced to translate four protein isoforms: AUF1p37, -p40, -p42, and -p45 [Wagner et al., 1998]. All bind AREs, but AUF1p40 and -p45 carry an mRNA stabilizing domain while the AUF1p37 and -p42 isoforms are associated with destabilized mRNAs [Sela-Brown et al., 2000; Loflin et al., 1999; Xu et al., 2001]. In vivo estradiol treatment increased AUF1p45 concentrations (probably nonphosphorylated) 6-fold in the uterine extracts [Ing et al., 2008]. Recombinant AUF1p45 stabilized ER mRNA in the mRNA stability assay with uterine extracts. A model for the mechanism of ER mRNA stabilization by estrogens is presented in Figure 1. In it, an estrogen up-regulates AUF1p45 which, along with other proteins and perhaps microRNAs, binds the two MEMSS within the 3'UTR of ER mRNA to form ribonucleoprotein complexes that stabilize ER mRNA.While this mechanism was discovered in sheep uteri, it is likely to operate in other species and tissues to up-regulate ER mRNA and protein concentrations.

The mechanistic work with the sheep uterus complements studies of other mRNAs. In rat uteri, estradiol treatment increased AUF1p40 expression by stabilizing the mRNA encoding it [Arao et al., 2002]. Estradiol treatment also up-regulated expression of two immediate early genes, encoding Immediate Early Response 2 (IER2) and TNFAIP3 Interacting protein 2 (TNIP2). These were initially identified by screening for mRNAs bound by AUF1 proteins [Arao et al., 2004].

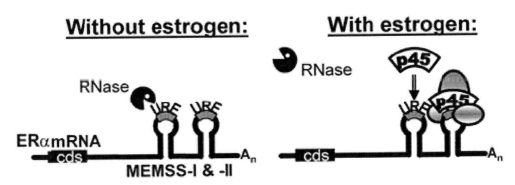

Figure 1. Model for E2 induction of a stabilizing ribonucleoprotein complex (RNP) on ER mRNA. The cartoon of ER mRNA shows the coding sequence (cds) 5' to the extensive 3'UTR. Within the 3'UTR, two 82 base Minimal Estradiol (E2) Modulated Stability Sequences (MEMSS) carry 10 base U-rich sequences (UREs) on stem loop structures. Estrogen induces binding of proteins to form a stabilizing RNP composed of proteins (AUF1p45 and shaded circles) on ER mRNA to protect it from ribonuclease (RNase).

This data implies that estradiol treatment stabilizes a set of mRNAs in rat uteri. This is a common theme in post-transcriptional gene regulation and some call these sets of coordinately regulated mRNAs "RNA regulons" [Keene, 2007]. It is expected that future experiments will identify groups of estrogen-stabilized mRNAs.

Most of the initial studies of estrogen effects on stabilities of mRNAs were performed in egg-laying animals [Dodson and Shapiro, 2002]. In the livers of male frogs treated with estradiol, the mRNA encoding vitellogenin is stabilized from having a half-life of 16 h to having one of 600 h. This is believed to occur in female frogs and other egg-laying species at the beginning of oogenesis. The mechanism is similar to the stabilization of ER mRNA in that estrogen induces a protein to bind vitellogenin mRNA on several sites within the 3' UTR. In the frog liver, estrogen induces production of the vigilin protein, which carries 15 RNA-binding domains [Cunningham et al., 2000]. The binding of vigilin to vitellogenin mRNA sterically blocks A(C/U)UGA sites that are susceptible to cleavage by an endonuclease named polysomal ribonuclease 1. Intriguingly, this endonuclease is also responsible for the concurrent destabilization of albumin mRNA in the livers of frogs treated with estradiol [Cunningham et al., 2001]. These estrogen-regulated posttranscriptional effects shift the gene expression program of the liver away from the production of serum proteins (such as albumin) and toward increased synthesis of proteins that will be packaged into eggs (such as vitellogenin).

There are numerous recent reports of estrogen-regulated stabilities of mRNAs in mammalian cells and tissues [Ing, 2005b]. One example with mechanistic information is the stabilization of luteinizing hormone receptor (LHR) in the granulosa cells within human and rat ovaries. Follicle stimulating hormone (FSH) followed by the preovulatory surge of estrogen increases LHR protein concentrations by stabilizing LHR mRNA [Wang et al., 2007a; Nair et al., 2008; Ikeda et al., 2008]. The stability of the LHR mRNA is inversely proportional to binding of a protein to sequence elements in the 3' UTR. The LHR mRNA binding protein was identified as mevalonate kinase, which was initially characterized for its role in cholesterol metabolism. The combination of FSH and estrogen influences down-regulate mevalonate kinase which, otherwise, would destabilize LHR mRNA. This action of mevalonate kinase in post-transcriptional regulation of LHR mRNA joins other examples of previously characterized metabolic enzymes that bind RNA and participate in post-transcriptional gene regulation [Kyrpides and Ouzounis, 1995; Ciesla, 2006]. These studies indicate the complexity of the interrelationships of gene products that participate in the regulation of vertebrate genes.

There are more and more intriguing examples of estrogens regulating of stabilities of mRNAs encoding critical gene products that await mechanistic information [Ing, 2005b]. The majority of examples of estrogen regulating mRNA stability involve stabilization in response to estrogen. In many cases, the regulation is tissue-specific. For example, in the mammalian pituitary, estrogen stabilizes some mRNAs, such as that encoding thyroid hormone releasing hormone receptor [el Meskini et al., 1997]. Other mRNAs, including that encoding the peptidylglycine alpha-amidating monooxygenase, are destabilized in the pituitary after estradiol treatment [Kimura et al., 1994]. Interestingly, one group reported that estradiol treatment destabilized ER mRNA in the MCF7 breast cancer cell line [Saceda et al., 1998]. The mechanism for destabilization of ER mRNA was not determined but could involve microRNAs, as discussed below. Comparisons of the molecular mechanisms by which estrogens regulate the stabilities of different mRNAs in various tissues may identify common or unique cis-elements and transacting factors. These could provide a way to predict and control the post-transcriptional effects of estrogen on gene regulation.

ESTROGENS AFFECT MRNA TRANSLATION

In many model systems, estrogens increase the rate of initiation of translation as well as the elongation rate of the nascent peptides. However, in most instances the effects are general and are not specific to one mRNA or a subset of mRNAs [Palmiter, 1972; Whelly and Barker, 1974]. There are a few exceptions that were published more than 20 years ago. In one example, estradiol treatment stimulated the synthesis of the ovalbumin protein in the chick oviduct while concentrations of ovalbumin mRNA decreased [Seaver, 1981]. This indicates that estradiol increased the rate of translation of ovalbumin mRNA and the effect appeared unique to that particular mRNA. The molecular mechanisms for the reported translational effects of estrogen are unknown. However, recent evidence demonstrates that estrogen effects on the expression of microRNA genes and microRNAs can alter the translation of individual mRNAs in positive as well as negative manners, as discussed below.

ESTROGEN ACTIONS AND MICRORNAS

MicroRNAs – Biogenesis and Regulation of Gene Expression

MicroRNAs are an abundant class of non-coding RNAs. There are hundreds of microRNAs (~600 confirmed in humans) which range from 21 to 23 bases in length [Maziere and Enright, 2007]. MicroRNA sequences are very highly conserved across species. There are families of microRNAs that have related sequences and, in some cases, their genes are clustered on chromosomes and share coordinate expression in animal tissues. Primary transcripts of microRNAs are synthesized by RNA polymerase II or III [Lodish et al., 2008]. Microprocessor complexes containing the RNAse III-type endonuclease Drosha cleave the primary microRNA transcript to an approximately 60 base long RNA with hairpin structure called a pre-microRNA, which is exported to the cytoplasm. Pre-microRNAs are then cleaved by another RNAse III endonuclease, Dicer, which produces a double-stranded approximately 22 base long product. One strand is the mature microRNA, which is incorporated into the RNA-induced silencing complex (RISC) so that the microRNA can direct the actions of this effector complex to specific mRNAs by hybridizing to them [Jackson and Standart, 2007]. Each microRNA is predicted to bind and regulate an average of about 200 mRNAs. In this way, microRNAs are proposed to regulate 30 to 90% of mammalian mRNAs [Ioshikhes et al., 2007; Lodish et al., 2008].

MicroRNAs bind sequences within the 3' UTRs of mRNAs to regulate mRNA stability or translation. The complementary binding of a microRNA to a target mRNA is usually not complete. The 5' ends of microRNAs generally bind most strongly to the mRNA, with a 6- to 8- base long "seed" sequence beginning at the second base of the microRNA. There are often mismatches in the next 9 to 12 bases of the microRNA, followed by more base-pairing between bases 13 to 16 of the microRNA and the mRNA [Grimson et al., 2007]. Algorithms (MiRanda, TargetScan and PicTar) have been developed to predict microRNAs binding sites on mRNAs [Ioshikhes et al., 2007; Maziere and Enright, 2007]. The PicTar algorithm also takes into account sequence conservation of the predicted site on the target mRNA across species as further indication of function of the putative microRNA binding sites [Ioshikhes et al., 2007]. However, some microRNA sites in nonconserved regions have been shown to very effectively bind microRNAs that regulate the function of that mRNA [Baek et al., 2008]. There is growing evidence for cooperativity between microRNAs (similar or different) bound to neighboring sites, resulting in enhanced effects on the function of the mRNA [Grimson et al., 2007]. The effectiveness of microRNAs is also enhanced by flanking AU-rich regions [Grimson et al., 2007]. With increased understanding of microRNA interactions with mRNAs, we will be better able to predict and potentially utilize microRNA effects on gene expression.

The molecular mechanisms by which microRNAs regulate mRNA stability or the rate of mRNA translation are being actively studied. After the microRNA binds to the mRNA, it generally has one of two actions: destabilization of the mRNA or inhibition of translation [Jackson and Standart, 2007; Wu and Belasco, 2008]. In cases of mRNA destabilization, the microRNA decreases the concentrations of both the mRNA targeted as well as of the protein encoded by the mRNA. MicroRNAs that destabilize mRNA targets were originally thought to direct endonucleolytic cleavage by the RISC, as do short interfering RNAs [Jackson and

Standart, 2007]. However, new evidence indicates that microRNAs can also induce rapid deadenylation of target mRNAs in animal cells, which then results in mRNA degradation [Wu and Belasco, 2008]. A recent study in animal cells indicated that the microRNAs that most powerfully down-regulate proteins also down-regulate the mRNAs encoding the proteins [Baek et al., 2008]. In the case of microRNAs inhibiting translation, the concentration of mRNA target remains unchanged while the concentration of the encoded protein decreases. MicroRNAs inhibit translation initiation by inhibiting recognition of the 5' cap of mRNA by ribosomes [Mathonnet et al., 2007]. The mode of microRNA binding to the mRNA may dictate whether the mRNA is destabilized or translation is repressed: more complete complementarity in binding may result in mRNA destabilization while less complementarity may lead to inhibition of translation [Saxena et al., 2003; Grimson et al., 2007; Maziere and Enright, 2007]. The actions of microRNAs are also dependent upon the proteins they associate with in ribonucleoprotein complexes [Jing et al., 2005; Bhattacharyya et al., 2006]. In addition, the binding of some proteins to mRNAs on sites near microRNA binding sites can prevent mRNA repression by the microRNA [Ketting, 2007; Kedde et al., 2007].

There is also a small but growing number of reports of microRNAs up-regulating gene expression, either by stabilizing mRNAs or by increasing mRNA translation rates [Krutzfeldt et al., 2005; Jackson and Standart, 2007]. The latter effect is apparent in quiescent cells and occurs by microRNAs recruiting Argonaut and other proteins to the mRNA [Jackson and Standart, 2007]. More experiments are required to discover the breadth of microRNA actions on gene expression in animal tissues.

Estrogens Regulate Expression of MicroRNA Genes in Responsive Tissues During Normal Physiology and Disease

Because of the importance of microRNAs in the regulation of gene expression, microRNA expression profiles are being studied in normal and pathological tissues [Landgraf et al., 2007; Liang et al., 2007; Hammond, 2006; Jiang et al., 2009]. For example, microRNA expression in human cancers is being evaluated to discern patterns for diagnostic and prognostic purposes [Jay et al., 2007; Lu et al., 2005; Lui et al., 2007; Boren et al., 2008]. Not surprisingly, since estrogens affect the expression of numerous protein-encoding genes, estrogens also affect the expression of numerous microRNA genes (Table 1). All of the studies listed in Table 1 initially used microarray analyses and then confirmed altered concentrations of the microRNAs with real time PCR and/or Northern blotting. This increases the validity of the experimental results reported in Table 1. However, it also means that the data summarized in Table 1 are only a subset of the microRNAs that are regulated by estrogen and that the data are biased by the investigators' selection of the microRNAs chosen for confirmation by the second technique. The data shown in Table 1 are compiled from studies in which animals or cells were treated directly with estradiol as well as from studies of animals and tissues that are known to have lesser or greater estrogen influence, as discussed below.

Estrogens regulate microRNA gene expression in tissue- and time-dependent manners. Regulation of microRNAs by estradiol treatment was directly tested in a fairly comprehensive study of zebrafish [Cohen et al., 2008]. Table 1 shows the four microRNAs up-regulated and four others down-regulated when whole fish were measured 24 h after treatment. The study went on to examine the microRNAs in specific tissues of the fish over a time course. Time dependent differences were found, such as down-regulation of dre-miR-26b concentrations by 40% at 12 h but return to baseline levels by 24 h after estradiol treatment. It is noteworthy that the ER protein concentrations peaked in the fish at 12 h post-treatment. It is also important to note that microRNAs are very highly conserved. For example, the zebrafish dre-miR-26b mentioned is identical to the human hsa-miR-26a except for two bases on the 3'end. In primary cultures of human endometrial stromal cells, estradiol treatment regulated the expression of three microRNA genes while concurrent treatment with the SERM ICI 182,780 blocked the effect [Pan et al., 2007]. In rats treated long-term with estradiol, expression of six microRNA genes was up-regulated and one other was down-regulated in the mammary tissue that was undergoing carcinogenesis [Kovalchuk et al., 2007]. For some of the other studies with results in Table 1, estrogen regulation of microRNAs was implied. For example, in the mouse uterus, tissues were compared between animals in estrogen-dominated vs. progesterone-dominated phases of the estrous cycle [Chakrabarty et al., 2007]. Since miR-101a and -199a-3p concentrations were greater in the latter, we infer that estrogens may down-regulate them. This study identified cyclooxygenase 2 (COX2) mRNA as a target of miR-101a and -199a-3p and that these microRNAs inhibited COX2 mRNA translation in the mouse uterus during early pregnancy [Chakrabarty et al., 2007]. Microarray data also indicated that expression of miR-31 and -368 genes was lower in myometrial tissue from women during the estrogen-dominated follicular phase of the menstrual cycle, implying down-regulation by estrogens [Wang et al., 2007b]. Together, these data are an indication of the long-reaching effects of estrogens on microRNA gene expression in normal vertebrate animal tissues.

Estrogens also regulate microRNA gene expression in estrogen-dependent diseases. The latter include endometriosis, endometrial cancer, leiomyoma (uterine fibroids) and breast cancer. Microarray data indicated that expression of miR-20a, -21, and -26a genes, previously mentioned as being estradiol-regulated in endometrial stromal cells, was also deregulated in endometriotic tissue compared to normal endometrium [Pan et al., 2007]. In matched samples of endometrial cancer and normal endometrium from women, microarray data indicated that miR-103, -107 and -185 concentrations were up-regulated while that of miR-let-7i was down-regulated [Boren et al., 2008]. In the case of leiomyomas, these tumors of the myometrium are well-known to have higher levels of estrogens as well as expression of estrogen-responsive genes compared to neighboring normal myometrium [Nowak, 2001]. Table 1 shows results from comparisions of matched leiomyoma and neighboring normal myometrial samples that detected several differentially expressed microRNAs [Wang et al., 2007b; Marsh et al., 2008]. Breast tumors are routinely characterized for treatment and prognosis by ER and progesterone receptor (PR) protein immunohistochemistry [Mattie et al., 2006]. Tumors containing significant immunoreactivity for ER and PR proteins (ER+/PR+) are considered to be more differentiated (normal) and responsive to treatment with SERMs. It is important to note that as estrogens up-regulate gene expression, they also down-regulate concentrations of ER protein in the responsive tissues/cells as part of their transcriptional activation [Chu et al., 2007]. Because estrogens up-regulate PR gene expression, PR+ status is probably the best indicator that estrogen influence is dominant in a breast tumor sample. Results from three studies that compared microRNAs in breast tumors and related the data to the ER and PR protein status of the tumors are summarized in Table 1 [Iorio et al., 2005; Mattie et al., 2006;

Foekens et al., 2008]. Microarray data from the first two studies agreed that microRNAs let-7c, 26a, 30a-5p, 30b and 30c were up-regulated in PR+ tumors compared to PR- tumors. The third study identified miR-22 and-34b as up-regulated in PR+ breast tumors compared to others [Foekens et al., 2008]. Intriguingly, all three studies identified only increases in microRNA concentrations (those named and others) in PR+ compared to PR- breast tumors, while no microRNA concentrations appeared to decrease. Differences in results between similar microRNA studies may be the result of individual differences in the women, the method or microarray platform used, or sampling technique (times of sampling, stage of the menstrual cycle, etc.). What is clear is that we need more good studies of microRNAs in human tissues, both during normal physiology and during the pathogenesis of estrogen-dependent diseases.

It is interesting that several microRNAs in Table 1 appear to be estrogen-regulated in more than one normal and/or pathological tissues. For example, miR-101 was down-regulated by estradiol treatment of zebrafish as well as during the estrogen-dominated phase of the estrous cycle in the uteri of mice [Chakrabarty et al., 2007; Cohen et al., 2008]. In addition, miR-26a concentrations were up-regulated by estradiol treatment in endometrial stromal cells in culture and were greater in leiomyoma compared to myometrium and in ER+/PR+ breast tumors compared to other tumors [Iorio et al., 2005; Pan et al., 2007; Wang et al., 2007b]. Concentrations of miR-21 appeared to be regulated by estrogens in five of the studies in Table 1. Concentrations of miR-21 were greater in leiomyomas and breast tumors compared to normal tissues as well as in mammary tissues of estradiol-treated rats [Yan et al., 2008; Kovalchuk et al., 2007; Wang et al., 2007b; Iorio et al., 2005]. It is noteworthy that high levels of miR-21 correlate to poor prognosis for breast cancer patients [Yan et al., 2008]. However, not all studies indicate the same direction of the microRNA gene's regulation by estrogen, either up or down. For example, estradiol treatment decreased concentrations of miR-21 in cultured endometrial stromal cells [Pan et al., 2007;]. Also, concentrations of miR-199a-3p were down-regulated in mouse uterus during the estrogen-dominated phase of the estrous cycle but were up-regulated in leiomyoma compared to normal myometrium. In conclusion, there are some consistent patterns for estrogen regulation of microRNA gene expression in vertebrate tissues. However, there are also some divergences that remain to be explained. Future research is needed to expand our understanding of how estrogens regulate the expression of microRNA genes in health and disease.

MicroRNAs Regulate ER Gene Expression and Estrogen Actions

While estrogens regulate the expression of microRNA genes, microRNAs also regulate the gene products that transduce the actions of estrogen in tissues: ER, cofactors, and kinase signaling cascades. Several studies demonstrated that microRNAs regulate expression of the ER gene. One study began with interest in microRNAs that had increased concentrations in ER- breast tumors compared to ER+ breast tumors [Iorio et al., 2005]. Among those microRNAs, the investigators focused on the few microRNAs that also had predicted binding sites within the 3'UTR of ER mRNA [Adams et al., 2007]. This lead to the discovery that experimentally altering the concentrations of miR-206 caused opposite effects on ER gene expression in MCF7 cells. The mechanism was by miR-206 destabilizing ER mRNA. Further studies demonstrated that estradiol treatment decreases concentrations of miR-206, leading to increased ER mRNA stability and ER protein concentrations. This describes a second mechanism by which estrogens post-transcriptionally up-regulate ER gene expression in addition to that described in the sheep uterus (Figure 1). Another group searching for microRNAs that regulate ER gene expression identified 12 microRNAs that were more highly expressed in ER- breast cancer cell lines compared to ER+ lines [Zhao et al., 2008]. These investigators predicted that miR-221 and -222 would target the 3'UTR of ER mRNA. By altering levels of the microRNAs in the MCF7 and T47D breast cancer cell lines, they demonstrated that miR-221 and -222 decreased ER protein but not mRNA concentrations, indicating that these two microRNAs inhibit translation of ER mRNA. Studies of a human pancreatic cancer cell line lead to the discovery that treatment with curcumin (a naturally occurring flavonoid that inhibits cancer growth) increased miR-22 and decreased miR-199a-3p concentrations [Sun et al., 2008]. Experimental up-regulation of miR-22 suppressed ER protein concentrations along with those of Sp1, which is a transcription factor that works with ER to activate transcription on many gene promoters that lack conventional estrogen-responsive elements. Thus, miR-22 appears to inhibit ER and ER/Sp1 responsive gene promoters by reducing the concentrations of those transcription factors. These examples show that several different microRNAs regulate the expression of the ER gene in a variety of tumor cells.

MicroRNAs can inhibit estrogen actions in ways other than decreasing ER protein concentrations. Phosphorylation of the ER protein is required for its function [Arnold et al., 1997; Vasudevan and Pfaff, 2008]. Phosphorylation is important for both the genomic actions of estrogens and the rapid, non-genomic effects of estrogens that involve kinase cascades [Watson et al., 2007]. We are just beginning to understand how microRNAs regulate phosphorylation pathways in cells. In fibroblast cell lines, miR-199a-3p down-regulates extracellular signal-regulated kinase 2 (ERK2) [Kim et al., 2008]. Although estradiol treatment was not a part of that study, concentrations of miR-199a-3p appeared to be down-regulated by estrogen in mouse uteri and were greater in leiomyomas compared to normal myometrium in women (Table 1). In another example, the expression of the miR-17-5p gene was inversely correlated to the expression of the Amplified In Breast cancer 1 (AIB1) gene in breast cancer cell lines [Hossain et al., 2006]. AIB1 is a coactivator that works with ER to activate transcription of responsive genes. AIB1 gene expression is up-regulated in many tumors including breast cancers. MiR-17-5p decreased the expression of the AIB1 gene by inhibiting AIB1 mRNA translation. Thus, miR-17-5p acts as a tumor suppressor by down-regulating AIB1 concentrations and estrogen signaling. Interestingly, miR-17-5p decreases both estrogen/ER-dependent and estrogen/ER-independent proliferation in the MCF7 breast cancer cell line [Hossain et al., 2006]. Increased knowledge about microRNA actions may lead to new therapies that utilize microRNAs to fight diseases such as cancer.

FUTURE THERAPEUTIC APPROACHES TO REGULATING POST-TRANSCRIPTIONAL ESTROGEN ACTIONS

Over the last decades, several therapies have been developed to control estrogen effects in humans in order to treat or prevent estrogen-dependent diseases. Because of that, there are several currently available drugs to modulate estrogen effects in vivo. From more general to specific, Gonadotrophin Releasing Hormone (GnRH) agonists block gonadal steroid production and thereby estrogen and progesterone effects. More recently, aromatase inhibitors were developed to block estrogen production in all tissues of the body. Both GnRH agonists and aromatase inhibitors reduce all estrogen effects in the body, including the desirable ones that preserve bone, brain and circulatory system health. The SERMs modulate ER actions in a more tissue-specific manner. For example, treatment with the SERM raloxifene interferes with ER+ breast cancer growth while providing healthful estrogenic effects in bone [Jordan and O'Malley, 2007]. Since SERMs block a subset of estrogen effects, they may be more desirable drugs than GnRH and aromatase inhibitors due to their greater specificity and fewer side effects.

It is likely that increased knowledge of the molecular mechanisms of the post-transcriptional effects of estrogen will be another step towards therapeutic targeting of known subsets of estrogen-responsive genes. It is possible to target proteins involved in post-transcriptional regulation either by using reagents like short interfering RNAs to down-regulate the protein and reverse its effects on mRNA stability. For the proteins that regulate mRNA stability by binding directly to cis-elements on the mRNA, oligonucleotide mimics of the cis-element (also called "RNA decoys") could be used to bind and sequester those RNA-binding proteins and interrupt the proteins' effect on mRNA lifespan and function [Makeyev et al., 2002]. For example, if uteri were treated with the ER mRNA regions responsible for its stabilization (such as MEMSS), the AUF1p45 and other binding proteins could be sequestered and unable to stabilize ER mRNA. This would block the up-regulation of the expression of the ER gene, as well as the genes it subsequently up-regulates. This approach is unique because it could spare basal influences of estrogens so that bone, brain and circulatory system health would be maintained.

The elucidation of the molecular mechanisms by which microRNAs regulate the stability and/or translation of target mRNAs opens even more new therapeutic avenues for altering both normal and abnormal physiology of animal tissues. Recent advances in oligonucleotide therapeutics and gene therapy approaches make it feasible to either up- or down-regulate key microRNAs in cell culture and, subsequently, in vivo [Esau and Monia, 2007; Ford and Cheng, 2008; Zhao et al., 2008]. One group called its antisense microRNAs "antagomirs" and used them successfully in mice to down-regulate an individual microRNA [Krutzfeldt et al., 2005]. A particular microRNA might be targeted by an antisense RNA to reverse the actions of the microRNA on expression of critical genes. In another example, down-regulation of miR-221 and -222 by antisense RNAs increased ER levels in breast cancer cells [Zhao et al., 2008]. This restored estrogen-dependent cell growth as well as sensitivity to growth inhibition by the SERM tamoxifen. This may be a useful approach to treat tamoxifen-resistant breast cancer in women. Alternatively, therapies might increase concentration of microRNAs that act as tumor suppressors, like miR-17-5p in breast cancer [Hossain et al., 2006]. As our knowledge of the post-transcriptional effects of estrogens on individual genes and sets of gene products grows, so will our ability to control physiology through them in order to enhance human and animal health.

Table 1. Estrogen regulation of microRNAs in normal and pathological tissues. Selected data from studies using microarrays followed by confirmation with real time PCR and/or Northern blotting

miRNA	Estrogen regulation	Estrogen treatment or influence and tissue or cultured cells	Reference
Let7h	Up	Estradiol-treated zebrafish (whole body)	Cohen et al., 2008
122	Up		
196b	Up		
29b*	Up		
101a	Down		
130c	Down		
19a	Down		
460-5p	Down		
101a	Down	Estrogen-dominated uterus (mouse)	Chakrabarty et al., 2007
199a-3p	Down		
26a	Up	Estradiol-treated primary cultures of endometrial stromal cells (human)	Pan et al., 2007
20a	Down		
21	Down		
323	Up	Leiomyoma compared to myometrium (human)	Marsh et al., 2008
34a	Up		
139	Down		
Let7b	Up	Leiomyoma compared to myometrium (human)	Wang, et al., 2007
145	Up		
199a-3p	Up		
21	Up		
23b	Up		
26a	Up		
99a	Down		
29b	Down		
203	Down		
220	Down		
106a	Up	Estradiol-treated rats - mammary tissue undergoing carcinogenesis	Kovalchuk et al, 2007
129-3p	Up		
17-5p	Up		
20a	Up		
21	Up		
92	Up		
127	Down		
21	Up	Breast tumors compared to normal mammary tissues (human)	Iorio et al., 2005
125b	Down		
145	Down		
21	Up	Breast tumors compared to normal mammary tissues (human)	Yan et al., 2008
497	Down		
206	Down	Estradiol-treated MCF7 breast cancer cell line (human)	Adams et al., 2007

CONCLUSION

As discussed above, the many post-transcriptional actions of estrogens are an important component of their effects on the expression of specific genes. Some of these estrogen actions appear to be central to aspects of normal physiology, as well as to the development and progression of estrogen-dependent diseases including uterine and breast cancers. These effects may occur by estrogen-induced increases in the transcription of genes encoding RNAbinding proteins and/or microRNAs. Estrogens could also activate RNA-binding proteins, such as AUF1p45, by post-translational mechanisms such as dephosphorylation [Proia et al., 2006]. As we elucidate the molecular mechanisms by which estrogens regulate gene expression post-transcriptionally, we will reveal new targets and approaches for therapeutic interventions in estrogen-dependent physiology.

REFERENCES

Adams, B.D., Furneaux, H., White, B.A., 2007. The micro-ribonucleic acid (miRNA) miR-206 targets the human estrogen receptor-alpha (ERalpha) and represses ERalpha messenger RNA and protein expression in breast cancer cell lines. *Mol. Endocrinol.* 21, 1132-1147.

Arao, Y., Kikuchi, A., Ikeda, K., Nomoto, S., Horiguchi, H., Kayama, F., 2002. A+U-rich-element RNA-binding factor 1/heterogeneous nuclear ribonucleoprotein D gene expression is regulated by oestrogen in the rat uterus. *Biochemical. Journal.* 361, 125-132.

Arao, Y., Kikuchi, A., Kishida, M., Yonekura, M., Inoue, A., Yasuda, S., Wada, S., Ikeda, K., Kayama, F., 2004. Stability of A+U-rich element binding factor 1 (AUF1)-binding messenger ribonucleic acid correlates with the subcellular relocalization of AUF1 in the rat uterus upon estrogen treatment. *Mol. Endocrinol* 18, 2255-2267.

Arnold, S.F., Melamed, M., Vorojeikina, D.P., Notides, A.C., Sasson, S., 1997. Estradiol-binding mechanism and binding capacity of the human estrogen receptor is regulated by tyrosine phosphorylation. *Mol. Endocrinol.* 11, 48-53.

Baek, D., Villen, J., Shin, C., Camargo, F.D., Gygi, S.P., Bartel, D.P., 2008. The impact of microRNAs on protein output.[see comment]. *Nature* 455, 64-71.

Bhattacharyya, S.N., Habermacher, R., Martine, U., Closs, E.I., Filipowicz, W., 2006. Relief of microRNA-mediated translational repression in human cells subjected to stress. *Cell* 125, 1111-1124.

Boren, T., Xiong, Y., Hakam, A., Wenham, R., Apte, S., Wei, Z., Kamath, S., Chen, D.T., Dressman, H., Lancaster, J.M., 2008. MicroRNAs and their target messenger RNAs associated with endometrial carcinogenesis. *Gynecol. Oncol.* 110, 206-215.

Chakrabarty, A., Tranguch, S., Daikoku, T., Jensen, K., Furneaux, H., Dey, S.K., 2007. MicroRNA regulation of cyclooxygenase-2 during embryo implantation. *Proc. Natl.Acad. Sci. USA* 104, 15144-15149.

Cheadle, C., Fan, J., Cho-Chung, Y.S., Werner, T., Ray, J., Do, L., Gorospe, M., Becker, K.G., 2005. Control of gene expression during T cell activation: alternate regulation of mRNA transcription and mRNA stability. *BMC Genomics* 6, 75.

Chen, C.Y., Shyu, A.B., 1995. AU-rich elements: characterization and importance in mRNA degradation. *Trends Biochem. Sci.* 20, 465-470.

Chen, C.Y., Xu, N., Shyu, A.B., 2002. Highly selective actions of HuR in antagonizing AU-rich element-mediated mRNA destabilization. *Mol. Cell Biol.* 22, 7268-7278.

Chu, I., Arnaout, A., Loiseau, S., Sun, J., Seth, A., McMahon, C., Chun, K., Hennessy, B., Mills, G.B., Nawaz, Z., Slingerland, J.M., 2007. Src promotes estrogen-dependent estrogen receptor alpha proteolysis in human breast cancer. *J. Clin. Invest* 117, 2205-2215.

Ciesla, J., 2006. Metabolic enzymes that bind RNA: yet another level of cellular regulatory network? *Acta Biochim. Pol.* 53, 11-32.

Cohen, A., Shmoish, M., Levi, L., Cheruti, U., Levavi-Sivan, B., Lubzens, E., 2008. Alterations in micro-ribonucleic acid expression profiles reveal a novel pathway for estrogen regulation. *Endocrinology* 149, 1687-1696.

Cramer, S.F., Patel, A., 1990. The frequency of uterine leiomyomas. *Am. J. Clin. Pathol.* 94, 435-438.

Cunningham, K.S., Dodson, R.E., Nagel, M.A., Shapiro, D.J., Schoenberg, D.R., 2000. Vigilin binding selectively inhibits cleavage of the vitellogenin mRNA 3'-untranslated region by the mRNA endonuclease polysomal ribonuclease 1. *Proc. Natl. Acad. Sci. USA* 97, 12498-12502.

Cunningham, K.S., Hanson, M.N., Schoenberg, D.R., 2001. Polysomal ribonuclease 1 exists in a latent form on polysomes prior to estrogen activation of mRNA decay. *Nucleic Acids Research* 29, 1156-1162.

Dodson, R.E., Shapiro, D.J., 2002. Regulation of pathways of mRNA destabilization and stabilization. *Progress in Nucleic Acid Research and Molecular Biology* 72, 129-164.

el Meskini, R., Delfino, C., Boudouresque, F., Hery, M., Oliver, C., Ouafik, L., 1997. Estrogen regulation of peptidylglycine alpha-amidating monooxygenase expression in anterior pituitary gland. *Endocrinology* 138, 379-388.

Esau, C.C., Monia, B.P., 2007. Therapeutic potential for microRNAs. *Adv. Drug Deliv. Rev.* 59, 101-114.

Farnell, Y.Z., Ing, N.H., 2003a. The effects of estradiol and selective estrogen receptor modulators on gene expression and messenger RNA stability in immortalized sheep endometrial stromal cells and human endometrial adenocarcinoma cells. *Journal ofSteroid Biochemistry and Molecular Biology* 84, 453-461.

Farnell, Y.Z., Ing, N.H., 2003b. Endometrial effects of selective estrogen receptor modulators (SERMs) on estradiol-responsive gene expression are gene and cell-specific. *Journal ofSteroid Biochemistry and Molecular Biology* 84, 513-526.

Farnell, Y.Z., Ing, N.H., 2003c. Myometrial effects of selective estrogen receptor modulators on estradiol-responsive gene expression are gene and cell-specific. *Journal of SteroidBiochemistry and Molecular Biology* 84, 527-536.

Foekens, J.A., Sicuwerts, A.M., Smid, M., Look, M.P., de Weerd, V., Boersma, A.W., Klijn, J.G., Wiemer, E.A., Martens, J.W., 2008. Four miRNAs associated with aggressiveness of lymph node-negative, estrogen receptor-positive human breast cancer. *Proc. Natl.Acad. Sci. USA* 105, 13021-13026.

Ford, L.P., Cheng, A., 2008. Using synthetic precursor and inhibitor miRNAs to understand miRNA function. *Methods Mol. Biol.* 419, 289-301.

Friend, K.E., Resnick, E.M., Ang, L.W., Shupnik, M.A., 1997. Specific modulation of estrogen receptor mRNA isoforms in rat pituitary throughout the estrous cycle and in response to steroid hormones. *Mol. Cell Endocrinol.* 131, 147-155.

Grimson, A., Farh, K.K., Johnston, W.K., Garrett-Engele, P., Lim, L.P., Bartel, D.P., 2007. MicroRNA targeting specificity in mammals: determinants beyond seed pairing. *Mol.Cell* 27, 91-105.

Gurates, B., Bulun, S.E., 2003. Endometriosis: the ultimate hormonal disease. *Semin. ReprodMed.* 21, 125-134.

Hammond, S.M., 2006. RNAi, microRNAs, and human disease. *Cancer Chemother Pharmacol.* 58 Suppl 1, s63-68.

Hossain, A., Kuo, M.T., Saunders, G.F., 2006. Mir-17-5p regulates breast cancer cell proliferation by inhibiting translation of AIB1 mRNA. Mol. *Cell Biol.* 26, 8191-8201.

Ikeda, S., Nakamura, K., Kogure, K., Omori, Y., Yamashita, S., Kubota, K., Mizutani, T., Miyamoto, K., Minegishi, T., 2008. Effect of estrogen on the expression of luteinizing hormone-human chorionic gonadotropin receptor messenger ribonucleic acid in cultured rat granulosa cells. *Endocrinology* 149, 1524-1533.

Ing, N.H., 2005a. Steroid hormones regulate gene expression post-transcriptionally by altering the stabilities of messenger RNAs. *Biol. Reprod* 72, 1290-1296.

Ing, N.H., 2005b. Steroid hormones regulate gene expression posttranscriptionally by altering the stabilities of messenger RNAs. *Biol. Reprod.* 72, 1290-1296.

Ing, N.H., Massuto, D.A., Jaeger, L.A., 2008. Estradiol up-regulates AUF1p45 binding to stabilizing regions within the 3'-untranslated region of estrogen receptor alpha mRNA. *J.Biol. Chem.* 283, 1764-1772.

Ing, N.H., Ott, T.L., 1999. Estradiol up-regulates estrogen receptor-alpha messenger ribonucleic acid in sheep endometrium by increasing its stability. *Biol. Reprod* 60, 134-139.

Ing, N.H., Robertson J. A., 1999. *Regulation of hormone receptor gene expression.* Springer-Verlag, New York.

Iorio, M.V., Ferracin, M., Liu, C.G., Veronese, A., Spizzo, R., Sabbioni, S., Magri, E., Pedriali, M., Fabbri, M., Campiglio, M., Menard, S., Palazzo, J.P., Rosenberg, A., Musiani, P., Volinia, S., Nenci, I., Calin, G.A., Querzoli, P., Negrini, M., Croce, C.M., 2005. MicroRNA gene expression deregulation in human breast cancer. *Cancer Res.* 65, 7065-7070.

Ioshikhes, I., Roy, S., Sen, C.K., 2007. Algorithms for mapping of mRNA targets for microRNA. *DNA Cell Biol.* 26, 265-272.

Jackson, R.J., Standart, N., 2007. How do microRNAs regulate gene expression? *Sci. STKE* 2007, re1.

Jay, C., Nemunaitis, J., Chen, P., Fulgham, P., Tong, A.W., 2007. miRNA profiling for diagnosis and prognosis of human cancer. *DNA Cell Biol.* 26, 293-300.

Jiang, Q., Wang, Y., Hao, Y., Juan, L., Teng, M., Zhang, X., Li, M., Wang, G., Liu, Y., 2009. miR2Disease: a manually curated database for microRNA deregulation in human disease. *Nucleic Acids Res.* 37, D98-D104.

Jing, Q., Huang, S., Guth, S., Zarubin, T., Motoyama, A., Chen, J., Di Padova, F., Lin, S.C., Gram, H., Han, J., 2005. Involvement of microRNA in AU-rich element-mediated mRNA instability. *Cell* 120, 623-634.

Jordan, V.C., O'Malley, B.W., 2007. Selective estrogen-receptor modulators and antihormonal resistance in breast cancer. *J. Clin. Oncol.* 25, 5815-5824.

Kedde, M., Strasser, M.J., Boldajipour, B., Oude Vrielink, J.A., Slanchev, K., le Sage, C., Nagel, R., Voorhoeve, P.M., van Duijse, J., Orom, U.A., Lund, A.H., Perrakis, A., Raz, E., Agami, R., 2007. RNA-binding protein Dnd1 inhibits microRNA access to target mRNA.[see comment]. *Cell* 131, 1273-1286.

Keene, J.D., 2007. RNA regulons: coordination of post-transcriptional events. *Nat. Rev.Genet.* 8, 533-543.

Kenealy, M.R., Flouriot, G., Sonntag-Buck, V., Dandekar, T., Brand, H., Gannon, F., 2000. The 3'-untranslated region of the human estrogen receptor alpha gene mediates rapid messenger ribonucleic acid turnover. *Endocrinology* 141, 2805-2813.

Ketting, R.F., 2007. A dead end for microRNAs.[comment]. *Cell* 131, 1226-1227.

Kim, S., Lee, U.J., Kim, M.N., Lee, E.J., Kim, J.Y., Lee, M.Y., Choung, S., Kim, Y.J., Choi, Y.C., 2008. MicroRNA miR-199a* regulates the MET proto-oncogene and the downstream extracellular signal-regulated kinase 2 (ERK2). *J. Biol. Chem.* 283, 18158-18166.

Kimura, N., Arai, K., Sahara, Y., Suzuki, H., 1994. Estradiol transcriptionally and posttranscriptionally up-regulates thyrotropin-releasing hormone receptor messenger ribonucleic acid in rat pituitary cells. *Endocrinology* 134, 432-440.

Kovalchuk, O., Tryndyak, V.P., Montgomery, B., Boyko, A., Kutanzi, K., Zemp, F., Warbritton, A.R., Latendresse, J.R., Kovalchuk, I., Beland, F.A., Pogribny, I.P., 2007. Estrogen-induced rat breast carcinogenesis is characterized by alterations in DNA methylation, histone modifications and aberrant microRNA expression. *Cell Cycle* 6, 2010-2018.

Krutzfeldt, J., Rajewsky, N., Braich, R., Rajeev, K.G., Tuschl, T., Manoharan, M., Stoffel, M., 2005. Silencing of microRNAs in vivo with 'antagomirs'. *Nature* 438, 685-689.

Kyrpides, N.C., Ouzounis, C.A., 1995. Nucleic acid-binding metabolic enzymes: living fossils of stereochemical interactions? *J. Mol. Evol.* 40, 564-569.

Landgraf, P., Rusu, M., Sheridan, R., Sewer, A., Iovino, N., Aravin, A., Pfeffer, S., Rice, A., Kamphorst, A.O., Landthaler, M., Lin, C., Socci, N.D., Hermida, L., Fulci, V., Chiaretti, S., Foa, R., Schliwka, J., Fuchs, U., Novosel, A., Muller, R.U., Schermer, B., Bissels, U., Inman, J., Phan, Q., Chien, M., Weir, D.B., Choksi, R., De Vita, G., Frezzetti, D., Trompeter, H.I., Hornung, V., Teng, G., Hartmann, G., Palkovits, M., Di Lauro, R., Wernet, P., Macino, G., Rogler, C.E., Nagle, J.W., Ju, J., Papavasiliou, F.N., Benzing, T., Lichter, P., Tam, W., Brownstein, M.J., Bosio, A., Borkhardt, A., Russo, J.J., Sander, C., Zavolan, M., Tuschl, T., 2007. A mammalian microRNA expression atlas based on small RNA library sequencing.[see comment]. *Cell* 129, 1401-1414.

Lasa, M., Mahtani, K.R., Finch, A., Brewer, G., Saklatvala, J., Clark, A.R., 2000. Regulation of cyclooxygenase 2 mRNA stability by the mitogen-activated protein kinase p38 signaling cascade. *Mol. Cell Biol.* 20, 4265-4274.

Liang, Y., Ridzon, D., Wong, L., Chen, C., 2007. Characterization of microRNA expression profiles in normal human tissues. *BMC Genomics* 8, 166.

Lodish, H.F., Zhou, B., Liu, G., Chen, C.Z., 2008. Micromanagement of the immune system by microRNAs.[erratum appears in Nat Rev Immunol. 2008 Mar;8(3):238]. *Nature RevImmunol.* 8, 120-130.

Loflin, P., Chen, C.Y., Shyu, A.B., 1999. Unraveling a cytoplasmic role for hnRNP D in the in vivo mRNA destabilization directed by the AU-rich element. *Genes and Development* 13, 1884-1897.

Lu, J., Getz, G., Miska, E.A., Alvarez-Saavedra, E., Lamb, J., Peck, D., Sweet-Cordero, A., Ebert, B.L., Mak, R.H., Ferrando, A.A., Downing, J.R., Jacks, T., Horvitz, H.R., Golub, T.R., 2005. MicroRNA expression profiles classify human cancers.[see comment]. *Nature* 435, 834-838.

Lui, W.O., Pourmand, N., Patterson, B.K., Fire, A., 2007. Patterns of known and novel small RNAs in human cervical cancer. *Cancer Res.* 67, 6031-6043.

Makeyev, A.V., Eastmond, D.L., Liebhaber, S.A., 2002. Targeting a KH-domain protein with RNA decoys. *RNA* 8, 1160-1173.

Marsh, E.E., Lin, Z., Yin, P., Milad, M., Chakravarti, D., Bulun, S.E., 2008. Differential expression of microRNA species in human uterine leiomyoma versus normal myometrium. *Fertil Steril* 89, 1771-1776.

Mathonnet, G., Fabian, M.R., Svitkin, Y.V., Parsyan, A., Huck, L., Murata, T., Biffo, S., Merrick, W.C., Darzynkiewicz, E., Pillai, R.S., Filipowicz, W., Duchaine, T.F., Sonenberg, N., 2007. MicroRNA inhibition of translation initiation in vitro by targeting the cap-binding complex eIF4F. *Science* 317, 1764-1767.

Mattie, M.D., Benz, C.C., Bowers, J., Sensinger, K., Wong, L., Scott, G.K., Fedele, V., Ginzinger, D., Getts, R., Haqq, C., 2006. Optimized high-throughput microRNA expression profiling provides novel biomarker assessment of clinical prostate and breast cancer biopsies. *Mol. Cancer* 5, 24.

Maziere, P., Enright, A.J., 2007. Prediction of microRNA targets. *Drug Discov. Today* 12, 452-458.

McKean-Cowdin, R., Feigelson, H.S., Pike, M.C., Coetzee, G.A., Kolonel, L.N., Henderson, B.E., 2001. Risk of endometrial cancer and estrogen replacement therapy history by CYP17 genotype. *Cancer Res.* 61, 848-849.

Miller, B.G., Wild, J., Stone, G.M., 1979. Effects of progesterone on the oestrogen-stimulated uterus: a comparative study of the mouse, guinea pig, rabbit and sheep. *Aust. J. Biol. Sci.* 32, 549-560.

Mitchell, D.C., Ing, N.H., 2003. Estradiol stabilizes estrogen receptor messenger ribonucleic acid in sheep endometrium via discrete sequence elements in its 3'-untranslated region. *Mol. Endocrinol.* 17, 562-574.

Moore, N.W., Miller, B.G., Trappl, M.N., 1983. Transport and development of embryos transferred to the oviducts and uteri of entire and ovariectomized ewes. *J. Reprod. Fertil* 68, 129-135.

Mukherjee, D., Gao, M., O'Connor, J.P., Raijmakers, R., Pruijn, G., Lutz, C.S., Wilusz, J., 2002. The mammalian exosome mediates the efficient degradation of mRNAs that contain AU-rich elements. *EMBO Journal* 21, 165-174.

Nagy, E., Maquat, L.E., 1998. A rule for termination-codon position within intron-containing genes: when nonsense affects RNA abundance. *Trends Biochem. Sci.* 23, 198-199.

Nair, A.K., Young, M.A., Menon, K.M., 2008. Regulation of luteinizing hormone receptor mRNA expression by mevalonate kinase--role of the catalytic center in mRNA recognition. *FEBS J.* 275, 3397-3407.

Nowak, R.A., 2001. Identification of new therapies for leiomyomas: what in vitro studies can tell us. *Clin. Obstet. Gynecol.* 44, 327-334.

Palmiter, R.D., 1972. Regulation of protein synthesis in chick oviduct. II. Modulation of polypeptide elongation and initiation rates by estrogen and progesterone. *J. Biol. Chem.* 247, 6770-6780.

Pan, Q., Luo, X., Toloubeydokhti, T., Chegini, N., 2007. The expression profile of micro-RNA in endometrium and endometriosis and the influence of ovarian steroids on their expression. *Mol. Hum. Reprod.* 13, 797-806.

Parker, R., Song, H., 2004. The enzymes and control of eukaryotic mRNA turnover. *Nature Structural and Molecular Biology* 11, 121-127.

Proia, D.A., Nannenga, B.W., Donehower, L.A., Weigel, N.L., 2006. Dual roles for the phosphatase PPM1D in regulating progesterone receptor function. *J. Biol. Chem.* 281, 7089-7101.

Robertson, J.A., Zhang, Y., Ing, N.H., 2001. ICI 182,780 acts as a partial agonist and antagonist of estradiol effects in specific cells of the sheep uterus. *Journal of SteroidBiochemistry and Molecular Biology* 77, 281-287.

Rodriguez-Pinon, M., Meikle, A., Tasende, C., Sahlin, L., Garofalo, E.G., 2005. Differential estradiol effects on estrogen and progesterone receptors expression in the oviduct and cervix of immature ewes. *Domestic Animal Endocrinology* 28, 442-450.

Saceda, M., Lindsey, R.K., Solomon, H., Angeloni, S.V., Martin, M.B., 1998. Estradiol regulates estrogen receptor mRNA stability. *Journal of Steroid Biochemistry andMolecular Biology* 66, 113-120.

Saxena, S., Jonsson, Z.O., Dutta, A., 2003. Small RNAs with imperfect match to endogenous mRNA repress translation. Implications for off-target activity of small inhibitory RNA in mammalian cells. *J. Biol. Chem.* 278, 44312-44319.

Seaver, S.S., 1981. The effects of sequential hormone treatment on ovalbumin synthesis in chick oviduct: a possible example of translation regulation. *J. Steroid Biochem.* 14, 949-957.

Sela-Brown, A., Silver, J., Brewer, G., Naveh-Many, T., 2000. Identification of AUF1 as a parathyroid hormone mRNA 3'-untranslated region-binding protein that determines parathyroid hormone mRNA stability. *J. Biol. Chem.* 275, 7424-7429.

Sengupta, S., Jang, B.C., Wu, M.T., Paik, J.H., Furneaux, H., Hla, T., 2003. The RNA-binding protein HuR regulates the expression of cyclooxygenase-2. *J. Biol. Chem.* 278, 25227-25233.

Sun, M., Estrov, Z., Ji, Y., Coombes, K.R., Harris, D.H., Kurzrock, R., 2008. Curcumin (diferuloylmethane) alters the expression profiles of microRNAs in human pancreatic cancer cells. *Mol. Cancer Ther* 7, 464-473.

Tsai, M.J., Clark, J.H., Schrader, W.T., O'Malley, B.W., 1998. Mechanisms of action of hormones that act as transcription factors. Saunders, Philadelphia.

Vasudevan, N., Pfaff, D.W., 2008. Non-genomic actions of estrogens and their interaction with genomic actions in the brain. *Front Neuroendocrinol* 29, 238-257.

Wagner, B.J., DeMaria, C.T., Sun, Y., Wilson, G.M., Brewer, G., 1998. Structure and genomic organization of the human AUF1 gene: alternative pre-mRNA splicing generates four protein isoforms. *Genomics* 48, 195-202.

Wang, L., Nair, A.K., Menon, K.M., 2007a. Ribonucleic acid binding protein-mediated regulation of luteinizing hormone receptor expression in granulosa cells: relationship to sterol metabolism. *Mol. Endocrinol.* 21, 2233-2241.

Wang, T., Zhang, X., Obijuru, L., Laser, J., Aris, V., Lee, P., Mittal, K., Soteropoulos, P., Wei, J.J., 2007b. A micro-RNA signature associated with race, tumor size, and target gene activity in human uterine leiomyomas. *Genes Chromosomes Cancer* 46, 336-347.

Watson, C.S., Alyea, R.A., Jeng, Y.J., Kochukov, M.Y., 2007. Nongenomic actions of low concentration estrogens and xenoestrogens on multiple tissues. *Mol. Cell Endocrinol.* 274, 1-7.

Whelly, S.M., Barker, K.L., 1974. Early effect of estradiol on the peptide elongation rate by uterine ribosomes. *Biochemistry* 13, 341-346.

Wilson, G.M., Sutphen, K., Bolikal, S., Chuang, K.Y., Brewer, G., 2001. Thermodynamics and kinetics of Hsp70 association with A + U-rich mRNA-destabilizing sequences. *J.Biol. Chem.* 276, 44450-44456.

Wu, L., Belasco, J.G., 2008. Let me count the ways: mechanisms of gene regulation by miRNAs and siRNAs. *Mol. Cell* 29, 1-7.

Xu, N., Chen, C.Y., Shyu, A.B., 2001. Versatile role for hnRNP D isoforms in the differential regulation of cytoplasmic mRNA turnover. *Mol. Cell Biol.* 21, 6960-6971.

Yan, L.X., Huang, X.F., Shao, Q., Huang, M.Y., Deng, L., Wu, Q.L., Zeng, Y.X., Shao, J.Y., 2008. MicroRNA miR-21 overexpression in human breast cancer is associated with advanced clinical stage, lymph node metastasis and patient poor prognosis. *RNA* 14, 2348-2360.

Zhao, J.J., Lin, J., Yang, H., Kong, W., He, L., Ma, X., Coppola, D., Cheng, J.Q., 2008. MicroRNA-221/222 negatively regulates estrogen receptor alpha and is associated with tamoxifen resistance in breast cancer. *J. Biol. Chem.* 283, 31079-31086.

Zhao, Z., Chang, F.C., Furneaux, H.M., 2000. The identification of an endonuclease that cleaves within an HuR binding site in mRNA. *Nucleic. Acids Research* 28, 2695-2701.

In: Biological Aspects of Human Health and Well-Being
Editor: Tsisana Shartava

ISBN: 978-1-61209-134-1
© 2011 Nova Science Publishers, Inc.

Chapter X

EPIGENETICS OF GESTATIONAL TROPHOBLASTIC DISEASE: GENOMIC IMPRINTING AND X CHROMOSOME INACTIVATION

*Pei Hui**

Programs in Women's Health and Molecular Diagnostics
Department of Pathology, Yale University School of Medicine,
310 Cedar Street, New Haven, CT 06520-8023

ABSTRACT

Genomic imprinting, the selective suppression of one of the two parental alleles of various genes, has been proposed to play an important regulatory role in the development of the placenta of eutherian mammals. The "parental conflict hypothesis" views that parents of opposite sex have conflicting interests in allocating resources to their offspring by the mother, proposing that growth-promoting genes are mainly expressed from the paternally inherited genome and are silent in the maternally inherited counterparts. X chromosome inactivation plays a central role in compensating for the double dose of X-linked genes in cells of the female relative to cells of the male.

In the placenta of some species, X inactivation represents a special form of epigenetic imprinting. The paternal X chromosome is preferentially imprinted and silent in mouse trophectoderm, a tissue type from which gestational trophoblastic diseases arise. Growing body of evidence has suggested that the pathogenesis of gestational trophoblastic diseases involves altered genomic imprinting. Abnormal expressions of imprinted genes, such as IGF2 and H19, have been implied in the development of molar pregnancies and gestational choriocarcinoma. Moreover, unique genetic modes have long been established in hydatidiform moles: all complete moles have diandric diploid or tetraploid paternal-only genome and partial moles have triploid diandric and monogynic genome. Consistent with parental imprinting theory, partial mole occurs only with diandroid but digynic troploidy. Recent studies have found that all human placental site trophoblastic tumors arose from a female conceptus, suggesting that a functional paternal

---

* E-mail: Pei Hui pei.hui@yale.edu

X chromosome is important for the neoplastic transformation, likely through inappropriate expression of paternal X-linked genes.

As epigenetic regulation of genomic imprinting and X chromosome inactivation are important for the genesis of gestational trophoblastic diseases, hydatidiform mole and placental site trophoblastic tumor may provide model systems with which genomic imprinting regulation of placenta development and the proliferative advantage conferred by the paternal X chromosome can be studied.

INTRODUCTION

As a transient organ during reproduction, the placenta consists of unique tissue type in eutherian mammals in that it is fetal in origin and shares half of the genome with that of the maternal uterus. Though existing for a short life span, there are constant morphological and biological fluctuations within the placenta proper and at the interface between the placenta and the maternal gestational endomyometrium. It is at the latter interface, where unique tissue interactions, hormonal regulations and immunological modulations are controlled in such a delicate balance so that appropriate maternal support can sustain the embryo development as well as rapid growth of the placenta, and in the mean time the fetus does not impost undue burden onto the mother and the mother does not illicit an immunologic rejection against the growing gestational tissues. In biology, it should be an extraordinary event when tumors of the placental origin occur within the maternal uterus, even long time after the pregnancy has accomplished. Tumors of these types are also remarkable in that they have to overcome additional growth and immunological barriers in order to develop and progress. Recent findings of genomic imprinting including imprinted X chromosome inactivation in the placenta and their biological roles in the pathogenesis of gestational trophoblastic diseases have raised some fundamental questions in mammalian biology and oncology. In this review, extraembryonic placenta and gestational trophoblastic diseases are placed onto the discussion platform to reflect our recent understanding of their intriguing biology and pathogenesis, respectively.

PLACENTA DEVELOPMENT AND TROPHOBLASTIC CELLS

After fertilization, the ovum rapidly proliferates within the fallopian tube into a blastocyst, which is then transported into the endometrial cavity and implanted into the receptive gestational endometrium to form the trophoblastic cell mass by the 10th post ovulatory day (Kurman 2002). During the implantation, the outer cell layers of the blastocyst differentiate into trophoblasts, and the inner cell mass develops into the ultimate embryo. The trophoblastic mass grows circumferentially and peripheral trophoblasts invade the stroma of endometrium at the implantation site. Trophoblastic columns are formed with blood filled spaces in between and covered by a syncytiotrophoblastic outer layer and an immature cytotrophoblastic inner layer. Extraembryonic mesenchyme then invades trophoblastic columns and forms fetal vascular channels, which eventually connect with each other and with those of the body stalk/allantois to establish the fetoplacental circulation. At the implantation site, the trophoblasts invade endometrial vasculatures including the spiral arteries to create maternal blood filled lacunae to form the precursor of the intervillous space. However, the ultimate fetal-maternal circulation is not established until the 12th week of gestation. Solid trophoblastic columns remain at the periphery of the stem villi, anchoring them to the basal plate to eventually form a complete shell, which continues to grow and expand. As chorion protrudes into the endometrial cavity, the villi toward the uterine surface undergo regression to form chorionic laeve or fetal membranes with eventual obliteration of the uterine cavity, while the villi on the embryonic site continue to proliferate and mature with additional septae formation and structural modification to form the ultimate placenta.

The main functional cell types in the placenta are various trophoblasts that are generally classified into three categories: syncytiotrophoblasts, cytotrophoblasts and intermediate trophoblasts (Shih 2001). Syncytiotrophoblasts are fully matured cells that are in direct contact with the maternal circulation and produce most of the placental hormones. These are multinucleated cells with large amount of cytoplasm and located externally to the immature cytotrophoblasts on the surface of the chorionic villi. Syncytiotrophoblasts do not have detectable proliferative activity or mitosis. They can be found to produce hCG and hPL as early as 12 days of gestation. Cytotrophoblasts are primitive cell type and likely have stem cell profile in the placenta. They are medium size cells with polygonal to oval shape and may rapidly proliferate and are mitotically active. They are not hormonally functional cells, however. Intermediate trophoblasts are large mononuclear cells with abundant cytoplasm and may exist in single or sheet. They show remarkable infiltrative growth patterns involving decidua, myometrium and maternal vasculatures. Three types of intermediate trophoblasts have been described: the implantation site intermediate trophoblasts are found at the implantation sites and responsible for establishment and maintenance of the placenta proper; chorionic laeve intermediate trophoblasts are found within the chorion membrane and their function are unknown; and villous intermediate trophoblasts are found mainly in the anchoring villi.

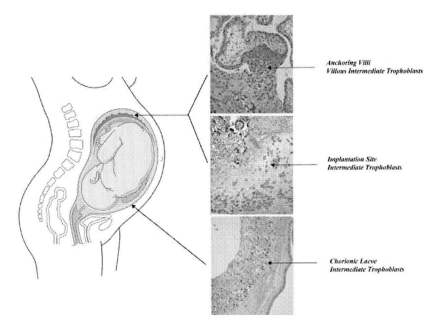

Figure 1. Schematic presentation of human gestation, placenta and various intermediate trophoblasts.

The intermediate trophoblasts play a key role in the fetomaternal interactions (Shih 2001, Figure 1). These cells are functional derivatives of the trophoblastic stem cells, i.e. cytotrophoblasts, present within the anchoring villi. At the base of the anchoring villi, implantation site intermediate trophoblasts come in contact with endomyometrium where they infiltrate the decidua and myometrium.

With the main function to establish fetomaternal circulation, implantation site intermediate trophoblasts invade and replace spiral arteries at the implantation site, essentially replacing maternal cells of the vasculatures including the endothelial cells while maintaining the vascular patency. The histological and cytological features of such invasion by these intermediate trophoblasts are reminiscent to those involved in the tissue invasion by malignant tumor cells.

GENOMIC IMPRINTING AND PLACENTAL DEVELOPMENT

Epigenetic imprinting, the selective suppression of various genes derived from one parent or the other, has been proposed to play an important regulatory role in the evolution and development of the placenta in eutherian mammals. Significant body of work has demonstrated that, in different mammalian species, genomic imprinting incurred during mammalian evolution and is linked to the placental development and function (Ferguson-Smith 2001, Reik 2005, Morison 2005, Coan 2005, Figure 2). The most remarkable example is that in mice, many imprinted genes are found in extraembryonic tissue and many of which are subject to imprinting only in the placenta (Table I, Sandell, 2003, Ono 2003, Higashimoto 2002, Caspary 1998, Mizuno 2002, Zwart 2001). Some of the imprinted genes important for the placental development are highly conserved in eutherian mammals (Ono 2006). It has been speculated that ancestral mammals might have evolved with the placenta, from newly acquired, retrotransposon-derived genes, or endogenous versions of the genes present in oviparous animals, some time after the divergence of mammals and birds, more than 90 to 130 million years ago (Hedges, 2002).

Approximately half of the known 80 or so imprinted genes are related to cellular proliferation and growth, and many of which are involved in placental development and function (Tycko 2002). Remarkably, all imprinted genes specific to the placenta are functionally expressed only from their maternal alleles in mouse (Table 1, Feil 2006).

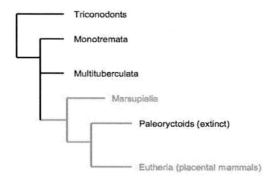

Figure 2. Evolution tree related to mammalian animals. Red lines indicate the appearance of genomic imprinting.

Table I. Selected Imprinted Genes Involved in Placenta Development and Function

Gene Name	Gene Location	Functional Aspect or gene product	Expression Tissue	Species	Imprint Alleles	References
Esx 1	X	X-linked homeobox gene	Placenta, testis	Mouse	Paternal	Li 1997,
Pem	X	X-linked homeobox gene	Placenta	Mouse	Paternal	Lin 1994
Xist	X	X dosage counting	Placenta	Mouse	Paternal	Mak 2004, Nolen 2005
BEX1(REX3)/ Bex1(Rex 3)	Xq22	X-linked	Blasstocyst, placenta	Human.mouse	Patern al	Brown 1999,Williams 2002
Eed	X	X-linked polycomb group gene	Fetus, placenta	Mouse	Paternal	Wang 2001, Monk 2008
Tsix	X	Regulating Xist	Fetus, placenta	Mouse	Paternal	Lee 2000
TFPI2	7q21/6		Placenta	Human/mouse	Paternal	Monk 2008
Rtl1/Peg11	Distal 12	Aspartyl protease motif/maintenance of fetal capillaries	Fetus, placenta	Mouse	Maternal	Seitz 2003
IGF-II/Igf-2	11p15.5	Growth factor	Fetus, placenta	Human/mouse	Maternal	Constancia 2002, Tilghman 1999, Ferguson-Smith 2000
H19/H19	11p15.5		Fetuas placenta	Human/mouse	Paternal	Tilghman 1999, Ferguson-Smith 2000
PEG10/Peg10	7q216	DNA/RNA binding/cell cycle regulator	Embryo, placenta	Human/mouse	Maternal	Ono 2006
Igf2-r	Distal 17		Fetus, placenta	mouse	Paternal	Wutz 1998, Sleutels 2002
INS/Ins2	11p15.5		Fetus, placenta	Human/mouse/marsupial	Maternal	Ager 2007
PEG1(MEST)/ Peg1(mest)	7q32.2	Hydrolase enzyme/Angiogenesis	Fetus, placenta	Human, mouse	Maternal	Mayer 2000,Kaneko-Ishino 1995, McMinn 2006
Gatm	Central 2	L-arginie:glycine amidinotranserase	Placenta	mouse	Paternal	Sandell 2003
Ppp1r9a	Proximal 6	Neural tissue-specific F-actin binding protein	Placenta	mouse	Paternal	Ono 2003
Pon 2 and 3	Proximal 6	Paraoxonases	Placenta	mouse	Paternal	Ono 2003

Table 1. (Continued)

Gene Name	Gene Location	Functional Aspect or gene product	Expression Tissue	Species	Imprint Alleles	References
Osbpl5	Distal 7	Oxysterol binding protein-like 5	Placenta	mouse	Paternal	Engemann 2000
Tssc4	Distal 7	Tumor suppressor	Placenta	mouse	Paternal	Paulsen 2000
Tspan32	Distal 7	AML1 regulated tranmembrance protein	Placenta	mouse	paternal	Umlauf 2004
Ascl2	Distal 7	Achaet-scute homolog 2	Placenta	mouse	Paternal	Guillemot 1995
Cd81	Distal 7	Cd81 antigen	Placenta	mouse	Paternal	Umlauf 2004
Dcn	Distal 10	Decorin	Placenta	mouse	Paternal	Mizuno 2002
Slc22a2/Slc22a3	Proximal 17	Solure carrier family 22	Placenta	mouse	Paternal	Sleutels 2002, Zwart 2001.
CTNNA3	10q21	Catenin, alpha3	Placenta	Human	Paternal	Van Dijk 2004
HERC4	10q21	Ubiquitin ligase domain	Placenta	Human	Paternal	Oudejans 2004
MAWBP	10q21	MAP Activator	Placenta	Human	Paternal	Oudejans 2004
STOX1	10q22	Storkead Box 1	Placenta	Human	Paternal	Van Dijk 2005
KCNMA1	10q22	Calcium-activated potassium channel alpha subunit 1	Placenta	Human	Paternal	Oudejans 2004
OSBPL5	11p15	Oxysterol binding protein-like	Placenta	Human	Paternal	Higashimoto 2002
P57kip2(CDKN1C)	11p15		Fetus, placenta	Human	Paternal	Takahashi 2000
KCNQ1OT1/Kcnq1ot1	11p15		Fetus, placenta	Human/mouse	Paternal	Mancini-DiNardo D 2003, Lewis 2004, Umlauf D 2004
Mash2	Distal 7	Helix-turn-helix transcription factor	Placenta, fetus	Mouse	Paternal	Guillemot 1995.

In human placenta, although limited in the number of studies, it has been shown that the patterns of genomic imprinting are similar to those of the mouse. Human PHLDA2 gene is expressed only from the maternal allele and is not expressed in paternal unidisomy complete hydatidiform moles (Salas 2004). P57kip1 is another paternally imprinted gene that is expressed only from the maternal allele in the cytotrophoblasts and villous stromal cells of human placenta (Caspary 1998, Takahashi 2000). Placenta specific imprinted genes such as CTNNA3/alpha3 catenin, HERC4/ubiquitin ligase, MAWBP/MAP activator, STOX1, KCNKMA1/calcium-activated potassium channel alpha subunit are all paternally imprinted in human placenta (Oudejans 2004, Van Dijk2004). Encoding a Forkhead/winged helix protein family, STOX1 is expressed in extravillous trophoblasts and possibly involved in the transition of invasive to non-invasive behaviors of these trophoblasts (Argnrimsson 2005).

Imprinted genes are frequently clustered in the genome. At DNA level, the maintenance of genomic imprinting in extraembryonic tissue is mainly dependant on non-DNA methylation mechanisms including histone deacetylation and methylation, unlike those imprinting genes found in somatic embryonic tissues (Lewis 2004, Feil 2006). Convincing molecular data confirms that in placenta, mechanisms in initiation and maintenance of imprinting in mouse involve imprinting initiation center from which a noncoding RNA is produced to coat the imprinting cluster in-cis, followed by histone H3 deacetylation at K4 and acquisition of H3 K9 and K27 methylation, and eventually leading to recruitment of the polycomb complex proteins to the repressed paternal chromosome region (Silva 2003, Cao 2004, Montgomery 2005, Mager 2003, Umlauf 2004). Similar molecular mechanism is well established for the imprinted X chromosome inactivation in the placenta (see below).

The "parental conflict hypothesis" is the dominant theory of genomic imprinting in placenta biology. It views that parents of opposite sex have conflicting interests in allocating resources to their offspring by the mother (Barlow 1995, Haig 2006, Abu-Amero 2006, Solter 1998). The interest of the paternal genome would be to maximize resources for the father's own progeny. In contrast, the one of the maternal genome would be to distribute resources equally among her offsprings. This implies that growth-promoting genes are mainly expressed from the paternally inherited genome and are silent in the maternally inherited counterparts. Analyses of many imprinted genes in mammals support this theory (Tilghman 1999, Constancia 2004) and many known imprinted genes are involved in cellular proliferation and growth and are important for placental development and function (Tycko 2002). Pronuclear transfer experiments showed that unipaternal disomy in mouse led to placental overgrowth along with early fetal lethality, and unimaternal disomy resulted in hypoplasia of the placenta (McGrath 1984). Phlda2 (Ipl/Tssc3) is a maternally expressed and paternally imprinted gene in the placenta. Homozygous knockout mice lacking the active maternal allele have markedly enlarged placenta (Frank 2002, Salas 2004). Transgenic overexpression of Phlda2 leads to a marked reduction of the placental size (Salas 2004). P57Kip2/CDKN1C is another paternally imprinted gene in human placenta. Loss of its expression is associated with trophoblastic hypertrophy and placentaomegaly in mice (Caspary 1998, Takahashi 2000), and hyperplasia of trophoblasts in human diandric hydatidiform mole (Fukunaga 2002, Castrillon DH 2001). Several mouse imprinted genes (Table 1) are found in placenta only and are paternally silenced, although their functions are not clear.

One other theoretical speculation for silencing paternal alleles in placenta by imprinting may involve prevention of allogeneic graft rejection potentially mounted by the maternal immune system. There has been no evidence that the maternal immune system is altered during the pregnancy and no serum antibodies or cellular immune response have been detected against trophoblastic cells regardless of the intimate cellular interactions at the placental-endomyometrial interface (Bulmer 1990, Divers 1995). The placenta is a fetal organ with half of its genome of paternal origin, therefore containing paternal allelic antigens that are foreign to the mother. It is clear that immunological interactions between placental and maternal tissues are important for normal placental function and a successful pregnancy. Suppression of the expression of many paternal imprinted genes would fit nicely with the inability of the mother to elicit an immune response against the placenta as an allogeneic graft, targeting particularly the intermediate trophoblasts at the chroiodecidual interface. Whether this speculation holds any truth requires future investigations. Currently, it is believed that not any single factor but rather a complex interplay among various players between trophoblastic antigenicity and maternal immune response factors. NK cells are emphasized to have important functions in mitigating the immunological balance (Le Bouiteiller 2008) and expression of HLA-G molecule by the trophoblasts is also implied in this immunologic silence (Kuroki 2007, Hunt 2007).

GESTATIONAL TROPHOBLASTIC DISEASES: ABNORMAL GENOMES AND GENOMIC IMPRINTING

Gestational trophoblastic diseases are a fascinating group of placental proliferative disorders because of their unique clinical settings, genetic compositions and varying biological behaviors. Arising from various trophoblastic cell types within the placenta, six distinct human diseases have been classified under gestational trophoblastic diseases by WHO (Genest 2003): complete hydatidiform mole, partial hydatidiform mole, invasive hydatidiform mole, choriocarcinoma, placental site trophoblastic tumor and more recently, epithelioid trophoblastic tumor. All hydatidiform moles are proliferative lesions of the villous trophoblasts. The most virulent choriocarcinoma is a fully malignant tumor of the trophoblasts recapitulating the primitive cells of the previllous stage of the placenta. The intermediate trophoblast at the implantation site is the putative cell type that gives rise to placental site trophoblastic tumor, whereas the intermediate trophoblast at the chorionic leave is considered the cell type found in epithelioid trophoblastic tumor.

Molecular and genetic investigations in the 1970's were pivotal to the understanding the etiology of hydatidiform moles (Kajii 1977, Szulman 1984). Essentially all complete hydatidiform moles have diandric, paternal-only genomes, with either 46, XX diploid karyotypes arising from the fertilization of an empty egg by one spermatozoon followed by duplication (homozygous, 80%), or 46 XX or XY karyotypes arising from the simultaneous fertilization of an empty egg by two spermatozoa (heterozygous, 20%). The genetic profile of partial hydatidiform moles is triploid with a diandric, monogynic genome arising from the fertilization of a haploid egg by either two heterozygous spermatozoa (heterozygous, 90%) or one spermatozoon with duplication (homozygous, 10%). With their obvious lacking of or relatively deficient in maternal genome, abnormal genomic imprinting has long been implicated in the development of hydatidiform moles. Consistent with the "parents of conflict interest" theory, the lack of maternal genome together with the global genome demethylation and abnormal paternal imprinting gene expression are key molecular features of complete hydatidiform moles).

IGF2 and H19 tightly linked on human chromosome 11 are of special interest because of their reciprocal imprinting and possible association with certain malignancy and congenital abnormalities. Normally, the IGF2 gene expressed from the paternally derived allele (Ohlsson 1993, Giannoukalis 1993), whereas the H19 gene is expressed from the maternally derived allele (Rachmilewitz 1992, Zhang 1993, Ferguson-Smith1993). Although twice expression of paternal genome may be a prerequisite for the pathogenesis of complete hydatidiform moles, trophoblasts of these androgenetic moles express abnormally both IGF2 (normally expressed only from paternal alleles) and H19 (normally expressed only from maternal alleles) (Mutter 1993, Wake 1998, Ariel 2000, Bestor 2006), suggesting a failure of allele-specific gene expression in CHM. It has been suggested that the mutated promoter is responsible for overcoming transcriptional suppression by CpG methylation in the H19 gene (Arima 1997). However, such imprinting alteration is selective because retention of other paternal imprinted genes, i.e. p57, exists (Fukunaga 2002, Castrillon 2001). Such abnormal expression of imprinted genes in trophoblasts may be connected to the global genome demethylation (Mann MR 2004) that occurs only in the trophoblastic cells of complete hydatidiform moles (Figure 3).

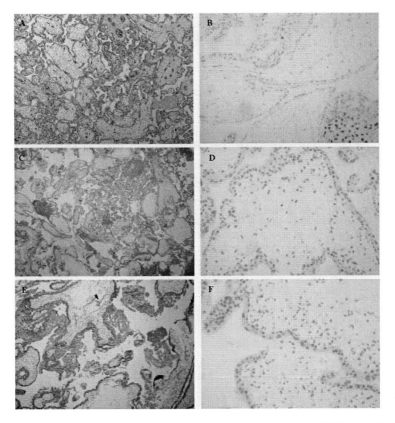

Figure 3. Imprinting patterns of P57kip2 gene in normal (A, B), partial hydatidiform mole (C, D) and complete hydatidiform mole (E, F). Note absence of nuclear staining of p57 immunohistochemistry of cytotrophoblasts in complete hydatidiform mole (F).

Therefore, the pathogenesis of androgenic complete hydatidiform mole is likely to involve, in addition to a loss of maternal genome, the combination of the gain of paternal expression of imprinted genes as well as additional alterations of epigenetic gene silencing due to the global genome demethylation in trophoblastic cells.

Alterations of genomic imprinting are not unique to molar gestations and have been implicated in the development of other gestational trophoblastic diseases including gestational choriocarcinoma, a bona fide malignant tumor. Choriocarcinomas, which may develop from complete hydatidiform moles, showed similar expression of IGF2 and enhanced expressing H19, although this biallelic expression of IGF2 or H19 was not consistently found in the tumors (Arima 1997, Wake 1998, Ariel 2000). Limited data has suggested that there is a lack of correlation between IGF2 and H19 imprinting status, and in fact, the imprinting status of H19 and IGF2 was found differentially modulated, indicating that allele-specific expression of IGF2 operates in the absence of a parental imprint in this type of trophoblastic disease (He 1998).

X Chromosome Inactivation and Placental Development

X chromosome inactivation plays a central role in balancing the gene dosage in a female cell. Growing body of evidence confirmed some similarities between placental imprinting of autosomal genes and X chromosomal inactivation in placenta in female mice (Feil 2006, Hemberger 2007). The paternal X chromosome is preferentially imprinted and silent in mouse trophectoderm and genes on X chromosome are expressed only from the maternal alleles. Therefore, in mouse placenta, X chromosome inactivation represents a special form of genomic imprinting. Such imprinted X inactivation is established early in pre-implantation development and in all cells of the embryo. However, the imprinting is maintained in the extra-embryonic tissue only. The embryonic tissues switch X imprinting pattern to random later in the development (Mak 2004, Okamoto 2004). At the genetic regulatory level, there are ontogenetic similarities of imprinting between autosomal genes and those of X chromosome. Best illustrated by Kcnq1 domain on distal chromosome 7 and Xist on X chromosome, both autosomal and X imprinting in mouse placenta are installed in the early pre-implantation embryo followed by somatic maintenance of imprinting only in the extraembryonic tissue (Umlauf 2004). Both imprinting conditions are independent of DNA methylation (Sado 2000) and involve a non-translated RNA gene: Kcnq1 in somatic imprinting and Xist in X imprinting. Both nontranslated RNA molecules interact with intronic regions through a histone-based mechanism involving deacetylation and hypomethylation of K4 combined with methylation of K9 and K27 of H3 protein (Fitzpatrick 2002, Lewis 2004). Xist RNA carpets the entire X chromosome in cis, followed by the recruitment of various chromatin-modifying molecules (Avner 2001, Brockdorff 2002). Similar mechanism for Kcnq1 non-translated RNA has been speculated (Fitzpatrick 2002). In sharp contrast, DNA methylation is the main mechanism involved in the random X inactivation and genomic imprinting in embryo proper (Li 1993, Sado 2000, Lewis 2004).

Paternal imprinting expression of autosomal genes and paternal X inactivation in mouse placenta raise an important biological question: are they co-evolutionarily established in placental animals (Lee 2003, Reik 2005)? If this is true, it is reasonable to argue that in primates including human, imprinting of X chromosome in placenta is possible. The paternal X chromosome is preferentially imprinted and silent in mouse trophectoderm (Salido 1992, Lin 1994, Li 1997, Huynh 2003,Sado 2005), a tissue type corresponding to that from which the placental site trophoblastic tumor and hydatidiform moles arise in humans. In marsupial mammals, the paternal X chromosome is preferentially inactivated in both embryonic and extraembryonic tissues (Cooper 1971, Graves 1996). Unexpectedly, the mode of X inactivation appears random in human extraembryonic tissue by recent data (Looijenga 1999, Migeon 2002, Zeng 2003), although still controversial (Goto 1997, Huynh 2003, Lee 2003), particularly because now we understand that the maintenance of X imprinting in placenta are largely dependent on histone deacetylation and methylation pathways, rather than through direct DNA methylation. Additionally, imprinting patterns of some autosomal genes appear cell type specific among various placental trophoblasts (unpublished data).

PATERNAL X CHROMOSOME IN THE PATHOGENESIS OF HUMAN TROPHOBLASTIC DISEASES

Tumors arising from the extraembryonic tissue are rare and include gestational chorionic carcinoma and intermediate trophoblastic tumors, i.e. placental site trophoblastic tumor and epithelioid trophoblastic tumor. In oncology, these three tumors are extraordinary due to their fetal origin and maternal tissue matrix that supports the tumor growth. These tumors behave quite variably clinically with chorionic carcinoma represents one of the most malignant tumors in human if untreated (Genest 2003). Placental site trophoblastic tumor is a rare neoplastic proliferation of extravillous intermediate trophoblasts at the implantation site (Shih 2001). The clinical presentation is generally that of a young woman who has a history of term pregnancy, an abortion or uncommonly, a complete mole.

At cytogenetic level, most of placental site trophoblastic tumors were diploid (Kotylo 1992, Fukunaga 1993) and rarely showed genetic imbalances analyzed by comparative genomic hybridization(Hui 2004). We have previously noted through literature reviews that a majority of the tumors were preceded by a female gestational event. In our pilot investigation, the presence of a paternal X chromosome was seen in 4 of 5 cases, and the absence of a Y chromosomal element was found in all 5 tumors by the semi-nested PCR amplification of SRY (human sex-determining region on Y chromosome, Hui 2000). Since then, additional case reports of placental site trophoblastic tumor have been published and again a female antecedent pregnancy was documented in 12 out of 14 new cases (Hassadia 2005, Khan 2006). One cell line from a placental site trophoblastic tumor was established and confirmed by karyotyping to have XX genome (Kobel 2005).

In our recent investigation (Hui 2006), the presence of a haploid set of paternal chromosomes without a Y chromosomal element was seen in all 20 cases (Figure 4), confirming a requirement of paternal X chromosome by placental site trophoblastic tumor. These results imply that the presence of paternal X chromosome is important for the neoplastic proliferation of placental site trophoblastic tumor.

One unlikely explanation is that placental site trophoblastic tumor might derive from an antecedent complete hydatidiform mole, a comparatively more common gestational trophoblastic disease with uniparental disomy of the paternal genome and an exclusive paternal X chromosome. However, only a small percentage of placental site trophoblastic tumors have been reported to follow the development of complete mole (Shih 2001, Baergen 2006). Alternatively, available data suggest an active role for the paternal X chromosome in the pathogenesis placental site trophoblastic tumor (Hui 2000).

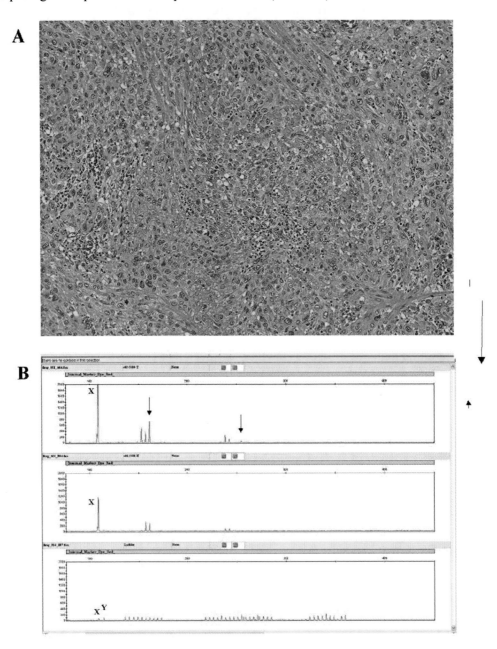

Figure 4. Placental site trophoblastic tumor consists of neoplastic proliferation of implantation site trophoblasts infiltrating maternal myometrium (A). STR analysis at three loci of human genome of the tumor tissue (top) and corresponding maternal tissue (middle) along with allelic distribution ladder (bottom). Note the presence of unique paternal alleles in the tumor (arrows) and absence of Y chromosomal allele in both the tumor and the maternal tissue.

The definite reason for the requirement of paternal X chromosome in placental site trophoblastic tumor is unclear at this time. Relaxation of an imprinted paternal X in placentawould explain such requirement as the paternal imprinted X chromosome will bring growth advantage according to the conflict of interest theory. However, currently there is no evidence of X imprinting in human placenta. Nevertheless, it is possible that a relaxation of the X inactivation (random or imprinted) will result in double dosing of X chromosomal genes in trophectoderm leading to trophoblast hyperplasia and eventually placental site trophoblastic tumor.

CONCLUSION

The introduction of imprinting genes during the evolution coincided with the appearance of eutherian mammals, leading to, for the first time, that the fetal tissue became in direct contact with maternal endomyometrial structures. Not only did such contact create a safer house, i.e. the womb, for the eutherial offspring to develop, but also brought in a spectrum of complex biological interactions between the fetal and maternal organs, where the placenta is at the center stage. Genomic imprinting and preferential X inactivation in mouse placental trophoblasts likely regulate in a significant way of these biological interactions, and perhaps play important roles in the pathogenesis of some reproductive disorders, such as gestational trophoblastic diseases. Imprinting alterations leading to over representation of the paternal genes in trophoblasts underline the development of hydatidiform moles. The propensity to malignancy of complete hydatidiform mole is likely associated with their genetic compositions and the growth advantage conferred by homozygosity and the selective inheritance of paternal genome. Homozygosity would lead to the inactivation of a tumor suppressor gene by paternal transmission of imprinted allele. In-depth investigations of these aspects are important to understand the pathogenesis of hydatidiform moles, a rather common disease of human reproduction. As issues of epigenetic regulation of imprinting in X chromosome inactivation are of significant biological implications, placental site trophoblastic tumor may provide an important tumor model with which the sex chromosome biology and the proliferative advantage conferred by the paternal X chromosome can be studied.

REFERENCES

Abu-Amero S, Monk D, Apostolidou S, Stanier P, Moore G. Imprinted genes and their role in human fetal growth. *Cytogenet. Genome Res.* 2006; 113:262-70.

Ager E, Suzuki S, Pask A, Shaw G, Ishino F, Renfree MB. Insulin is imprinted in the placenta of the marsupial, Macropus eugenii. *Dev. Biol.* 2007; 15;309:317-28.

Ariel I, de Groot N, Hochberg A. Imprinted H19 gene expression in embryogenesis and human cancer: the oncofetal connection. *Am. J. Med. Genet.* 2000;91:46-50.

Arima T, Matsuda T, Takagi N, Wake N. Association of IGF2 and H19 imprinting with choriocarcinoma development. *Cancer Genet. Cytogenet.* 1997;93:39-47.

Arngrímsson R. Epigenetics of hypertension in pregnancy. *Nat. Genet.* 2005; 37:460-1.

Avner P, Heard E. X-chromosome inactivation: counting, choice and initiation. *Nat. Rev. Genet.* 2001; 2:59-67.

Baergen RN, Rutgers JL, Young RH, Osann K, Scully RE. Placental site trophoblastic tumor: A study of 55 cases and review of the literature emphasizing factors of prognostic significance. *Gynecol. Oncol.* 2006;100:511-20.

Barlow D. Gametic imprinting in mammals. *Science* 1995; 270:1610-3.

Bestor TH, Burchis D. Genetics and epigenetics of hydatidiform moles. *Nat. Genet.* 2006; 38:274-6.

Brockdorff N. X-chromosome inactivation: closing in on proteins that bind Xist RNA. *Trends Genet.* 2002;18:352-8.

Brown AL, Kay GF. Bex1, a gene with increased expression in parthenogenetic embryos, is a member of a novel gene family on the mouse X chromosome. *Hum. Mol.* Genet. 1999;8:611-9.

Bulmer JN, Johnson PM. Immunology of human placental trophoblast membrane antigens. *Arch. Immunol. Ther. Exp.*1990;38:103-10.

Cao R, Zhang Y. The functions of E(Z)/EZH2-mediated methylation of lysine 27 in histone H3. *Curr. Opin. Genet. Dev.* 2004;14:155-64.

Caspary T, Cleary MA, Baker CC, Guan XJ, Tilghman SM. Multiple mechanisms regulate imprinting of the mouse distal chromosome 7 gene cluster. *Mol. Cell Biol.* 1998;18:3466-74.

Castrillon DH, Sun D, Weremowicz S, Fisher RA, Crum CP, Genest DR. Discrimination of complete hydatidiform mole from its mimics by immunohistochemistry of the paternally imprinted gene product p57KIP2. *Am. J. Surg. Pathol.* 2001;25:1225-30.

Coan PM, Burton GJ, Ferguson-Smith. Imprinted genes in the placenta – a review. *Placenta* 2005, 26:S10-20.

Constância M, Hemberger M, Hughes J, Dean W, Ferguson-Smith A, Fundele R, Stewart F, Kelsey G, Fowden A, Sibley C, Reik W. Placental-specific IGF-II is a major modulator of placental and fetal growth. *Nature.* 2002;417:945-8.

Cooper DW, VandeBerg JL, Sharman GB, Poole WE. Phosphoglycerate kinase polymorphism in kangaroos provides further evidence for paternal X inactivation. *Nature New Biol.* 1971; 230:155-7.

Divers MJ, Miller D, Bulmer JN, Vail A, Lilford RJ. Maternal levels of serum-soluble CD8 and IL-2R are not significantly elevated in idiopathic preterm labour. *Eur. J. Obstet. Gynecol. Reprod. Biol.* 1995; 62:209-12.

Engemann S, Strödicke M, Paulsen M, Franck O, Reinhardt R, Lane N, Reik W, Walter J. Sequence and functional comparison in the Beckwith-Wiedemann region: implications for a novel imprinting centre and extended imprinting. *Hum. Mol. Genet.* 2000; 9:2691-706.

Wagschal A, Feil AWR. Genomic imprinting in the placenta. *Cytogent. Genome Res.* 2006;113:90-8.

Ferguson-Smith AC, Sasaki H, Cattanach BM, Surani MA.Parental-origin-specific epigenetic modification of the mouse H19 gene. *Nature.* 1993;362:751-5.

Ferguson-Smith AC, Surani MA. Imprinting and the epigenetic asymmetry between parental genomes. Science. 2001;293:1086-9.

Ferguson-Smith AC. Genetic imprinting: silencing elements have their say. *Curr. Biol.* 2000; 10:R872-5.

Fitzpatrick GV, Soloway PD, Higgins MJ. Regional loss of imprinting and growth deficiency in mice with a targeted deletion of KvDMR1. *Nat. Genet.* 2002; 32:426-31.

Frank D, Fortino W, Clark L, Musalo R, Wang W, Saxena A, Li CM, Reik W, Ludwig T, Tycko B. Placental overgrowth in mice lacking the imprinted gene Ipl. *Proc. Natl. Acad. Sci. USA*. 2002; 99:7490-5.

Fukunaga M, Ushigome S. Malignant trophoblastic tumor: immunohistochemical and flow cytometric comparison of choriocarcinoma and placental site trophoblastic tumors. *Hum. Pathol.* 1993; 24:1098-106.

Fukunaga M. Immunohistochemical characterization of p57(KIP2) expression in early hydatidiform moles. *Hum. Pathol.* 2002; 33:1188-92.

Genest DR, Berkowitz RS, Fisher RA, Newlands ES, Fehr M. Gestational trophoblastic disease. In: Tavassoli FA, Devilee P. (Eds.): World Health Organization Classification of Tumours. Pathology and Genetics of Tumours of the Breast and Female Genital Organs. IARC Press: Lyon 2003. p250-254.

Giannoukakis N, Deal C, Paquette J, Goodyer CG, Polychronakos C.Parental genomic imprinting of the human IGF2 gene. *Nat. Genet.* 1993;4:98-101.

Goto T, Wright E, Monk M. Paternal X-chromosome inactivation in human trophoblastic cells. *Mol. Hum. Reprod.* 1997;3:77-80.

Graves JA. Mammals that break the rules: Genetics of marsupials and monotremes. *Annu. Rev. Genet.* 1996; 30:233-6.

Guillemot F, Caspary T, Tilghman SM, Copeland NG, Gilbert DJ, Jenkins NA, Anderson DJ, Joyner AL, Rossant J, Nagy A. Genomic imprinting of Mash2, a mouse gene required for trophoblast development. *Nat. Genet.* 1995; 9:235-42.

Haig D, Westoby M. An earlier formulation of the genetic conflict hypothesis of genomic imprinting. *Nat. Genet.* 2006; 38:271.

Hassadia A, Gillespie A, Tidy J, Everard RGNJ, Wells M, Coleman R, Hancock B. Placental site trophoblastic tumour: Clinical features and management. *Gynecol. Oncol.* 2005;99:603-7.

He L, Cui H, Walsh C, Mattsson R, Lin W, Anneren G, Pfeifer-Ohlsson S, Ohlsson R. Hypervariable allelic expression patterns of the imprinted IGF2 gene in tumor cells. *Oncogene*. 1998; 16:113-9.

Hedges SB. The origin and evolution of model organisms. *Nat. Rev. Genet.* 2002; 3:838-49.

Hemberger M.Epigenetic landscape required for placental development. *Cell Mol. Life Sci.* 2007;64:2422-36.

Higashimoto K, Soejima H, Yatsuki H, Joh K, Uchiyama M, Obata Y, Ono R, Wang Y, Xin Z, Zhu X, Masuko S, Ishino F, Hatada I, Jinno Y, Iwasaka T, Katsuki T, Mukai T. Characterization and imprinting status of OBPH1/Obph1 gene: implications for an extended imprinting domain in human and mouse. *Genomics*. 2002;80:575-84.

Hui P, Parkash V, Perkins AS, Carcangiu ML. Pathogenesis of placental site trophoblastic tumor may require the presence of a paternally derived X chromosome. *Lab. Invest.* 2000; 80:965.

Hui P, Riba A, Pejovic T, Johnson T, Baergen RN, Ward D. Comparative genomic hybridization study of placental site trophoblastic tumour: a report of four cases. *Mod. Pathol.* 2004;17:248-51.

Hui P, Wang HL, Chu P, Yang B, Huang J, Baergen RN, Sklar J, Yang XJ, Soslow RA.Absence of Y chromosome in human placental site trophoblastic tumor. *Mod. Pathol.* 2007; 20:1055-60.

Hunt JS, Morales PJ, Pace JL, Fazleabas AT, Langat DK. A commentary on gestational programming and functions of HLA-G in pregnancy. *Placenta*. 2007; 28 Suppl A:S57-63.

Huynh KD, Lee JT. Inheritance of a pre-inactivated paternal X chromosome in early mouse embryos. *Nature* 2003;426:857-62.

Isles AR, Holland AJ. "Imprinted genes and mother-offspring interactions". *Early Hum. Develop.* 2005; 81: 73–7.

Kajii T, Ohama K. Androgenetic origin of hydatidiform mole. *Nature* 1977; 268:633-4.

Kaneko-Ishino T, Kuroiwa Y, Miyoshi N, Kohda T, Suzuki R, Yokoyama M, Viville S, Barton SC, Ishino F, Surani MA. Peg1/Mest imprinted gene on chromosome 6 identified by cDNA subtraction hybridization. *Nat. Genet* 1995;11:52-9.

Khan S, Dancey G, Lindsay I, et al. Placental site trophoblastic tumour derived from an oocyte donation pregnancy. *BJOG* 2006;113:344-6.

Kobel M. Pohl G, Schmitt WD, Hauptmann S, Wang TL, Shih IeM. Activation of mitogen-activated protein kinase is required for migration and invasion of placental site trophoblastic tumor. *Am. J. Pathol.* 2005;167:879-85.

Kotylo PK, Michael H, Davis TE, Sutton GP, Mark PR, Roth LM. Flow cytometric DNA analysis of placental-site trophoblastic tumors. *Int. J. Gynecol. Pathol.* 1992;11:245-52.

Kurman R. Blaustein' Pathology of the Female Genital Tract. Springer (5th ed) 2002.

Kuroki K, Maenaka K. Immune modulation of HLA-G dimer in maternal-fetal interface. *Eur. J. Immunol.* 2007;37:1727-9.

Le Bouteiller P, Piccinni MP. Human NK cells in pregnant uterus: why there? *Am. J. Reprod. Immunol.* 2008; 59:401-6.

Lee JT.. Disruption of imprinted X inactivation by parent-of-origin effects at Tsix. *Cell* 2000;103, 17 – 27.

Lee JT. Molecular links between X-inactivation and autosomal imprinting: X-inactivation as a driving force for the evolution of imprinting. *Curr. Biol.* 2003; 13:R242-54.

Lefebvre S, Antoine M, Uzan S, McMaster M, Dausset J, Carosella ED, Paul P. Specific activation of the non-classical class I histocompatibility HLA-G antigen and expression of the ILT2 inhibitory receptor in human breast cancer. *J. Pathol.* 2002;196:266-74.

Lewis A, Mitsuya K, Umlauf D, Smith P, Dean W, Walter J, Higgins M, Feil R, Reik W. Imprinting on distal chromosome 7 in the placenta involves repressive histone methylation independent of DNA methylation. *Nat. Genet.* 2004;36:1291-5.

Li E, Beard C, Jaenisch R. Role for DNA methylation in genomic imprinting. Nature. 1993;366:362-5.

Li Y, Lemaire P, Behringer RR. Esx1, a novel X chromosome-linked homeobox gene expressed in mouse extraembryonic tissues and male germ cells. *Developmental Biology* 1997; 188:85-95.

Lin TP, Labosky PA, Grabel LB, Kozak CA, Pitman JL, Kleeman J, MacLeod CL. The Pem homeobox gene is X-linked and exclusively expressed in extraembryonic tissues during early murine development. *Develop. Biol.* 1994; 166:170-9.

Looijenga L, Gillis AJM, Verkerk AJ, van Putten WL, Oosterhuis JW. Heterogeneous X inactivation in trophoblastic cells of human full-term female placentas. *American J Hum. Genet.*1999;64:1445-52.

Mager J, Montgomery ND, de Villena FP, Magnuson T. Genome imprinting regulated by the mouse Polycomb group protein Eed. *Nat. Genet.* 2003; 33:502-7.

Mak W, Nesterova TB, de Napoles M, Appanah R, Yamanaka S, Otte AP, Brockdorff N. Reactivation of the paternal X chromosome in early mouse embryos. *Science.* 2004;303:666-9.

Mancini-DiNardo D, Steele SJ, Ingram RS, Tilghman SM. A differentially methylated region within the gene Kcnq1 functions as an imprinted promoter and silencer. *Hum. Mol. Genet.* 2003, 12:283-94.

Mann MR, Lee SS, Doherty AS, Verona RI, Nolen LD, Schultz RM, Bartolomei MS. Selective loss of imprinting in the placenta following preimplantation development in culture. *Development.* 2004;131:3727-35.

Mayer W, Hemberger M, Frank HG, Grümmer R, Winterhager E, Kaufmann P, Fundele R. Expression of the imprinted genes MEST/Mest in human and murine placenta suggests a role in angiogenesis. *Dev. Dyn.* 2000;217:1-10.

McGrath J, Solter D. "Completion of mouse embryogenesis requires both the maternal and paternal genomes". *Cell* 1984;37: 179–83.

McMinn J, Wei M, Sadovsky Y, Thaker HM, Tycko B. Imprinting of PEG1/MEST isoform 2 in human placenta. *Placenta* 2006;27:119-26.

Migeon BR, Lee CH, Chowdhury AK, Carpenter H. Species differences in TSIX/Tsix reveal the roles of these genes in X-chromosome inactivation. *Am. J. Hum. Genet.* 2002; 71:286-93.

Mizuno Y, Sotomaru Y, Katsuzawa Y, Kono T, Meguro M, Oshimura M, Kawai J, Tomaru Y, Kiyosawa H, Nikaido I, Amanuma H, Hayashizaki Y, Okazaki Y. Asb4, Ata3, and Dcn are novel imprinted genes identified by high-throughput screening using RIKEN cDNA microarray. *Biochem. Biophys. Res. Commun.* 2002;290:1499-505.

Monk D, Wagschal A, Arnaud P, Müller PS, Parker-Katiraee L, Bourc'his D, Scherer SW, Feil R, Stanier P, Moore GE. Comparative analysis of human chromosome 7q21 and mouse proximal chromosome 6 reveals a placental-specific imprinted gene, TFPI2/Tfpi2, which requires EHMT2 and EED for allelic-silencing. *Genome Res.* 2008;18:1270-81.

Montgomery ND, Yee D, Chen A, Kalantry S, Chamberlain SJ, Otte AP, Magnuson T. The murine polycomb group protein Eed is required for global histone H3 lysine-27 methylation. *Curr. Biol.* 2005;15:942-7.

Moore T, Haig D. "Genomic imprinting in mammalian development: a parental tug-of-war". *Trends in Genetics* 1991; 7: 45–9.

Morison IM, Ramsay JP, Spencer HG. A census of mammalian imprinting. Trends *Genet.* 2005;21:457-65.

Mutter GL, Stewart CL, Chaponot ML, Pomponio RJ. Oppositely imprinted genes H19 and insulin-like growth factor 2 are coexpressed in human androgenetic trophoblast. *Am. J. Hum. Genet.* 1993;53:1096-102.

Nolen LD, Gao S, Han Z, Mann MR, Gie Chung Y, Otte AP, Bartolomei MS, Latham KE. X chromosome reactivation and regulation in cloned embryos. *Dev. Biol.* 2005;279:525-40.

Ohlsson R, Nyström A, Pfeifer-Ohlsson S, Töhönen V, Hedborg F, Schofield P, Flam F, Ekström TJ.IGF2 is parentally imprinted during human embryogenesis and in the Beckwith-Wiedemann syndrome. *Nat. Genet.* 1993;4:94-7.

Okamoto I, Otte AP, Allis CD, Reinberg D, Heard E. Epigenetic dynamics of imprinted X inactivation during early mouse development. *Science* 2004;303:644-9.

Ono R, Nakamura K, Inoue K, Naruse M, Usami T, Wakisaka-Saito N, Hino T, Suzuki-Migishima R, Ogonuki N, Miki H, Kohda T, Ogura A, Yokoyama M, Kaneko-Ishino T, Ishino F. Deletion of Peg10, an imprinted gene acquired from a retrotransposon, causes early embryonic lethality. *Nat. Genet.* 2006;38:101-6.

Ono R, Shiura H, Aburatani H, Kohda T, Kaneko-Ishino T, Ishino F.Identification of a large novel imprinted gene cluster on mouse proximal chromosome 6. *Genome Res.* 2003;13:1696-705.

Oudejans CB, Mulders J, Lachmeijer AM, van Dijk M, Könst AA, Westerman BA, van Wijk IJ, Leegwater PA, Kato HD, Matsuda T, Wake N, Dekker GA, Pals G, ten Kate LP, Blankenstein MA. The parent-of-origin effect of 10q22 in pre-eclamptic females coincides with two regions clustered for genes with down-regulated expression in androgenetic placentas. *Mol. Hum. Reprod.* 2004;10:589-98.

Paulsen M, El-Maarri O, Engemann S, Strödicke M, Franck O, Davies K, Reinhardt R, Reik W, Walter J. Sequence conservation and variability of imprinting in the Beckwith-Wiedemann syndrome gene cluster in human and mouse. *Hum. Mol. Genet.* 2000;9:1829-41.

Rachmilewitz J, Goshen R, Ariel I, Schneider T, de Groot N, Hochberg A.Parental imprinting of the human H19 gene. *FEBS Lett.* 1992;309:25-8.

Reik W, Ferguson-Smith AC. Developmental biology: the X-inactivation yo-yo. *Nature.* 2005;438:297-8.

Reik W, Lewis A. "Co-evolution of X-chromosome inactivation and imprinting in mammals". *Nat. Rev. Genet.* 2005; 6: 403–10.

Sado T, Fenner MH, Tan SS, Tam P, Shioda T, Li E. X inactivation in the mouse embryo deficient for Dnmt1: distinct effect of hypomethylation on imprinted and random X inactivation. *Dev. Biol.* 2000;225:294-303.

Sado T, Ferguson-Smith AC. Imprinted X inactivation and reprogramming in the preimplantation mouse embryo. *Hum. Mol. Genet.* 2005;14:R59-64.

Salas M, John R, Saxena A, Barton S, Frank D, Fitzpatrick G, Higgins MJ, Tycko B. Placental growth retardation due to loss of imprinting of Phlda2. *Mech. Dev.* 2004;121:1199-210.

Salido EC, Yen PH, Mohandas TK, et al. Expression of the X-inactivation-associated gene XIST during spermatogenesis. *Nat. Genet.* 1992; 2:196-9.

Sandell LL, Guan XJ, Ingram R, Tilghman SM. Gatm, a creatine synthesis enzyme, is imprinted in mouse placenta. *Proc. Natl. Acad. Sci. USA.* 2003;100:4622-7.

Seitz H, Youngson N, Lin SP, Dalbert S, Paulsen M, Bachellerie JP, Ferguson-Smith AC, Cavaillé J. Imprinted microRNA genes transcribed antisense to a reciprocally imprinted retrotransposon-like gene. *Nat. Genet.* 2003;34:261-2.

Shih IM, Kurman RJ. The pathology of intermediate trophoblastic tumors and tumor-like lesions. Int J Gynecol. Pathol. 2001; 20:31-47.

Silva J, Mak W, Zvetkova I, Appanah R, Nesterova TB, Webster Z, Peters AH, Jenuwein T, Otte AP, Brockdorff N. Establishment of histone h3 methylation on the inactive X chromosome requires transient recruitment of Eed-Enx1 polycomb group complexes. *Dev. Cell.* 2003;4:481-95.

Sleutels F, Zwart R, Barlow DP. The non-coding Air RNA is required for silencing autosomal imprinted genes. *Nature.* 2002;415:810-3.

Solter D. Imprinting. International Journal of Developmental Biology 1998;42:951-4.

Szulman A. Syndromes of hydatidiform moles, partial vs complete. *J. Reprod. Med.* 1984;29:788-91.

Takahashi K, Kobayashi T, Kanayama N. p57(Kip2) regulates the proper development of labyrinthine and spongiotrophoblasts. *Mol. Hum. Reprod.* 2000;6:1019-25.

Tilghman SM The sins of the fathers and mothers: genomic imprinting in mammalian development. *Cell* 1999; 96:185-93.

Tycko B, Efstratiadis A. Genomic imprinting: piece of cake. *Nature.* 2002;417:913-4.

Tycko B, Morison IM. Physiological functions of imprinted genes. *J. Cel. Physiol.* 2002;192:245–58.

Umlauf D, Goto Y, Cao R, Cerqueira F, Wagschal A, Zhang Y, Feil R. Imprinting along the Kcnq1 domain on mouse chromosome 7 involves repressive histone methylation and recruitment of Polycomb group complexes. *Nat. Genet.* 2004;36:1296-300.

van Dijk M, Mulders J, Könst A, Janssens B, van Roy F, Blankenstein M, Oudejans C.Differential downregulation of alphaT-catenin expression in placenta: trophoblast cell type-dependent imprinting of the CTNNA3 gene. *Gene. Expr. Patterns.* 2004;5:61-5.

van Dijk M, Mulders J, Poutsma A, Könst AA, Lachmeijer AM, Dekker GA, Blankenstein MA, Oudejans CB. Maternal segregation of the Dutch preeclampsia locus at 10q22 with a new member of the winged helix gene family. *Nat. Genet.* 2005;37:514-9.

Wake N, Arima T, Matsuda T. Involvement of IGF2 and H19 imprinting in choriocarcinoma development. *Int. J. Gynaecol. Obstet.* 1998;60 Suppl 1:S1-8.

Wang J, Mager J, Chen Y, Schneider E, Cross JC, Nagy A, Magnuson T. Imprinted X inactivation maintained by a mouse Polycomb group gene. *Nat. Genet.* 2001;28:371-5.

Williams JW, Hawes SM, Patel B, Latham KE. Trophectoderm-specific expression of the X-linked Bex1/Rex3 gene in preimplantation stage mouse embryos. *Mol. Reprod. Dev.* 2002;61:281-7.

Wutz A, Barlow DP. Imprinting of the mouse Igf2r gene depends on an intronic CpG island. *Mol. Cell Endocrinol.* 1998;140:9-14.

Zeng SM, YankowizJ: X-inactivation patterns in human embryonic and extra-embryonic tissues. *Placenta* 2003; 24:270-5.

Zhang Y, Shields T, Crenshaw T, Hao Y, Moulton T, Tycko B.Imprinting of human H19: allele-specific CpG methylation, loss of the active allele in Wilms tumor, and potential for somatic allele switching. *Am. J. Hum. Genet.* 1993;53:113-24.

Zwart R, Sleutels F, Wutz A, Schinkel AH, Barlow DP. Bidirectional action of the Igf2r imprint control element on upstream and downstream imprinted genes. *Genes. Dev.* 2001;15:2361-6.

In: Biological Aspects of Human Health and Well-Being ISBN: 978-1-61209-134-1
Editor: Tsisana Shartava © 2011 Nova Science Publishers, Inc.

Chapter XI

THE ROLE OF SUPRASPINAL GABA AND GLUTAMATE IN THE MEDIATION AND MODULATION OF PAIN

Kieran Rea and David P. Finn[*]

Department of Pharmacology and Therapeutics,
NCBES Neuroscience Cluster and Centre for Pain Research,
University Road, NationalUniversity of Ireland,
Galway, Ireland

ABSTRACT

Gamma-aminobutyric acid (GABA) and glutamate play critical roles in the mediation and modulation of nociception at peripheral, spinal and supraspinal levels. Supraspinally, these amino acid neurotransmitters, and their receptors, are present in key brain regions involved in the sensory-discriminative, affective and cognitive dimensions of pain perception.

Modulation of central GABAergic and glutamatergic neurotransmission underlies both activation of the endogenous analgesic system and the therapeutic effects of a number of analgesics. Enhancement or suppression of firing of GABAergic and glutamatergic neurons, and associated changes in neurotransmitter release, have been reported in supraspinal sites associated with nociception in animal models of acute, inflammatory and neuropathic pain. Moreover, pharmacological modulation of central GABAergic and glutamatergic signaling results in altered nociceptive behaviour.

Here we review recent evidence in this area. We consider how this research has enhanced our understanding of the neurochemical mechanisms underpinning nociception and discuss its implications for the development of novel analgesic agents.

[*] Corresponding author: David P. Finn, Address: Department of Pharmacology and Therapeutics, University Road, National University of Ireland, Galway, Ireland. E-mail: David.Finn@Nuigalway.ie Tel. +353 91 495280; Fax. +353 91 525700; URL address: http://www.nuigalway.ie/pharmacology/Dr_David_Finn.html

Keywords: Pain; Brain; GABA; Glutamate; Neurotransmission; Nociception.

ABBREVIATIONS

↑ Increase;
↓ Decrease;
↔ no change;
AMPA: α-amino-3-hydroxy-5-methylisoxazole-4- propionic acid;
BLA: basolateral amygdala;
CeA: central nucleus of the amygdala;
CCI: chronic constriction injury;
COX: cyclooxygenase;
EPSPs: excitatory postsynaptic potentials;
GABA: γ-aminobutyric acid;
GAD: glutamic acid decarboxylase;
GiA: nucleus reticularis gigantocellularis pars alpha;
ICV: intracerebroventricular;
IPSPs: inhibitory postsynaptic potentials;
LA: lateral amygdala;
LOX: lipoxygenase;
mGluR: metabotropic glutamate receptors;
NMDA: N-methyl-D-aspartate;
NMR: nucleus raphe magnus;
NSAID: non-steroidal anti-inflammatory drug;
PAG: periaqueductal gray;
PLC: Phospholipase C;
RVM: rostral ventromedial medulla;
SIA: stress-induced analgesia;
SSRI: selective serotonin reuptake inhibitor

INTRODUCTION

Pain is mediated by a complex network of sensory pathways, and while many of the critical neuroanatomical loci involved in pain have been identified, the precise neural mechanisms underlying the perception and modulation of pain are poorly understood. The cost of pain to health, well-being, society and the economy is immense. Elucidation of the neurochemical mechanisms seems a necessary prerequisite for a detailed understanding of pain and development of novel analgesics with improved efficacy and safety profiles. Nociception is the process by which noxious stimuli activate a select population of primary afferent sensory neurons (nociceptors), which innervate the skin, muscle, joints or organs. Nociceptors may be activated or sensitized by mechanical, thermal or chemical stimuli or by molecules released at a site of inflammation or injury (bradykinin, prostaglandins, proteases, histamine, substance P, serotonin and others). The nociceptive signals are propagated along the primary afferent neurons to the dorsal horn of the spinal cord where they synapse on ascending spinal relay neurons. The main supraspinal projection sites for these ascending pain pathways include the thalamic nuclei, relay stations in the midbrain (gigantocellular nucleus (GiA) and periaqueductal gray matter (PAG)), hypothalamus and the amygdaloid complex. Nociceptive information is then further relayed to areas in the cortex to be decoded and processed as a 'painful' sensation. The transfer of information from the spinal cord to higher centres in the brain may be suppressed or enhanced via a series of interconnecting anatomical projections known collectively as the descending pain pathway. Neurons of the descending pain pathway originate in higher brain regions including the amygdala and hypothalamus and project to the lower brainstem (including the A5, A6/A7 noradrenergic neurons) and spinal cord, via the periaqueductal gray (PAG) and rostral ventromedial medulla (RVM). Interestingly, the anatomical regions involved in the descending facilitatory and inhibitory pain pathways overlap, and as such, a single transmitter may both suppress and enhance nociception through the activation of different receptor types within these brain regions (Millan, 1997). There is an accumulating body of neurochemical, pharmacological, electrophysiological and behavioural evidence for a key role of GABA, glutamate and their receptors in modulating supraspinal pain pathways (Bleakman et al. 2006, Enna and McCarson 2006). Indeed, GABAergic and glutamatergic neurons at most, if not all, supraspinal components of the ascending and descending pain pathways mediate facilitatory and/or inhibitory effects on pain perception, and as such are the main focus of this chapter.

THE GABAERGIC AND GLUTAMATERGIC SYSTEMS

Billions of neurons exist in the mammalian brain which can communicate with each other via specialized junctions called synapses. The vast majority of these synapses occur at contacts between presynaptic axon terminals and postsynaptic dendrites, and many use glutamate and GABA as the excitatory and inhibitory neurotransmitter, respectively. Glutamate and GABA released from the presynaptic terminal act upon pre- and post-synaptic receptors to modulate neuronal activity. The strength of synaptic transmission can be modified (synaptic plasticity) by various endogenous and exogenous factors that affect ligand-receptor binding.

Where synapses occur, a variety of receptor tyrosine kinases, metabotropic (G protein–coupled) receptors, ionotropic (ion-linked receptor channels) receptors, and cell adhesion molecules exist, which mediate physical linkage and/or functional communication with the presynaptic specialization as well as function in postsynaptic signaling. Ionotropic and metabotropic receptors are also found just outside the synaptic junction at perisynaptic sites, or extrasynaptically where they modulate the transmission of neuronal impulses.

The various classes of ionotropic and metabotropic glutamate and GABA receptors can show differential distribution in the pre- and post-synaptic membrane, depending on their neuroanatomical location. This phenomenon greatly complicates the interpretation of data derived from supraspinal activation of specific receptors by pharmacological means – GABA and glutamate are the natural endogenous ligands for these receptors, and their pharmacological manipulation may result in a net excitatory or inhibitory response depending on which receptor subtype is activated, and its synaptic location. While considerable progress has been made, the precise functional significance and the differential distribution of each receptor subtype in the control of nociceptive transmission still remains unclear.

Throughout the central nervous system, and in neuroanatomical areas involved in pain, GABA and glutamate receptors are found in high density suggesting a role for GABA and glutamate in the mediation and modulation of nociception. GABA is synthesised in vivo by a metabolic pathway known as the GABA shunt (figure 1). Glutamate is converted to GABA by the enzyme glutamic acid decarboxylase (GAD), an enzyme expressed solely in cells that use GABA as a neurotransmitter.

Following receptor activation, GABA is taken back into the presynaptic nerve terminal by active transport where it can be stored in vesicles and reutilised, or into surrounding glial cells where it is degraded by GABA-transferase to succinic semialdehyde and can eventually be reconstituted into GABA as illustrated in Figure 1.

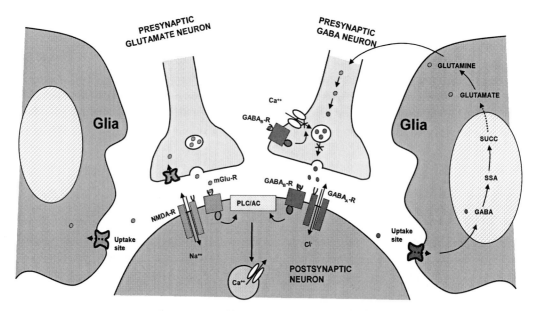

Figure 1. Representation of a GABAergic/ glutamatergic synapse. The activation of GABA and glutamate receptors can lead to a depolarization or hyperpolarisation of the neuron upon which they are located depending on the receptor subtype activated. Activation of NMDA receptors by glutamate opens post-synaptic ion channels that allow an increase in cytosolic sodium (Na^+) leading to the depolarization of the neuron, while activation of $GABA_A$ receptors open chloride ion channels leading to a hyperpolarisation. These changes in membrane potential due to depolarisations or hyperpolarisations are referred to as excitatory (EPSPs) or inhibitory (IPSPs) post-synaptic potentials respectively, and determine the firing rate or activity of the neuron. The activation of G-proteins by group I mGlu receptors result in the activation of phospholipase C, which produces second messengers that mobilise calcium release from intracellular stores. Group II and group III mGlu receptors as well as $GABA_B$ receptors are negatively coupled to adenylyl cyclase and inhibit cAMP production thus decreasing the intracellular Ca^{++} concentration. These effects also impact on the susceptibility of the cell to activation, as they affect the potential of the cell membrane. Following activation of postsynaptic receptors, GABA and glutamate are either taken back up into the neurons or taken into the nearby glial cells. In the glial cells GABA is metabolized by GABA transferase to succinic semialdehyde and further to succinic acid by succinate semialdehyde dehydrogenase. The mitochondrial enzymes then use this succinic acid in the TCA cycle resulting in α-ketoglutarate as a by-product, which can be converted to glutamate by GABA transaminase. Glutamate is further converted to glutamine by glutamate synthase and can be transferred to neighbouring neurons. In the neurons, glutamine is metabolized first to glutamate and then further to GABA by the enzymes glutaminase and glutamic acid decarboxylase. The neurotransmitter can then be stored in vesicles or released again as needed.

GABA RECEPTORS

GABA$_A$ Receptor Subunit Composition

Within the central nervous system, GABA mediates its effects by acting on ionotropic GABA$_A$ receptors and G-protein coupled GABA$_B$ receptors. Fast-acting inhibitory post-synaptic currents are mediated by GABA$_A$ receptors (Barnard et al. 1998). GABA$_A$ receptors are comprised of five subunits from seven different subunit families (α(1–6), β(1–3), γ(1–3), ρ(1-3), δ, ϵ and θ) (numbers in brackets indicate known isoforms) (Whiting et al. 1999). The most common GABA$_A$ configuration is two α(1–6), two β(1–3) and one γ(1–3) subunit (Farrar et al., 1999; Sieghart et al., 1999; Whiting, 1999). Over 50 different subtypes of GABA$_A$ receptor result from these subunits alone. However, most of the GABA$_A$ receptor subtypes cannot be easily distinguished from one another on the basis of their electrophysiological properties or pharmacological profile (Costa, 1998). The location of GABA$_A$ heteroreceptors at pre- or post-synaptic sites, the various subunit compositions and the regional specificity of these subunits complicate elucidation of the exact roles of the GABA$_A$ receptor. It is postulated that different GABA$_A$ receptor subunit compositions are integrated in distinct neuronal circuits and may serve specific cellular signaling functions. Neural networks in the CNS are largely controlled by the activity of interneurons – the majority of which are GABAergic (Buzsaki and Chrobak 1995; Paulsen and Moser 1998; Freund and Buzsaki 1996; Klausberger et al. 2002). GABAergic interneurons can innervate the input to, and output from, principal cells; and the responses to these interneuron connections are determined by the type of GABA$_A$ receptor expressed in the respective domain of the principal cell. For example, in the hippocampus, the soma of pyramidal cells form synapses with two distinct types of interneurons. The fast spiking parvalbumin-containing basket cells form synapses that contain α1 GABA$_A$ receptors (Nyiri et al. 2001; Pawelzik et al. 2002; Freund and Buzsaki 1996; Klausberger et al. 2002), while regular spiking CCK-positive basket cells form synapses containing α2 GABA$_A$ receptors (Brussaard et al. 1997; Hutcheon et al. 2000; Juttner et al. 2001), suggesting the existence of functionally specialized interneurons operating in tandem with the kinetically appropriate GABA$_A$ receptor subtype. Using point mutations of and immunostaining for the different GABA$_A$ receptor subtypes, it was shown that IPSCs in the lateral / basolateral amygdala were mediated by both α2- and α1-subunit containing GABA$_A$ receptors whereas those in the central amygdala were mediated only by α2-subunit-containing GABA$_A$ receptors (Marowski et al. 2004; Kaufmann et al. 2003), suggesting the differential expression of specific subtypes of GABA$_A$ receptors. Ultimately, the outcome of their activation however remains the same; the binding of active agonists to the GABA$_A$ receptor subunit complex induces a conformational change, which allows the conductance of chloride ions (Cl$^-$) through the receptor resulting in a rapid hyperpolarisation of the postsynaptic membrane (Dudel and Kuffler 1961, MacDermott et al. 1999).

The GABA$_A$ receptor is the primary inhibitory neurotransmitter receptor in the brain and the pharmacological exploitation of this receptor has led to the development of several therapeutic compounds such as anxiolytics, sedatives, hypnotics, or antiepileptics. As such, the vast majority of genetic studies with GABA$_A$ subunit specific knockout or knock-in mice, or agonists or antagonists with specificity for one type of GABA$_A$ receptor subunit have focused on the role of GABA$_A$ receptor subtypes in mediating these behavioural effects in various regions of the brain. Pharmacological studies and studies in transgenic mice have suggested a role for brain GABA$_{A\alpha1}$ receptors in sedation, GABA$_{A\alpha2}$/GABA$_{A\alpha3}$ receptors in anxiety, and GABA$_{A\alpha5}$ receptors in memory (Squires et al., 1980; Griebel et al., 1999a,b; Paronis et al., 2001; Rudolph and Mohler, 2006; Ballard et al. 2008). Two recent studies have demonstrated antinociception following systemic (Munro et al. 2008) or spinal (Knabl et al. 2008) administration ofnon-sedative, α1-sparingsubtype selective GABA$_A$ receptor agonists. However, there is a paucity of studies investigating the role of supraspinal GABA$_A$ receptor subtypes in pain.

GABA$_B$ Receptor Subunit Composition

GABA$_B$ receptors are G-protein coupled receptors and can be located presynaptically - on GABAergic neurons as autoreceptors or on other neurons as heteroreceptors; or postsynaptically mediating slow, long-lasting, inhibitory post synaptic potentials (IPSPs). Presynaptic GABA$_B$ receptors are negatively coupled to adenylyl cyclase (Kerr and Ong 1995, Mott and Lewis 1994), and decrease cell conductance to calcium ions (Ca^{++}) (Couve et al. 2000, Kerr and Ong 1995) resulting in a hyperpolarisation of the neuron. Post-synaptic GABA$_B$ receptors are linked via G-proteins to potassium (K$^+$) channels and can cause a slow hyperpolarisation response by opening these channels (Chieng and Christie 1995, Christie and North 1988, Gage 1992, Kerr and Ong 1995, Lagrange et al. 1996, Wagner et al. 1998). Recombinant studies and molecular cloning and pharmacological characterization has revealed that native GABA$_B$ receptors are heterodimers composed of two subunits, GABA$_{B(1)}$ and GABA$_{B(2)}$ (Marshall et al. 1999; Schuler et al. 2001; Gassmann et al. 2004; Prosser et al. 2001). The GABA$_{B1}$ subunit consists of two isoforms resulting in two independently regulated receptor subtypes, GABA$_{B(1a,2)}$ and GABA$_{B(1b,2)}$ (Schwarz et al. 2000; Wei et al. 2001a,b; Steiger et al. 2004). The two isoforms differ in their developmental regulation and regional distribution, with GABA$_{B1b}$ being more widely expressed in adulthood. Most strikingly, the GABA$_{B1b}$ isoform, in addition to its synaptic localization, was localized extrasynaptically, as shown in Purkinje cell dendrites and spines (Fritschy et al. 1999). This subcellular localization is unrelated to GABAergic input and points to GABA$_B$-receptors as being activated by spillover of GABA released from adjacent synapses. Structurally, the GABA$_{B1a}$ and GABA$_{B1b}$ isoforms differ solely in their N-terminal ectodomain by a tandem pair of sushi domains that are present in GABA$_{B1a}$ but not in GABA$_{B1b}$ (Hawrot et al. 1998; Blein et al. 2004). Sushi domains, also known as complement control modules or short consensus repeats, mediate protein interactions in adhesion molecules and in G-protein-coupled receptors binding to peptide hormones (Lehtinen et al. 2004; Grace et al. 2004). Pharmacological tools that distinguish between GABA$_{B(1a,2)}$ and GABA$_{B(1b,2)}$ receptors are lacking, but work with genetic knockout mice lacking either isoform of the GABA$_{B1}$ subunits has determined that GABA$_{B1a}$ isoform is mainly, although not exclusively, involved in heteroreceptor function regulating glutamate release, whereas both GABA$_{B1a}$ and GABA$_{B1b}$ contribute to autoreceptor function (Waldmeier et al. 2008). However, the role for supraspinal GABA$_B$ receptors of either receptor subtype in animal models of pain has yet to be investigated. Recent evidence also suggests a role for other GABA$_{B1}$ subunit isoforms formed from gene splicing in modulating presynaptic inhibition at glutamatergic terminals (Tiao et al. 2008).

IONOTROPIC GLUTAMATE RECEPTORS

The glutamatergic ionotropic receptors, N-methyl-D-aspartate (NMDA), α-amino-3-hydroxy-5-methylisoxazole-4-propionic acid (AMPA), and kainate receptors are responsible for rapid excitatory synaptic transmission, while metabotropic receptors (mGlu) mediate more prolonged responses following activation. Most glutamatergic mediated excitatory postsynaptic potentials (EPSPs) are comprised of a fast AMPA receptor-mediated component that depolarises the cell and a slower NMDA-receptor mediated component. As with the $GABA_A$ receptor, activation of the ionotropic glutamate receptors results in the opening of ion channels, but in this case the passage of sodium (Na^+) and Ca^{++} ions into the cell is permitted, while K^+ ions exit the cell (Mayer and Miller 1990, Nakanishi and Masu 1994).

NMDA Receptor Subunit Composition

The NMDA receptor family is composed of seven subunits, NR1, $NR2_{(A-D)}$ and $NR_{(3A, 3B)}$, which are all products of separate genes. Eight functional splice variants and one non-functional truncated splice variant of NR1 have been described (McBain and Mayer 1994). Expression of functional recombinant NMDA receptors in mammalian cells requires the co-expression of at least one NR1 and one NR2 subtype. The stoichiometry of NMDA receptors has not yet been fully established, but the consensus is that NMDA receptors are tetramers that most often incorporate two NR1 and two NR2 subunits of the same or different subtypes (Dingledine and Conn 2000). In cells expressing NR3, it is thought that this subunit co-assembles with NR1 and NR2 to form ternary NR1/NR2/NR3 tetrameric complexes (Sasaki et al. 2002). In the heteromeric NMDA receptors, the arrangement of the different subunit isoforms determine the ion permeability of the pore (Ciabarra et al. 1995; Nishi et al. 2001; Matsuda et al. 2002; Sasaki et al. 2003), as incorporating the NR3 subunit markedly decreases single-channel conductance, Ca^{++} permeability and Mg^{++} block (Wollmoth and Sobolevski 2004). The NMDA receptor complex has a number of sites where compounds can act to inhibit, activate or enhance the function of the receptor (see Kemp and McKernan 2002), including the glutamate/NMDA binding site, a glycine binding site and the channel pore itself. As well as mediating an excitatory component of EPSPs, NMDA receptors have also been shown to mediate a net inhibitory effect on postsynaptic neurons presumably through presynaptic activation of GABAergic neurons (Duguid and Smart, 2004; Glitsch and Marty, 1999; Liu and Lachamp, 2006).

Highly selective subtype-specific antagonists have recently been developed that act allosterically through an interaction with the extracellular N-terminal domain of the NR2 subunits. A large number of compounds have been derived from ifenprodil, the first NR2B subtype selective antagonist compound to be identified (Williams 1993). The Zn^{++} ion is also reported to show selectivity for the NR2A subtype (Traynelis et al. 1998), and is thought to be an endogenous allosteric modulator of NMDA receptors.

AMPA Receptor Subunit Composition

As with the NMDA receptor, the AMPA receptors are also thought to be heteromeric in nature. They are made up of a four subunit family ($GluR_{1-4}$) with AMPA receptors lacking the $GluR_2$ subunit being more permeable to Ca^{++}. At AMPA receptors, glutamate and AMPA act as full agonists and induce rapidly desensitizing responses, while kainate acts as a partial agonist and induces receptor responses that show little desensitisation. Several studies indicate that in the hippocampus and amygdala, AMPA receptors lacking $GluR_2$, and with high Ca^{++} permeability, predominate in inhibitory interneurons, whereas pyramidal cells have largely non-Ca^{++}- permeable, $GluR_2$-containing AMPA receptors (Geiger et al. 1995; Isa et al. 1996; Mahanty and Sah 1998). Similarly to NMDA receptor activation, AMPA receptors generally result in the excitation of post-synaptic neurons; however, activation of presynaptic AMPA receptors on GABAergic neurons has been demonstrated in the cerebellum (Rusakov et al. 2005; Satake et al. 2000; 2006).

Kainate Receptor Subunit Composition

Five different kainate receptor subunits exist, Glu $_{(K5, K6, K7)}$ and Glu $_{(K1, K2)}$ with the Glu_{K5} and Glu_{K6} subunits being more permeable to Ca^{++} in a manner similar to the AMPA $GluR_2$ subunit (Egebjerg and Heinemann, 1993; Burnashev et al., 1995, 1996). Cells from different regions of the nervous system express kainate receptors with distinct physiological and pharmacological properties (Wilding and Huettner, 2001). Knockout mice have provided valuable information about the composition of native receptors and the function of individual subunits, although their role in pain is yet to be determined.

Kainate autoreceptors in the hippocampus are thought to be responsible for short-term synaptic plasticity and associative learning mechanisms (Contractor et al. 2001; Pinheiro et al. 2007). Presynaptic kainate receptors are also reported on glutamatergic (Lauri et al. 2005) and GABAergic (Ali et al. 2001; Jiang et al. 2001) neurons resulting in a net excitatory or net inhibitory response in the postsynaptic neuron respectively.

METABOTROPIC GLUTAMATE RECEPTORS

The metabotropic glutamate receptors (mGlu) can be classed according to their sequence homologies and physiological activity (Nakanishi and Masu 1994). Group I (mGlu1 and mGlu5 receptor subtypes), group II (mGlu2 and mGlu3 receptor subtypes) and group III (mGlu4, mGlu6, mGlu7, mGlu8 receptor subtypes) receptors all display differences in their ability to couple to intracellular second messenger systems. Group I mGlu receptors are generally positively coupled to phospholipase C, resulting in the hydrolysis of phosphoinositide phospholipids in the cell membrane, which in turn leads to the activation of second messenger systems and ultimately the release of Ca^{++} from intracellular stores. Group II and III mGlu receptors are negatively coupled to adenylyl cyclase, inhibiting cAMP (cyclic adenosine monophosphate) production and decreasing the intracellular Ca^{++} concentration. Although mGlu receptor activation can mediate synaptic transmission via activation of slow

EPSPs, they generally exert a more modulatory role, regulating neuronal excitability, synaptic plasticity and can even affect the activity of other glutamate receptor types (Lea et al. 2002). These receptors may also regulate ion channel activity through the action of the $\beta\gamma$ subunits of the G-protein (Anwyl 1999). Alternative splicing of the mGlu genes can result in receptor variants for mGlu1, mGlu3 and mGlu5-mGlu8 (Pin and Duvoisin 1995; Laurie et al. 1996; Flor et al. 1997; Corti et al. 1998; Malherbe et al. 1999; Zhu et al.1999; Valerio et al. 2001; Schulz et al. 2002; Sartoriuset al. 2006). mGlu receptors function at both glutamatergic and GABAergic synapses (Desai and Conn 1991; Stefani et al. 1994; Poncer and Miles 1995; Kinoshita et al. 1998), and can be located at extrasynaptic sites on dendrites and axons. In order to comprehend the myriad of physiological roles of glutamate receptors, it is important to identify the subtypes of mGlu receptors expressed in distinct subpopulations of neurons, their synaptic location (pre- vs post- vs extra-synaptic) and the proximity of mGlu receptors to sites of glutamate release.

SOME IMPORTANT CONSIDERATIONS

In studying the effects of selective GABAergic and glutamatergic antagonist administration on nociceptive behaviour, we can ascertain whether the endogenous neurotransmitters are involved in the physiological manifestation of, or response to, pain. However, data can often be difficult to interpret, as many receptors can exist on both pre-and post-synaptic sites as well as extrasynaptically. Another difficulty in interpreting data arises from the knowledge that the endogenous compounds GABA and glutamate can act as agonists at a multitude of ionotropic and metabotropic GABA and glutamate receptors, some of which mediate membrane depolarisation and others hyperpolarisation. Thus, in any brain region, the net effect on neuronal activity may be either inhibitory or excitatory depending on which receptor type is activated and on which type of neuron the receptor is located. Furthermore, neuronal activity is subject to modulation at a number of sites due to the heterogenous expression of GABAergic and glutamatergic neurons and their receptors in most brain areas.

THE ROLE OF SUPRASPINAL GABAERGIC AND GLUTAMATERGIC SYSTEMS DURING PAIN

GABA and glutamate receptors are ubiquitously expressed in brain regions associated with nociception (Millan, 1997). The animal models (usually rats or mice) most commonly used to screen for antinociceptive activity can be divided into three groups: (1) acute pain (e.g. hot plate test; tail flick test; paw withdrawal test) (2) inflammatory pain involving tissue injury (e.g. administration of carrageenan/ capsaicin/ formalin/ complete Freund's adjuvant (CFA)/ kaolin/ acetic acid) (3) neuropathic pain involving peripheral nerve injury (e.g. models of sciatic nerve ligation/ spinal nerve ligation/ chronic constriction injury). Acute pain is thought to involve different neural mechanisms than chronic pain. Ultimately, the behavioural response to an acute noxious stimulus involves spinal reflexive responses and is modulated by a 'ON'/'OFF' gating mechanism believed to be mediated by glutamatergic/GABAergic neurons in the rostral ventromedial medulla (Fields et al. 1983a, Fields et al. 1983b). The neural responses to chronic inflammatory or neuropathic pain however are more complex and involve neuroplastic changes in a number of systems including GABAergic and glutamatergic systems.

PLASTICITY OF SUPRASPINAL GABAERGIC AND GLUTAMATERGIC SYSTEMS DURING PAIN

In this section, we look at changes in neuronal firing, extracellular GABA and glutamate concentrations, or receptor densities, in discrete brain regions in various animal models of pain.

Changes in Activity of Supraspinal GABA and Glutamate Neurons in Animal Models of Acute Pain

Exposure to an acute noxious stimulus, or the behavioural response that ensues, is often associated with discrete, site-specific alterations in neuronal firing. Following tail immersion in hot water, an immediate increase in neuronal firing of thalamic neurons of anaesthetised rats results (Salt and Binns 2000, Salt and Eaton 1994). Similarly, increases in thalamic neuronal firing were associated with noxious pressure (Bordi and Quartaroli 2000). Evidence of a role for GABA, and glutamate, in the mediation of these neuronal responses via activation of their ionotropic and metabotropic receptors is provided in Table 1.

The thalamic nuclei have widespread connections with many brain regions critically involved in the mediation and modulation of nociception including the amygdala, PAG and RVM. Nociceptive information reaches the amygdala from thalamic nuclei, and this information is relayed from the lateral (LA) and basolateral amygdala (BLA) to the central nucleus of the amygdala (CeA). In lightly anaesthetised rats, an increase in evoked CeA neuronal firing was reported following a brief noxious pinch stimulus (Li and Neugebauer 2004), and the latero-capsular part of the CeA has now been established as the 'nociceptive amygdala'. The RVM is another key brain area, associated with the relay of information from the PAG to the spinal cord as part of the descending pain pathway. RVM neurons can be classified according to changes in the activity evoked by noxious heat stimulation of the tail in anaesthetised rats: on-cell activity is increased, off-cell activity is decreased and neutral-cell activity is unaffected by such noxious stimulation, just before the tail is withdrawn (Fields et al. 1983a, Fields et al. 1983b). Evidence of a role for endogenous GABA in the modulation of RVM cell firing during nociception is provided by studies reporting that intra-RVM microinjection of the $GABA_A$ receptor antagonist bicuculline increased the firing of off-cells, suppressed firing of on-cells and concomitantly increased the latency to tail withdrawal in the rat tail-flick test (Heinricher and Tortorici 1994, Moreau and Fields 1986). These electrophysiological and behavioural effects were reversed by the microinjection of the $GABA_A$ agonist muscimol (Moreau and Fields 1986), indicating the involvement of $GABA_A$ receptors in RVM cell firing and nociception. Similarly, glutamate microinjection into the PAG increased tail-flick latency and blocked the response of noxious-evoked neuronal activity in cells of the RVM in the rat tail-flick test, thus providing evidence for the involvement of glutamate in the PAG-RVM relay in nociception (Hutchison et al. 1996).

Table 1. The effects of supraspinal injection of GABAergic and glutamatergic compounds on neuronal firing in animal models of pain

Compound	in	Model	Effect	Reference
GABA$_A$ receptor antagonists				
Bicuculline	RVM	TFT	↑ off-cell firing ↓ on-cell firing	Moreau & Fields 1986; Heinricher & Tortorici 1994
SR95531	Thal	SNL	↓ neuronal firing	Roberts *et al.* 1992
Metabotropic glutamatergic receptor agonists				
Glutamate	PAG	TFT	↓ on-cell firing	Hutchison *et al.* 1996
DHPG	Thal	TWT	↓ neuronal firing	Salt & Turner 1998
	CeA	Arth	↑ neuronal firing	Li & Neugebauer 2004b
	CeA slices	Arth	↑ neuronal firing	Neugebauer *et al.* 2003
CHPG	CeA	Arth	↑ neuronal firing ↑	Li & Neugebauer 2004b
	CeA slices	Arth	neuronal firing	Neugebauer *et al.* 2003
ACPD	Thal	TWT	↓ neuronal firing	Salt & Turner 1998
LY354740	CeA	Arth	↓ neuronal firing	Li & Neugebauer 2006
	CeA slices	Arth	↓ neuronal firing	Han *et al.* 2006
L-AP4	CeA	Arth	↓ neuronal firing ↓	Li & Neugebauer 2006
	CeA slices	Arth	neuronal firing	Han *et al.* 2004
AMN082	lPAG	TFT	Shortened onset of off-cell pause	* Marabese *et al.* 2007
DCPG	lPAG	TFT	Increased onset of off-cell pause	* Marabese *et al.* 2007
Metabotropic glutamatergic receptor antagonists				
MCPG	Thal	TWT	↓ neuronal firing	Eaton *et al.* 1993; Salt & Eaton 1994
CPCCOEt	CeA	Arth	↓ neuronal firing ↓	Li & Neugebauer 2004b
	CeA slices	Arth	neuronal firing	Neugebauer *et al.* 2003
MPEP	Thal	TWT	↓ neuronal firing ↓	Salt & Binns 2000
	CeA	Arth	neuronal firing	Li & Neugebauer 2004b
	CeA slices	Arth	↓ neuronal firing	Neugebauer *et al.* 2003
EGlu	CeA	Arth	↑ neuronal firing	Li & Neugebauer 2006
UPB1112	CeA	Arth	↑ neuronal firing	Li & Neugebauer 2006
Ionotropic glutamatergic receptor antagonists				
CPP	Thal	TWT	↓ neuronal firing	Eaton & Salt 1990
APV	CeA	Arth	↓ neuronal firing	Li & Neugebauer 2004a
NBQX	CeA	Arth	↓ neuronal firing	Li & Neugebauer 2004a

* All experiments were performed on rats except mouse studies denoted by asterix.

Abbreviations: ↑ Increase; ↓ Decrease; ↔ no change; Arth: the kaolin/carrageenan-induced monoarthritic model; CeA: central nucleus of the amygdala; lPAG: lateral periaqueductal gray; PAG: periaqueductal gray; RVM: rostral ventromedial medulla; SNL: sciatic nerve ligation model; TFT: tail-flick test; Thal: Thalamic nuclei; TWT: tail-withdrawal test.

Changes in Activity of Supraspinal GABA and Glutamate Neurons in Animal Models of Inflammatory and Neuropathic Pain

In rats with sciatic nerve ligation, the prolongation of thalamic neuronal firing observed following noxious pressure application was reversible by systemic administration of the NMDA receptor antagonists MK801 or GV196771A (Bordi and Quartaroli 2000) as well as by the GABA$_A$ receptor antagonist SR95531 (Roberts et al. 1992). These studies suggest that the normal glutamatergic and GABAergic mechanisms for coping with, or initiating, nociception are altered in animal models of chronic pain. They provide evidence for recruitment of additional receptor subtypes, including NMDA and GABA$_A$ receptors, during thalamic processing of nociception in nerve-injured animals.

The intra-articular injection of kaolin and carageenan into the knee of rats induces an acute-onset inflammation, which later develops into a more prolonged inflammation representing a model of chronic inflammatory pain. Using extracellular single-unit recordings in anaesthetised animals it was determined that in the latero-capsular CeA, neurons respond more strongly and more readily to noxious stimuli in arthritic animals as compared to controls (Neugebauer et al. 2003). Brain slices containing CeA were obtained from arthritic rats at the time of maximum sensitization of CeA neurons, and it was determined that lower thresholds were required for generation of excitatory post-synaptic potentials (EPSPs) as compared to controls (Neugebauer et al. 2003). The pharmacological effects of glutamatergic agonists and antagonists on neuronal firing in the CeA of control and arthritic animals have been examined extensively in vivo and ex vivo and are reviewed in Table 2. There is, however, a paucity of studies examining the role of GABA despite the evidence for GABA interneuron connectivity between the LA and BLA to the CeA (Pare et al. 2004). The apparent changes in functionality of the various glutamate receptors in the CeA during chronic inflammatory pain are also mirrored by data from ex vivo electrophysiological studies demonstrating differences in neuronal firing in CeA neurons from brain slices of arthritic versus control rats (Neugebauer et al. 2003, Han et al. 2004, Han et al. 2006). These data suggest that altered group I, group II and group III mGlu, and NMDA receptor-mediated modulation of neuronal activity in the amygdala may occur in the arthritic pain model, suggesting a role for the glutamatergic system in the mediation and modulation of chronic inflammatory pain.

Brain Regional Alterations in Levels of GABA and Glutamate or Their Receptors in Animal Models of Inflammatory and Neuropathic Pain

Surprisingly few studies have been performed examining changes in extracellular levels of supraspinal GABA or glutamate in animal models of pain by microdialysis (see Table 3). Changes in extracellular glutamate levels have been reported in rat (Silva et al. 2000) and mouse (Oliva et al. 2006) PAG, the rat ventral posterolateral thalamic nuclei (VPL) (Silva et al. 2001), the medial and lateral preoptic nuclei of the rat hypothalamus (Silva et al. 2004) and the rat BLA (Silva and Hernandez 2007) following formalin administration. Similarly alterations in GABA levels were reported in the mouse (Oliva et al. 2006) and rat (Maione et al. 1999) PAG.

Table 2. Effects of supraspinal injection of GABAergic compounds on nociceptive behaviour in animal models of pain

Compound	in	Model	Effect	Reference
GABA receptor agonists				
GABA	PAG	FT	No effect	Monhemius et al. 2001
	GiA	FT	No effect	Azami et al. 2001
		TFT	No effect	
		pSNL	No effect	
Muscimol	icv	TFT	No effect	Liebman & Pastor 1980
		FT	Antinociceptive	* Chung et al. 1999; Shafizadeh et al. 1997; Mahmoudi & Zarrindast 2002
	CeA	TFT	Antinociceptive	Manning et al. 2003
		SNL	Antinociceptive	Pedersen et al. 2006
	RVM	Cap PWT; FT	Antinociceptive	Gear et al. 1999
			No effect	Lane et al. 2005
Baclofen	icv	TFT	No effect	Liebman & Pastor 1980
		FT	Antinociceptive	Shafizadeh et al. 1997; Mahmoudi & Zarrindast 2002
		SNL	No effect	Zarrindast et al. 2000
	Thal	FT	Antinociceptive	Potes et al. 2006
	GiA	TFT	Anti-/Pro-nociceptive	Hammond et al. 1998;
	NRM	TFT	Anti-/Pro-nociceptive	Thomas et al. 1996
GABA receptor antagonists				
Picrotoxin	Cortex	TFT	Antinociceptive	Qu et al. 2006
	vlPAG		Antinociceptive	Moreau & Fields 1986
Bicuculline	icv	TFT	Antinociceptive	Yokoro et al 2001
		FT	Antinociceptive	Mahmoudi & Zarrindast 2002
	vlPAG	FT	Antinociceptive	Morgan et al 2003
		PWT	Antinociceptive	Morgan & Clayton 2005
		TFT	Antinociceptive	Budai et al. 1998; Sandkuhler et al. 1989; Moreau & Fields 1986
	RVM	PWT	No effect	Nason & Mason 2004
		TFT	Antinociceptive	Heinricher & Tortorici 1994; Nason & Mason 2004
	CeA	SNL	No effect	Pedersen et al. 2007
	Cortex	TFT	Antinociceptive	Qu et al. 2006
	Thal	TFT	Antinociceptive	Jia et al. 2004; Xiao et al. 2005
CGP35348	icv	FT	Antinociceptive	Mahmoudi & Zarrindast 2002
		SNL	No effect	Zarrindast et al. 2000
	Thal	FT	Antinociceptive	Potes et al. 2006
	NRM;GiA	TFT; PWT	No effect	Thomas et al. 1996

* All experiments were performed on rats except mouse studies denoted by asterix.

Abbreviations: ↑ Increase; ↓ Decrease; ↔ no change; Cap:capsaicin-induced inflammation test; CeA: central nucleus of the amygdala; CGP35348: P-(3-aminopropyl)-p-diethoxymethyl-phosphinic acid; FT: formalin test; PAG: periaqueductal gray; PWT:paw-withdrawal test; RVM: rostral ventromedial medulla; Thal: Thalamic nuclei; icv: intracerebroventricular; GiA: nucleus reticularis gigantocellularis pars alpha; NRM: nucleus raphe magnus; pSNL: partial sciatic nerve ligation model; SNL: sciatic nerve ligation model; TFT: tail-flick test; vlPAG: ventrolateral periaqueductal gray.

Table 3. Effects of supraspinal injection of glutamatergic compounds on nociceptive behaviour in animal models of pain

Compound	in	Model	Effect on behaviour	Reference
Glutamate receptor agonists				
Glutamate	RVM	TFT	Anti-/Pro-nociceptive	Zhuo & Gebhart 1997
	VLO PAG		Antinociceptive	Zhang et al. 1997
			Antinociceptive	Hutchison et al. 1996; Urca et al. 1980
NMDA	RVM	CFA	Anti-/Pro-nociceptive	Guan et al. 2002
	PAG	FT	Antinociceptive	* Berrino et al. 2006; * Miguel & Nunes De Souza 2005
AMPA	RVM	CFA	Antinociceptive	Guan et al. 2002
Kainate	vlPAG	PWT	Antinociceptive	Morgan et al. 2003
DHPG	Thal	FT	Pronociceptive	* Miyata et al. 2003
	PAG	PWT	Pronociceptive	* Maione et al. 1998
		FT	Antinociceptive	* Maione et al. 2000
L-CCG-I	PAG	PWT	Pronociceptive	* Maione et al. 1998
		FT	Antinociceptive	* Maione et al. 2000
L-SOP	PAG	PWT	Pronociceptive	* Maione et al. 1998
		FT	Pronociceptive	* Maione et al. 2000
ACPD	PAG	PWT	Pronociceptive	* Maione et al. 1998
	RVM	TFT	Antinociceptive	Kim et al. 2002
DCGIV	RVM	TFT	Antinociceptive	Kim et al. 2002
L-AP4	RVM	TFT	No effect	Kim et al. 2002
AMN082	PAG	TFT	Antinociceptive	Marabese et al. 2007b
DCPG	PAG	TFT	Pronociceptive	Marabese et al. 2007b
		FT;	Antinociceptive	* Marabese et al. 2007a
		Carag;	Antinociceptive	
		SNL	No effect	
Glutamate receptor antagonists				
APV	Icv; ACC RVM	FT	No effect	Lei et al. 2004
	NRM	CRD	Antinociceptive	Coutinho et al. 1998
	PAG	TFT	Antinociceptive	Aimone & Gebhart 1987
		PWT	Pronociceptive	Palazzo et al. 2001
MK801	Icv	TFT	No effect	* Suh et al. 1995
		FT	Antinociceptive	* Chung et al. 2000
	RVM	TFT	Pronociceptive	Spinella et al. 1996
	PAG	FT	No effect	* Berrino et al. 2006
AP7	RVM	TFT	No effect	Spinella et al. 1996
DNQX	Icv; ACC RVM	FT	No effect	Lei et al. 2004
			No effect	
	CRD		Pronociceptive	Coutinho et al. 1998
CNQX	Icv	TFT	No effect	* Suh et al. 1995
		FT	Antinociceptive	* Chung et al. 2000
	RVM	TFT	No effect	Spinella et al. 1996
CPCCOEt	Thal	FT	Antinociceptive	* Miyata et al. 2003
	PAG	PWT	No effect	Palazzo et al. 2001
	CeA	Arth/Voc	Antinociceptive	Han & Neugebauer 2005

Table 3. (Continued)

Compound	in	Model	Effect on behaviour	Reference
AIDA	Icv	FT	Antinociceptive	* Chung et al. 2000
MPEP	Icv	FT	Antinociceptive	* Chung et al. 2000
	PAG	PWT	Pronociceptive	Palazzo et al. 2001
		FT	Antinociceptive	* Berrino et al. 2006
	CeA	Arth/Voc	No effect	Han & Neugebauer 2005
EGlu	PAG	PWT	No effect	Palazzo et al. 2001
		FT	No effect	* Berrino et al. 2006
MSOP	PAG	PWT	Pronociceptive	Palazzo et al. 2001
		FT	No effect	* Berrino et al. 2006

* All experiments were performed on rats except mouse studies denoted by asterix.

Abbreviations: ↑ Increase; ↓ Decrease; ↔ no change; ACC: anterior cingulated cortex, Arth/Voc. vocalisations due to monoarthritic pain; Carag:carrageenan-induced inflammation test; CeA: central nucleus of the amygdala; CFA: complete Freund's adjuvant-induced inflammation test; CRD: colorectal distension model; FT:formalin test; GiA: nucleus reticularis gigantocellularis pars alpha; Icv: intracerebroventricular; NRM: nucleus raphe magnus; PAG: periaqueductal gray; PWT:paw-withdrawal test; RVM: rostral ventromedial medulla; SNL:sciatic nerve ligation model; Thal: Thalamic nuclei; TFT: the tail-flick test; VLO: ventrolateral orbital cortex.

Another model of chronic inflammatory pain (hindpaw injection of complete Freund's adjuvant) was associated with a decrease in PAG GABA levels 24 hours and 7 days post-injection (Renno and Beitz 1999), while an increase in PAG glutamate levels (Renno 1998) was reported 7 days post-injection. These data suggest changes in GABA and glutamate neurotransmission in areas of the brain associated with nociception. However, it must be noted that all of these microdialysis studies are correlative in nature and do not directly address the question of whether alterations in extracellular levels of GABA or glutamate are a cause, or consequence, of altered nociception. They do, nevertheless, provide a firm foundation upon which to directly test hypotheses concerning the role of these neurotransmitters in discrete brain regions during nociception.

In chronic models of inflammation and models of arthritis, researchers have reported changes in GABA and glutamate receptor expression with respect to time. Rats receiving intra-articular injection of CFA or kaolin/carrageenan and non-arthritic controls were sacrificed at identical time points and glutamate and GABA receptor mRNA was measured in discrete brain regions by in situ hybridisation. Thalamic mGlu1, mGlu4 and mGlu7 receptor mRNA expression was decreased in CFA-induced monoarthritic animals compared to control rats, while mGlu3 mRNA expression was increased (Lourenco Neto et al. 2000). Further studies by these researchers demonstrated that cortical mGlu1, mGlu4 and mGlu7 receptor mRNA expression was unaltered, while mGlu3 receptor expression in the cortex was significantly increased (Neto et al. 2001). Similar studies reported an increase in expression of mGlu1 and mGlu5 receptor protein in tissue from the central amygdala (CeA) of kaolin/carrageenan induced arthritic animals compared to non-arthritic controls (Neugebauer et al. 2003). In another study, intra-plantar injection of CFA resulted in upregulation of gluR1-AMPA receptor subunit expression in the RVM 1-3 days post-inflammation (Guan et al. 2003). Expression of NMDAR1 mRNA has been shown to be increased in the rat caudal and mid ventrolateral PAG and the caudal dorsal raphe nucleus 3 days post-CFA injection (Renno et al. 1998). These workers also reported similar alterations in the rat chronic constrictive injury (CCI) model of neuropathic pain. Fewer studies have assessed alterations in supraspinal GABA receptor protein or mRNA levels in animal models of pain, with the vast majority of studies focusing instead at the spinal level. Decreases in $GABA_{B1}$ and $GABA_{B2}$ receptor subtype mRNA expression have, however, been reported in thalamic nuclei of kaolin-induced arthritic rats compared to controls (Ferreira-Gomes et al. 2004, 2006). Collectively, these studies suggest altered expression of specific glutamate and GABA receptor subtypes throughout supraspinal pain circuitry following tissue injury or inflammation. However, the majority of studies have focused on measuring mRNA levels and the extent to which these alterations are manifested as a change in receptor protein expression remains largely unknown. This question, together with more detailed analysis of the temporal profile of altered receptor expression in animal models of chronic inflammatory and neuropathic pain should be a focus for further research in this area.

PHARMACOLOGICAL ELUCIDATION OF THE ROLES OF SUPRASPINAL IONOTROPIC AND METABOTROPIPC GABA AND GLUTAMATE RECEPTOR SUBTYPES IN ANIMAL MODELS OF ACUTE PAIN

The intracerebral administration of compounds modulates synaptic transmission in discrete brain regions and is a useful approach for isolating the contribution of supraspinal receptors to pain and its modulation. Pharmacological studies where specific agonists and antagonists have been administered intracerebrally provide strong evidence for the involvement of supraspinal GABA and glutamate, and their receptors, in nociception (see Tables 2 and 3).

Effects of Supraspinal Administration of Agents Acting at $GABA_A$ Receptors in Animal Models of Acute Pain

In the rat tail-flick test, blockade of supraspinal $GABA_A$ receptors following intracerebroventricular (icv) administration of bicuculline increased the latency to response (Yokoro et al. 2001), while activation of $GABA_A$ receptors by icv administration of muscimol had no effect (Liebman and Pastor 1980). Nociceptive responses in the tail-flick test were also reduced following bicuculline administration into the rat vlPAG (de Novellis et al. 2005, Moreau and Fields 1986, Sandkuhler et al. 1989), rat ventrolateral orbital cortex (Qu et al. 2006), thalamic nucleus submedius (Jia et al. 2004, Xiao et al. 2005) and rat RVM (Heinricher and Tortorici 1994, Nason and Mason 2004). Similarly, blockade of $GABA_A$ receptors by picrotoxin also exhibited antinociceptive responses in the tail-flick test when injected into rat vlPAG (Moreau and Fields 1986), rat ventrolateral orbital cortex (Qu et al. 2006) and thalamic nucleus submedius (Jia et al. 2004, Xiao et al. 2005). Together these data suggest that endogenous GABAergic tone at $GABA_A$ receptors in these brain regions may mediate nociceptive responding.

In the rat paw withdrawal test, bicuculline microinjection into the vlPAG increased latency to response (Morgan and Clayton 2005), while bicuculline microinjection into the rat RVM had no effect (Nason and Mason 2004). The finding that administration of the same dose of bicuculline into the same brain area had different effects in two models of acute pain suggests that $GABA_A$ receptors in the RVM may differentially modulate pain circuitry depending on the type of noxious stimulus (Nason and Mason 2004). $GABA_A$ receptor activation by intra-RVM muscimol injection was ineffective in the rat paw withdrawal test (Lane et al. 2005). These data suggest that endogenous GABA, acting at $GABA_A$ receptors, plays a critical role in modulating nociceptive responses in the tail-flick test, at a number of neuroanatomical loci throughout the pain pathways. The effects of administration of the antagonist (bicuculline) alone suggest a role for endogenous GABAergic tone at supraspinal $GABA_A$ receptors in the modulation of nociception, while the lack of response to muscimol in the tail-flick test may suggest that supraspinal $GABA_A$ receptors are maximally activated following acute noxious thermal stimulation. However, the lack of effect of either bicuculline or muscimol on nociceptive behaviour in the paw withdrawal test when injected into the RVM suggests that in this animal model, the transmission of descending inhibitory nociceptive information may not involve $GABA_A$ receptors.

Effects of Supraspinal Administration of Agents Acting at $GABA_B$ Receptors in Animal Models of Acute Pain

The $GABA_B$ receptor antagonist, CGP 35348, elicited antinociceptive responses in the mouse hot-plate test (Zarrindast et al. 2000) when administered icv, suggesting a role for endogenous GABA in the mediation of nociception via the activation of supraspinal $GABA_B$ receptors. However, activation of supraspinal $GABA_B$ receptors following icv baclofen microinjection was also antinociceptive in the mouse hot-plate test and yet reversed the effect of CGP 35348 (Zarrindast et al. 2000), but had no effect in the rat tail-flick test (Liebman and Pastor 1980). This conflicting evidence for antinociceptive effects of both the $GABA_B$ receptor antagonist and agonist is difficult to explain but the authors suggested that blockade of the presynaptic $GABA_B$ receptor by CGP 35348 may release endogenous GABA which, in turn would activate the postsynaptic $GABA_B$ receptor sites and cause antinociception. In another study, CGP35348 had no effect on nociception in the tail-flick test and the hot plate test, when administered into the rat NRM or GiA (Thomas et al. 1996), while baclofen resulted in a biphasic effect on the latency to response, which was dependent on the dose administered (Hammond et al. 1998, Thomas et al. 1995).

Effects of Supraspinal Administration of Agents Acting at Ionotropic Glutamate Receptors in Animal Models of Acute Pain

In the mouse tail-flick test, icv administration of the NMDA receptor antagonist, MK801, or the AMPA/kainate receptor antagonist, CNQX, had no effect on nociceptive behaviour (Suh et al. 1995). Microinjection of glutamate itself into the dlPAG (Hutchison et al. 1996, Urca et al. 1980) or ventrolateral orbital cortex (Zhang et al. 1997) had an antinociceptive effect in the tail-flick test, and demonstrated both anti- and pro-nociceptive effects dependant on concentration when administered into the RVM (Zhuo and Gebhart 1997). These results may be due to activation of ionotropic or metabotropic glutamate receptors, which may mediate a facilitatory or inhibitory effect on neuronal firing depending on the subtype activated (see Table 4).

Table 4. Summary of drugs utilised in the various studies reviewed, and the receptors which they activate

Group	GABAAergic		Ionotropic Glutamate Receptor		Metabotropic Glutamate Receptor			
					I		II	III
Subtype	GABA_A	GABA_B	NMDA	AMPA	mGluR5	mGluR1	mGlu2,3	mGlu4-8
Agonists	GABA	GABA	NMDA	AMPA	Glutamate	Glutamate	Glutamate	Glutamate
	muscimol	baclofen		Kainate	CHPG	DHPG	L-AP4	LY354740
							L-CCG-I	AMN082
							DCGIV	DCPG
								L-SOP
Antagonists	SR95531	CGP 35348	CPP	NBQX	MPEP	LY367385	MCPG	UBP1112
	picrotoxin		MK801	CNQX		CPCCOEt	EGlu	MSOP
	bicuculline		GV196771A	DNQX		AIDA		
			APV			ACPD		

Metabotropic drug key

Abbreviation	Chemical name
MPEP	6-methyl-2-phenylethynyl-pyridine
LY367385	(+)-2-methyl-4-carboxyphenylglycine
MCPG	(±)-α-methyl-4-carboxyphenylglycine
DHPG	S-3,5-Dihydroxyphenylglycine
LY354740	(+)-2-Aminobicyclo[3.1.0]hexane-2,6-dicarboxylicacid
CPCCOEt	7-(hydroxyimino)cyclopropa[b]chromen-1a-carboxylate ethyl ester
L-AP4	L-(+)-2-amino-4-phosphonobutyrate
AMN082	N,N(I)-dibenzhydrylethane-1,2-diamin dihydrochloride
DCPG	(S)-3,4-dicarboxyphenylglycine
MSOP	(RS)-alpha-methylserine-O-phoshate
AIDA	(RS)-1 aminoindan-1,5-dicarboxylicacid
L-CCG-I	2S,1'S,2'S)-2-(carboxycyclopropyl)glycine
DCGIV	(2S,2'R,3'R)-2-(2',3'-dicarboxycyclopropyl) glycine
CHPG	(RS)-2-chloro-5-hydroxy-phenyl-glycine
UBP1112	alpha-methyl-4-phosphonophenylglycine
EGlu	S)-alpha-ethylglutamic acid
L-SOP	L-serine-O-phosphate
ACPD	1S,3R-aminocyclopentane-1,3-dicarboxylate

Ionotropic and GABAergic drug key

Abbreviation	Chemical name
NMDA	N-methyl-D-aspartate
AMPA	α-amino-3-hydroxy-5-methylisoxazole-4-propionic acid
CPP	3-((+)-2-carboxypiperazin-4-yl)propyl-l-phosphonate
MK801	(+)-5-methyl-10,11- dihydro-5H-dibenzo[a,d]cyclohepten-5,10-imine maleate
GV196771A	E-4, 6-dichloro-3-(2-oxo-1-phenyl-pyrrolidin-3-ylidenemethyl)-1H- indole-2- carboxylic acid sodium salt
SR95531	2-(3'(carboxypropyl)-3-amino-6-(paramethoxyphenyl)pyridazinium bromide
APV	2-Amino-5-phosphonovalerate
NBQX	2,3-dioxo-6-nitro-1,2,3,4-tetrahydrobenzo[f]quinoxaline-7-sulfonamide disodium salt
CGP 35348	P-(3-aminopropyl)-p-diethoxymethyl-phosphinic acid
CNQX	6-Cyano-7-nitroquinoxaline-2,3-dione
DNQX	6,7-dinitroquinoxaline-2,3-dione
AP7	2-amino-7-phosphonoheptanoate

In the rat tail-flick test, microinjection of the NMDA receptor antagonist APV into the NRM (Aimone and Gebhart 1987) or RVM (Coutinho et al. 1998), respectively, suppressed nociceptive responses, but intra-PAG microinjection of APV decreased the latency to response in the rat paw withdrawal test (Palazzo et al. 2001). These data suggest differential roles for NMDA receptors in specific regions of the brain associated with the descending inhibitory pain pathway.

Effects of Supraspinal Administration of Agents Acting at Metabotropic Glutamate Receptors in Animal Models of Acute Pain

The two best studied brain regions with respect to the actions of drugs acting at metabotropic glutamate receptors are the PAG and RVM. In the mouse hot plate test, dlPAG microinjection of the group I mGlu receptor antagonist AIDA into the dlPAG decreased the latency to response, while blockade of group II mGlu receptors, by EGlu microinjection into the dlPAG, had no effect on the latency of the nociceptive response, yet reversed the pronociceptive effect of the group II mGlu agonist L-CCG-I (Maione et al. 1998). This study also showed that intra-dlPAG microinjection of MSOP (Group III mGlu receptor antagonist) increased the latency to response, and reversed the pronociceptive effects of L-SOP (group III mGlu receptor agonist) (Maione et al. 1998). In another acute animal model of pain, the paw withdrawal test, no effect was reported with intra-dlPAG microinjection of CPCCOEt (mGlu1 receptor antagonist), while MPEP (mGlu5 receptor antagonist) reduced the latency to response and reversed the antinociceptive effect of CHPG microinjection (Palazzo et al. 2001). Intra-dlPAG EGlu administration had no effect on latency to response, while intra-dlPAG MSOP microinjection decreased the response latency in the rat paw withdrawal test (Palazzo et al. 2001). These data suggest that several types of mGlu receptors are present in the dlPAG, all of which can be activated by endogenous glutamate and can play a role in nociception.

In the RVM, the activation of group III mGlu receptors by L-AP4 microinjection had no effect in the rat tail-flick test, while the intra-RVM microinjection of DCGIV (group II mGlu receptor agonist) increased the latency to nociceptive response in the rat tail-flick test (Kim et al. 2002).

PHARMACOLOGICAL ELUCIDATION OF THE ROLES OF SUPRASPINAL IONOTROPIC AND METABOTROPIPC GABA AND GLUTAMATE RECEPTOR SUBTYPES IN ANIMAL MODELS OF INFLAMMATORY AND NEUROPATHIC PAIN

Effects of Supraspinal Administration of Agents Acting at GABA$_A$ Receptors in Animal Models of Inflammatory and Neuropathic Pain

Blockade of supraspinal GABA$_A$ receptors following icv administration of bicuculline demonstrated antinociceptive activity in the rat formalin test (Mahmoudi and Zarrindast 2002), while activation of GABA$_A$ receptors by icv administration of muscimol was also antinociceptive in the mouse (Chung et al. 1999) and rat (Mahmoudi and Zarrindast 2002, Shafizadeh et al. 1997) formalin test. It was postulated that blockade of the $GABA_A$ receptor at presynaptic sites may release endogenous GABA which, in turn would activate the postsynaptic $GABA_B$ receptor sites and cause antinociception.

Site specific administration of drugs targeting GABA receptors has aided elucidation of their role in discrete brain regions during persistent inflammatory of neuropathic pain. Similarly, bicuculline microinjection into the vlPAG decreased formalin-evoked nociceptive behaviour in the rat (Morgan et al. 2003). In the chronic constriction injury model of pain (CCI), no effect was observed when bicuculline was microinjected into the central amygdala (Pedersen et al. 2007), while intra-CeA muscimol injection reduced the hyperalgesic response of rats to von Frey hair stimulation in a chronic constriction injury model of neuropathic pain (Pedersen et al. 2007). $GABA_A$ receptor activation by intra-RVM muscimol injection was ineffective in the formalin test (Lane et al. 2005), yet decreased the jaw-opening reflex in the capsaicin-induced model of inflammatory pain in rats (Gear et al. 1999). The inconsistency in behavioural responses with intra-RVM administration of GABAergic compounds again suggests the existence of multiple circuits in this region, and that their activation is dependent on noxious stimulus modality.

Microinjection of GABA itself into the gigantocellular nucleus (GiA) had no effect on nociception in the rat tail-flick test but further enhanced nociceptive behaviour in the formalin test and in rats administered formalin into the hindpaw following partial ligation of the sciatic nerve (Azami et al. 2001) as compared to non sciatic-nerve-ligated controls. Moreover, intra-PAG microinjection of GABA had no effect in the rat formalin test in control animals but attenuated the decrease in formalin-evoked nociceptive behaviour observed following sciatic nerve ligation (Azami et al. 2001, Monhemius et al. 2001). Interestingly, while administration of GABA or muscimol directly into the rostral anterior cingulate cortex had no effect on mechanical allodynia in the rat L5 spinal nerve ligation model of neuropathic pain, these pharmacological interventions did attenuate place escape/avoidance behaviour in nerve-injured animals suggesting a role for $GABA_A$ receptors in this region in higher order supraspinal nociceptive processing (LaGraize and Fuchs 2007). However, with GABA microinjection it is not certain which receptor type, and/or subtype is being activated nor its synaptic location. As already mentioned, the activation of GABA receptors leads to a depolarisation of the membrane of the neuron upon which it is located. Care must be taken when interpreting data involving supraspinal administration of GABAergic modulators, as the synaptic location of the receptor can determine the overall effect at this synapse, which in turn will affect downstream processes.

Effects of Supraspinal Administration of Agents Acting at $GABA_B$ Receptors in Animal Models of Inflammatory and Neuropathic Pain

The $GABA_B$ receptor antagonist, CGP 35348, elicited antinociceptive responses in the rat formalin test (Mahmoudi and Zarrindast 2002) when administered icv, suggesting a role for endogenous GABA in the mediation of nociception via the activation of supraspinal $GABA_B$ receptors. Nociceptive behaviour in the mouse (Chung et al. 1999) and rat (Mahmoudi and Zarrindast 2002, Shafizadeh et al. 1997) was reduced by icv administration of the $GABA_B$ receptor agonist baclofen microinjection in the formalin test. Further studies determined that there was no significant difference between the antinociceptive effect of icv baclofen or CGP35348 in the mouse hot-plate test in sciatic nerve ligated animals compared to non-ligated controls (Zarrindast et al. 2000). This would suggest that there is little or no change in supraspinal GABA$_B$ receptor activity following sciatic nerve ligation. Both CGP35348 and baclofen demonstrated antinociceptive effects when injected into the ventrobasal complex of the thalamic nuclei in the rat formalin test and rat model of arthritic pain (Potes et al. 2006). These apparently paradoxical findings of antinociceptive activity for intracerebrally administered GABA$_B$ receptor agonists and antagonists in various animal models of pain may be due to differential inhibitory/disinhibitory effects of GABA$_B$ receptor activation dependant on neuroanatomical and synaptic location. Depending on the localisation of the receptor (pre- vs post- vs extra-synaptic), and onto which type of neuron this neuron synapses with, the administration of GABA receptor agonists or antagonists could lead to inhibitory or disinhibitory effects on pain pathways in the brain.

Effects of Supraspinal Administration of Agents Acting at Ionotropic Glutamate Receptors in Animal Models of Inflammatory and Neuropathic Pain

Icv administration of the NMDA receptor antagonist, MK801, or the AMPA/kainate receptor antagonist, CNQX, reduced formalin-evoked nociceptive behaviour in mice (Chung et al. 2000). However, DNQX (AMPA/kainite antagonist) and APV (NMDA receptor antagonist) demonstrated no antinociceptive effects when administered icv in the rat formalin test (Lei et al. 2004). The contrasting results may be due to species differences or discrepancies associated with the different antagonists employed and their pharmacokinetics. The icv administration of the NR2B subtype selective NMDA receptor antagonist (CP-101,606) was shown to suppress the mechanical allodynia associated with chronic constriction injury (Nakazato et al. 2005). However, further studies investigating the supraspinal administration of NMDA receptor subtype-specific compounds have not yet been performed in animal models of pain.

Pharmacological activation of ionotropic glutamate receptors in the RVM was examined in the CFA model of inflammatory hyperalgesia, and it was determined that NMDA could facilitate, as well as suppress, nociceptive responses depending on the dose and time of administration post-inflammation (Guan et al. 2002). These results mirror the biphasic behavioural effects seen with the intra-RVM administration of glutamate, suggesting that these behavioural effects may be attributable to the activation of supraspinal NMDA receptors by endogenous glutamate. Formalin-evoked nociceptive responses were also reduced by intra-dlPAG NMDA microinjection in mice (Berrino et al. 2001, Miguel and Nunes-de-Souza 2006). Meanwhile, intra-RVM microinjection of AMPA resulted in a dose-dependent inhibition of nociceptive responses in the rat CFA model (Guan et al. 2002). AMPA/kainate receptors were also shown to be involved in nociception following demonstration of the antinociceptive effects of intra-vlPAG kainate microinjection in the rat paw-withdrawal test (Morgan et al. 2003).

Antagonists of ionotropic glutamate receptor subtypes also demonstrate antinociceptive effects when administered into supraspinal sites associated with nociception. The NMDA receptor antagonist APV was antinociceptive in the rat colonic inflammation test when microinjected into the NRM (Aimone and Gebhart 1987), or RVM (Coutinho et al. 1998), respectively. APV had no effect in the rat formalin test when microinjected into the anterior cingulate cortex (Lei et al. 2004). However, there is good evidence from other studies to suggest an important role for NMDA receptors in forebrain regions including the anterior cingulate cortex and insular cortex in nociception (Zhuo 2002). It was also demonstrated in an elegant study that transgenic mice overexpressing the NR2B subunit of the NMDA receptor exhibited enhanced nociceptive behaviour in the formalin and models of inflammatory pain (Wei et al. 2001). Moreover, intra-RVM microinjection of the AMPA/kainate receptor antagonist, DNQX facilitated pronociceptive responses in the colonic inflammatory test (Coutinho et al. 1998). These studies provide good evidence for glutamate-mediated modulation of nociception via activation of ionotropic NMDA and AMPA/kainate receptors in these discrete brain regions.

Effects of Supraspinal Administration of Agents Acting at Metabotropic Glutamate Receptors in Animal Models of Inflammatory and Neuropathic Pain

In mice, icv administration of mGlu1 (AIDA) and mGlu5 (MPEP) (Miyata et al. 2003) receptor antagonists reduced nociceptive behaviour in the formalin test (Chung et al. 2000). The role of mGlu receptors in a number of discrete brain regions has been examined using the mouse formalin test (Berrino et al. 2001) and assessment of vocalisations in the kaolin/carageenan-induced arthritic rat model (Han and Neugebauer 2005) (see table 2). Despite displaying pronociceptive properties in the paw withdrawal test, intra-dlPAG microinjection of MPEP decreased formalin-evoked nociceptive behaviour in mice (Berrino et al. 2001). Meanwhile, intra-dlPAG microinjection of the mGlu1 receptor antagonist, CPCCOEt, reversed the antinociceptive response to intra-dlPAG DHPG microinjection (Maione et al. 2000), suggesting differential roles for mGlu5 and mGlu1 receptors in the PAG depending on the nature of the noxious stimulus. Similarly, both AIDA and CPCCOEt microinjection into the ventroposterior lateral thalamic nuclei decreased the second phase of formalin-evoked nociceptive behaviour in mice, while DHPG increased the behavioural response in both phases of the mouse formalin response (Miyata et al. 2003). Following kaolin/carageenan-induced arthritis in rats, pain-evoked vocalisations were reversed by intra-CeA microinjection of CPCCOEt but not MPEP (Han and Neugebauer 2005). Taken together these data suggest distinct roles for supraspinal mGlu1 and mGlu5 receptors in response to different kinds of noxious stimuli.

Intra-dlPAG microinjection of the mGlu2 receptor antagonist, EGlu, had no effect on formalin-evoked nociceptive behaviour (Berrino et al. 2001), but prevented the antinociceptive effects of local microinjection of the mGlu2 receptor agonist, L-CCG-I (Maione et al. 2000). In the mouse formalin test, intra-dlPAG MSOP (group III mGlu receptor antagonist) microinjection had no effect on nociceptive response (Berrino et al. 2001), but reversed the hyperalgesic response to intra-dlPAG administration of the mGlu group III agonist, L-SOP (Maione et al. 2000). Similarly, intra-lPAG DCPG (mGlu8 receptor agonist) reversed the formalin- and carageenan-evoked nociceptive responses as well as the hyperalgesic responses in the sciatic nerve ligation model in mice (Marabese et al. 2007). As with the animal models of acute pain, these studies suggest a modulatory role for metabotropic glutamate receptors in pain. The diversity of receptor subtypes likely facilitates differential effects of the endogenous ligand, glutamate, on nociceptive processing dependent on brain region and neuronal localisation.

There has been considerable progress made in the identification and development of receptor subtype-specific ligands for ionotropic and metabotropic GABA and glutamate receptors. With the help of genetic knockout and knock-in models, as well as pharmacological studies, our understanding of the physiological and pathophysiological role of receptor subfamilies and particular receptor subtypes has grown. Knockout mice lacking specific glutamatergic (Cheng et al. 2008; Youn et al. 2008; Garraway et al. 2007; Hizue et al. 2005; Ko et al. 2005; Hartmann et al. 2004; Fundytus et al. 2002) and GABAergic (Knabl et al. 2008) receptor subtypes or their transporters (Xu et al. 2008; Gomeza et al. 2006; Moechars et al. 2006) have demonstrated a role for these receptors in nociception. Studies employing conditional knockouts (Quintero et al. 2007), intracerebral administration of siRNA or antisense oligonucleotides against these receptors are now required to further investigate the contribution of these receptors at the supraspinal level.

Further studies examining the role of GABAergic and glutamatergic receptor subtypes in pain need to be performed. Moreover, a more comprehensive anatomical map of the various receptor GABA and glutamate receptor subtypes in regions associated with pain is required. It remains to be seen whether the investigation of the GABAergic and glutamatergic systems will be productive in identifying a target for therapeutic intervention in pain. It is possible that these neurotransmitters, their receptors and receptor subtypes may simply be expressed too ubiquitously and/or be too important for overall brain function to be exploited as analgesic targets. Nevertheless, considerable scope remains both for the characterisation and further development of novel subtype selective ligands which might act either as agonists or antagonists, or allosterically as positive or negative modulators of the various GABAergic and glutamatergic receptors. A combination of genetic and pharmacological approaches should facilitate full elucidation of the roles of these receptors in nociceptive processing at the supraspinal level.

THE ROLE OF CENTRAL GABA AND GLUTAMATE IN THE ANTINOCICEPTIVE EFFECTS OF ANALGESICS

Behavioural studies have indicated that supraspinal GABAergic and glutamatergic systems are involved in the mediation of pharmacological effects of classical analgesics. Current analgesic treatments in common use include non-steroidal anti-inflammatory drugs (NSAIDs), paracetamol, opioids, cannabinoids, antidepressants and anticonvulsants. We will now consider the evidence for a role of supraspinal GABA and/or glutamate in contributing to the analgesic effects of these agents.

NSAIDs and Paracetamol

Evidence for the involvement of the supraspinal GABAergic system in the antinociceptive effects of NSAIDs is provided by the observation that intra-RVM microinjection of muscimol abolished the antinociceptive effects of intra-vlPAG dipyrone microinjection in a rat model of carrageenan-induced hyperalgesia (Vazquez et al. 2007). This suggests that dipyrone administration locally into the PAG is sufficient to induce an antinociception, which is mediated, in part, by GABA neurons in the descending PAG-RVM-spinal cord pathway.

Icv microinjection of dipyrone was shown to reverse both capsaicin- and formalin-evoked nociceptive behaviour in mice (Beirith et al. 1998). In vitro receptor binding studies determined that [^{3}H] Glutamate binding in rat and mouse cortical membranes were decreased after bath application of dipyrone, suggesting a role for the cortical glutamatergic system in the mechanism of action of NSAIDs (Beirith et al. 1998). Further evidence for the involvement of glutamate in the mechanism of action of NSAIDs is provided by the observed increase in glutamate-induced ion current in rat PAG neurons following COX-2 inhibitor (celecoxib) bath application (Shin et al. 2003). Whilst the primary mechanism of action of the NSAIDs remains inhibition of COX enzymes, the studies above suggest that centrally-mediated COX-independent mechanisms may also play a key role in the pharmacological action of this class of analgesics, possibly through the modulation of supraspinal GABAergic and glutamatergic neurons.

Paracetamol also impacts on cyclooxygenase activity by reducing the oxidised form of cyclooxygenase enzymes, thus preventing the formation of pro-inflammatory chemicals (Aronoff et al. 2006, Boutaud et al. 2002). It has also been demonstrated that paracetamol is metabolised to AM404 (Hogestatt et al. 2005), a compound that inhibits the uptake of the endogenous cannabinoid anandamide, which may partially account for the effectiveness of paracetamol in analgesia. However, no studies have been performed investigating the supraspinal administration of paracetamol in animal models of pain, or the potential role of GABA or glutamate in the manifestation of paracetamol's effects.

Opioids

μ-opioid receptors are located on GABAergic neurons in almost all brain areas associated with nociception (Huo et al. 2005, Kalyuzhny and Wessendorf 1998) and κ-opioid receptors have been reported on GABA-containing off-cells and neutral cells but to a lesser extent on the on-cells of the RVM (Winkler et al. 2006). δ-opioid receptors, however, have not been reported on GABA-containing neurons, yet were reported to be in close proximity to GABA-containing neurons, as assessed by immunohistochemistry combined with retrograde tracing, and as such are well-positioned to indirectly modulate GABAergic neuronal firing (Commons et al. 2001, Kalyuzhny and Wessendorf 1998).

There are a number of studies reporting changes in GABAergic or glutamatergic neuronal firing (Chieng and Christie 1994, Chiou and Huang 1999, Coutinho-Netto et al. 1980, Finnegan et al. 2005, Hahm et al. 2004, Sugita and North 1993, Tanaka and North 1994, Vaughan 1998, Vaughan and Christie 1997, Vaughan et al. 1997), receptor expression (Inoue et al. 2003) or synaptic release (Bie and Pan 2005, Coutinho-Netto et al. 1980, Ma et al. 2006, Nicol et al. 1996, Sbrenna et al. 1999, Stiller et al. 1995, Tanaka and North 1994) following administration of opioidergic compounds, in a number of brain areas associated with nociception. A full review of these studies is beyond the scope of this chapter, but for extensive review see (Williams et al. 2001, Zhu et al. 1998). Fewer studies have examined the effects of opioidergic compounds on GABA- and glutamate-mediated nociceptive responses, in these brain areas, in animal models of pain. In the rat tail-flick test, the antinociceptive effects of icv administration of morphine were reversed by pre-injection of icv muscimol (Hough et al. 2001). Similarly, the antinociceptive effects of intra-thalamic (nucleus submedius thalamic neuron) morphine microinjection was prevented by muscimol and enhanced by bicuculline co-injection (Jia et al. 2004). However, while bicuculline microinjection itself was antinociceptive, muscimol had no effect in animals not receiving morphine (Jia et al. 2004). Furthermore, the antinociceptive effect of intra-thalamic bicuculline microinjection was prevented by preinjection of the opioid receptor antagonist naloxone (Jia et al. 2004). These results indicate that the antinociceptive effects of morphine may be mediated by a decrease in supraspinal $GABA_A$ receptor activation, presumably resulting from a morphine-induced decrease in GABA release. Correspondingly, in the rat hot-plate test, intra-vlPAG microinjection of either morphine or bicuculline resulted in the increased latency to paw withdrawal, and the antinociceptive effect of morphine was further augmented in a subset of animals with bicuculline co-injection (Morgan et al. 2005).

The antinociceptive effects of icv DAMGO (μ-opioid receptor agonist) administration in the mouse hot-plate test were enhanced by the icv administration of dizocilpine (NMDA receptor antagonist) and CHPG (group I mGlu receptor antagonist), and reversed by NMDA and DHPG (Suzuki et al. 2000). In contrast, the antinociceptive effects of intra-PAG administration of morphine in the rat tail-flick and hot plate tests were reversed by intra-RVM administration of the NMDA receptor antagonists APV, AP-7 and MK801 (Spinella et al. 1996, van Praag and Frenk 1990), suggesting that the antinociceptive effect of supraspinal μ-opioid receptor activation may be mediated in part via NMDA receptors in brain areas involved in the descending pain system.

Using microdialysis it was demonstrated that GABA and glutamate levels in the rat vlPAG were decreased following formalin administration, and local vlPAG naloxone infusion was capable of reversing this effect (Maione et al. 1999). Moreover, icv muscimol and baclofen microinjection were shown to decrease formalin-evoked nociceptive behaviour in rats and this effect was reversed by naloxone (Mahmoudi and Zarrindast 2002), suggesting an opioid receptor-mediated modulation of GABA-mediated antinociception. The effect of systemic morphine administration on formalin-evoked nociceptive responses in rats was also reversed by the intra-CeA microinjection of both muscimol and NMDA (Manning 1998), implicating a role for both GABAergic and glutamatergic receptors in the CeA in the antinociceptive effect of morphine. In the mouse formalin test, the co-injection of baclofen or muscimol had no effect on the antinociception resulting from icv administration of morphine, but the antinociceptive effects of icv administration of beta endorphin were reversed by baclofen, MK801 and CNQX co-injection (Chung et al. 1999, Chung et al. 2000). It was suggested that morphine and beta-endorphin may stimulate different opioid receptors, which activate separate descending pain control systems, possibly explaining the results observed (Chung et al. 1999, Chung et al. 2000).

Interestingly, a number of studies have reported that opioids and NSAIDs may mediate their central antinociceptive effects via a common pathway, possibly involving GABA or glutamate (Bjorkman 1995, Tortorici et al. 1996, Vasquez and Vanegas 2000, Vaughan 1998, Vaughan et al. 1997, Vazquez et al. 2007). Evidence suggests that opioid receptor activation reduces GABA release by stimulating the formation of arachidonic acid metabolites via the non-selective lipoxygenase inhibition (12-LOX) pathway, which open voltage-dependent

K^+channels (Christie et al. 2000). NSAIDs are believed to synergize with opioids by blocking the cyclooxygenase enzymes, thereby making more arachidonic acid available for the 12-LOX pathway and thus further decrease GABA release. This mechanism would also explain the reversal of NSAID antinociception which results with co-injection of naloxone in the same site (Tortorici et al. 1996), or injection of opioid antagonists (Vasquez and Vanegas 2000) and GABAergic compounds (Vazquez et al. 2007) downstream from the site of administration of NSAIDs.

Cannabinoids

We (Rea et al. 2007) have recently reviewed the involvement of the endocannabinoid system in the modulation of nociception, with particular reference to the role of supraspinal GABA and glutamate. The anatomical and behavioural studies reported in that review strongly implicate a role for supraspinal GABA and glutamate in the antinociceptive effects of cannabinoids.

Anticonvulsants and Antidepressants

The precise mechanisms underlying the antinociceptive effects of anticonvulsants and antidepressants remain unclear but these two drug classes are widely prescribed for the treatment of neuropathic pain (Dickenson and Ghandehari 2007). Indeed, the proposed mechanisms of anticonvulsants and antidepressants are quite diverse. Anticonvulsants fall into one of the three following categories; those which functionally block voltage gated sodium and/or calcium channels, those which enhance GABA-mediated events via interaction with GABAergic receptors, inhibition of its metabolism or reduction of its neuronal uptake, and those which inhibit excitatory glutamatergic neurotransmission via blockade of receptors or alteration of glutamate metabolism or release. Few studies have been performed examining the supraspinal involvement of GABA and glutamate receptors in the antinociceptive effects of anticonvulsant administration. In the rat tail-flick test, the icv administration of diazepam had no effect but icv muscimol microinjection reversed the increase in tail-flick latency observed following systemic administration of diazepam (Zambotti et al. 1991). Similarly, antinociceptive effects of systemic tiagabine administration in the mouse hot-plate test, the mouse abdominal constriction test and the rat noxious pressure test, were reversed by supraspinal $GABA_B$ receptor antagonism with CGP35348. Vigabatrin microinjection into the rostral agranular insular cortex was analgesic in the rat paw-withdrawal test, an effect which was reversible by both bicuculline and saclofen ($GABA_B$ receptor antagonist) microinjection. Conversely, systemic pentobarbital administration was hyperalgesic in the tail-flick test, the hot-plate test, the formalin test and the abdominal–constriction test in mice, and this was reversible by icv bicuculline microinjection (Yokoro et al. 2001). Finally, icv phenobarbital microinjection demonstrated hyperalgesic effects in the tail-flick test when administered alone (Yokoro et al. 2001). Taken together, these studies demonstrate a role for supraspinal GABA receptors in the effects of anticonvulsant administration on nociception, however, further studies are required to investigate the effects of systemically- and locally-administered anticonvulsants on supraspinal glutamatergic and GABAergic systems.

Antidepressants, in particular the tricyclics, are also used in the treatment of neuropathic pain. Although the selective serotonin reuptake inhibitors (SSRIs) have a better tolerated profile than tricyclic antidepressants, their effectiveness in pain remains to be determined. As most current antidepressant treatments involve the manipulation of monoaminergic systems, it seems likely that the effectiveness of antidepressants in the treatment of neuropathic pain is mediated via central effects on these systems. It has been proposed that they mediate their effects by enhancing descending inhibitory serotonergic and noradrenergic modulation of nociception (Millan 1999). Recent studies at the spinal level suggest that modulation of glutamatergic plasticity may underlie the attenuation of tolerance to the antinociceptive effects of morphine produced by the antidepressant amitriptyline (Tai et al. 2007) and that the antinociceptive effects of amitriptyline in an animal model of neuropathic pain may be related to its effects on GABA$_B$ receptor expression in the spinal cord (McCarson et al. 2005). There is, however, a paucity of studies investigating the possible involvement of supraspinal GABA and glutamate in the analgesic effects of antidepressants. Given the evidence for indirect secondary effects of antidepressants on a variety of neurotransmitter and neuropeptide signalling systems, further study in this area is warranted.

CONCLUSION

The data currently available and reviewed in this chapter suggest key roles for supraspinal GABA and glutamate in the mediation and modulation of pain. We have discussed the neuroanatomical localisation of GABAergic and glutamatergic neurons in areas associated with pain, the changes in GABA and glutamate levels and receptor expression in response to painful stimuli, the effects of supraspinal administration of selective GABAergic or glutamatergic compounds in animal models of pain, and the role of GABA and glutamate in analgesic treatment.

Owing to methodological difficulties and ethical constraints, the vast majority of studies examining the role of supraspinal GABA or glutamate in pain have been carried out in laboratory animals and there is a paucity of translational clinical studies examining the extent to which these mechanisms and/or alterations may apply in humans. Recent advances in functional imaging methodologies, however, should facilitate increased research in this area. Similarly, advances in molecular biology have revealed the existence of multiple GABA and glutamate receptor subtypes in the brain playing a role in pain.

The challenge now is to harness this wealth of knowledge with a view to developing novel therapeutic approaches characterised by improved analgesic efficacy and increased safety profiles. These could take the form of new drugs targeting GABA and/or glutamate receptor subtypes, polypharmacy approaches, or even targeted delivery of GABAergic and glutamatergic agents to discrete brain regions.

ACKNOWLEDGMENTS

This work was supported by a grant from Science Foundation Ireland.

REFERENCES

Aimone, L. D. and Gebhart, G. F. (1987). Spinal monoamine mediation of stimulation-produced antinociception from the lateral hypothalamus. *Brain Res. 403, 2,* 290-300.

Ali, A. B., Rossier, J., Staiger, J. F. and Audinat, E. (2001). Kainate receptors regulate unitary IPSCs elicited in pyramidal cells by fast-spiking interneurons in the neocortex. *J. Neurosci. 21, 9,* 2992-9.

Anwyl, R. (1999). Metabotropic glutamate receptors: electrophysiological properties and role in plasticity. *Brain Res. Brain Res. Rev. 29, 1,* 83-120.

Aronoff, D. M., Oates, J. A. and Boutaud, O. (2006). New insights into the mechanism of action of acetaminophen: Its clinical pharmacologic characteristics reflect its inhibition of the two prostaglandin H2 synthases. *Clin. Pharmacol Ther. 79,* 1, 9-19.

Azami, J., Green, D. L., Roberts, M. H. and Monhemius, R. (2001). The behavioural importance of dynamically activated descending inhibition from the nucleus reticularis gigantocellularis pars alpha. *Pain. 92, 1-2,* 53-62.

Ballard, T. M., Knoflach, F., Prinssen, E., Borroni, E., Vivian, J. A., Basile, J., Gasser, R., Moreau, J. L., Wettstein, J. G., Buettelmann, B., Knust, H., Thomas, A. W., Trube, G. and Hernandez, M. C. (2008). RO4938581, a novel cognitive enhancer acting at GABA(A) alpha5 subunit-containing receptors. *Psychopharmacology (Berl).*

Barnard, E. A., Skolnick, P., Olsen, R. W., Mohler, H., Sieghart, W., Biggio, G., Braestrup, C., Bateson, A. N. and Langer, S. Z. (1998). International Union of Pharmacology. XV. Subtypes of gamma-aminobutyric acidA receptors: classification on the basis of subunit structure and receptor function. *Pharmacol. Rev. 50, 2,* 291-313.

Beirith, A., Santos, A. R., Rodrigues, A. L., Creczynski-Pasa, T. B. and Calixto, J. B. (1998). Spinal and supraspinal antinociceptive action of dipyrone in formalin, capsaicin and glutamate tests. Study of the mechanism of action. *Eur. J. Pharmacol.* 345, 3, 233-45.

Berrino, L., Oliva, P., Rossi, F., Palazzo, E., Nobili, B. and Maione, S. (2001). Interaction between metabotropic and NMDA glutamate receptors in the periaqueductal grey pain modulatory system. *Naunyn Schmiedebergs Arch. Pharmacol. 364, 5,* 437-43.

Bie, B. and Pan, Z. Z. (2005). Increased glutamate synaptic transmission in the nucleus raphe magnus neurons from morphine-tolerant rats. *Mol. Pain.* 1, 1, 7.

Bjorkman, R. (1995). Central antinociceptive effects of non-steroidal anti-inflammatory drugs and paracetamol. Experimental studies in the rat. *Acta Anaesthesiol. Scand Suppl. 103,* 1-44.

Bleakman, D., Alt, A. and Nisenbaum, E. S. (2006). Glutamate receptors and pain. *Semin. Cell Dev. Biol. 17, 5,* 592-604.

Blein, S., Ginham, R., Uhrin, D., Smith, B. O., Soares, D. C., Veltel, S., McIlhinney, R. A., White, J. H. and Barlow, P. N. (2004). Structural analysis of the complement control protein (CCP) modules of GABA(B) receptor 1a: only one of the two CCP modules is compactly folded. *J. Biol. Chem. 279, 46,* 48292-306.

Bordi, F. and Quartaroli, M. (2000). Modulation of nociceptive transmission by NMDA/glycine site receptor in the ventroposterolateral nucleus of the thalamus. *Pain. 84, 2-3,* 213-24.

Boutaud, O., Aronoff, D. M., Richardson, J. H., Marnett, L. J. and Oates, J. A. (2002). Determinants of the cellular specificity of acetaminophen as an inhibitor of prostaglandin H(2) synthases. *Proc. Natl. Acad. Sci. USA. 99, 10,* 7130-5.

Brussaard, A. B., Kits, K. S., Baker, R. E., Willems, W. P., Leyting-Vermeulen, J. W., Voorn, P., Smit, A. B., Bicknell, R. J. and Herbison, A. E. (1997). Plasticity in fast synaptic inhibition of adult oxytocin neurons caused by switch in GABA(A) receptor subunit expression. *Neuron. 19, 5,* 1103-14.

Burnashev, N., Villarroel, A. and Sakmann, B. (1996). Dimensions and ion selectivity of recombinant AMPA and kainate receptor channels and their dependence on Q/R site residues. *J. Physiol. 496 (Pt 1),* 165-73.

Burnashev, N., Zhou, Z., Neher, E. and Sakmann, B. (1995). Fractional calcium currents through recombinant GluR channels of the NMDA, AMPA and kainate receptor subtypes. *J. Physiol. 485 (Pt 2),* 403-18.

Buzsaki, G. and Chrobak, J. J. (1995). Temporal structure in spatially organized neuronal ensembles: a role for interneuronal networks. *Curr. Opin. Neurobiol. 5, 4,* 504-10.

Cheng, H. T., Suzuki, M., Hegarty, D. M., Xu, Q., Weyerbacher, A. R., South, S. M., Ohata, M. and Inturrisi, C. E. (2008). Inflammatory pain-induced signaling events following a conditional deletion of the N-methyl-D-aspartate receptor in spinal cord dorsal horn. *Neuroscience. 155, 3,* 948-58.

Chieng, B. and Christie, M. J. (1994). Inhibition by opioids acting on mu-receptors of GABAergic and glutamatergic postsynaptic potentials in single rat periaqueductal gray neurones in vitro. *Br. J. Pharmacol. 113, 1,* 303-9.

Chieng, B. and Christie, M. J. (1995). Hyperpolarization by GABAB receptor agonists in mid-brain periaqueductal gray neurones in vitro. *Br. J. Pharmacol. 116, 1,* 1583-8.

Chiou, L. C. and Huang, L. Y. (1999). Mechanism underlying increased neuronal activity in the rat ventrolateral periaqueductal grey by a mu-opioid. *J. Physiol. 518 (Pt 2),* 551-9.

Christie, M. J., Connor, M., Vaughan, C. W., Ingram, S. L. and Bagley, E. E. (2000). Cellular actions of opioids and other analgesics: implications for synergism in pain relief. *Clin. Exp. Pharmacol. Physiol. 27, 7,* 520-3.

Christie, M. J. and North, R. A., 1988. Agonists at mu-opioid, M2-muscarinic and GABAB-receptors increase the same potassium conductance in rat lateral parabrachial neurones. *Br. J. Pharmacol. 95, 3,* 896-902.

Chung, K. M., Kim, Y. H., Song, D. K., Huh, S. O. and Suh, H. W. (1999). Differential modulation by baclofen on antinociception induced by morphine and beta-endorphin administered intracerebroventricularly in the formalin test. *Neuropeptides. 33, 6,* 534-41.

Chung, K. M., Song, D. K., Huh, S. O., Kim, Y. H., Choi, M. R. and Suh, H. W. (2000). Supraspinal NMDA and non-NMDA receptors are differentially involved in the production of antinociception by morphine and beta-endorphin administered intracerebroventricularly in the formalin pain model. *Neuropeptides. 34, 3-4,* 158-66.

Ciabarra, A. M., Sullivan, J. M., Gahn, L. G., Pecht, G., Heinemann, S. and Sevarino, K. A. (1995). Cloning and characterization of chi-1: a developmentally regulated member of a novel class of the ionotropic glutamate receptor family. *J. Neurosci. 15, 10,* 6498-508.

Commons, K. G., Beck, S. G., Rudoy, C. and Van Bockstaele, E. J. (2001). Anatomical evidence for presynaptic modulation by the delta opioid receptor in the ventrolateral periaqueductal gray of the rat. *J. Comp. Neurol. 430, 2,* 200-8.

Contractor, A., Swanson, G. and Heinemann, S. F. (2001). Kainate receptors are involved in short- and long-term plasticity at mossy fiber synapses in the hippocampus. *Neuron. 29, 1,* 209-16.

Corti, C., Restituito, S., Rimland, J. M., Brabet, I., Corsi, M., Pin, J. P. and Ferraguti, F. (1998). Cloning and characterization of alternative mRNA forms for the rat metabotropic glutamate receptors mGluR7 and mGluR8. *Eur. J. Neurosci. 10, 12,* 3629-41.

Costa, E. (1998). From GABAA receptor diversity emerges a unified vision of GABAergic inhibition. *Annu. Rev. Pharmacol Toxicol. 38,* 321-50.

Coutinho-Netto, J., Abdul-Ghani, A. S. and Bradford, H. F., 1980. Suppression of evoked and spontaneous release of neurotransmitters in vivo by morphine. *Biochem. Pharmacol. 29, 20,* 2777-80.

Coutinho, S. V., Urban, M. O. and Gebhart, G. F. (1998). Role of glutamate receptors and nitric oxide in the rostral ventromedial medulla in visceral hyperalgesia. *Pain. 78, 1,* 59-69.

Couve, A., Moss, S. J. and Pangalos, M. N. (2000). GABAB receptors: a new paradigm in G protein signaling. *Mol. Cell Neurosci. 16, 4,* 296-312.

de Novellis, V., Mariani, L., Palazzo, E., Vita, D., Marabese, I., Scafuro, M., Rossi, F. and Maione, S. (2005). Periaqueductal grey CB1 cannabinoid and metabotropic glutamate subtype 5 receptors modulate changes in rostral ventromedial medulla neuronal activities induced by subcutaneous formalin in the rat. *Neuroscience. 134, 1,* 269-81.

Desai, M. A. and Conn, P. J. (1991). Excitatory effects of ACPD receptor activation in the hippocampus are mediated by direct effects on pyramidal cells and blockade of synaptic inhibition. *J. Neurophysiol. 66, 1,* 40-52.

Dickenson, A. H. and Ghandehari, J. (2007). Anti-convulsants and anti-depressants. *Handb Exp. Pharmacol. 177,* 145-77.

Dingledine, R. and Conn, P. J. (2000). Peripheral glutamate receptors: molecular biology and role in taste sensation. *J. Nutr. 130, 4S Suppl,* 1039S-42S.

Dudel, J. and Kuffler, S. W., 1961. Presynaptic inhibition at the crayfish neuromuscular junction. *J. Physiol. 155,* 543-62.

Duguid, I. C. and Smart, T. G. (2004). Retrograde activation of presynaptic NMDA receptors enhances GABA release at cerebellar interneuron-Purkinje cell synapses. *Nat. Neurosci. 7, 5,* 525-33.

Egebjerg, J. and Heinemann, S. F. (1993). Ca2+ permeability of unedited and edited versions of the kainate selective glutamate receptor GluR6. *Proc. Natl. Acad. Sci. USA. 90, 2,* 755-9.

Enna, S. J. and McCarson, K. E. (2006). The role of GABA in the mediation and perception of pain. *Adv. Pharmacol. 54,* 1-27.

Farrar, S. J., Whiting, P. J., Bonnert, T. P. and McKernan, R. M. (1999). Stoichiometry of a ligand-gated ion channel determined by fluorescence energy transfer. *J. Biol. Chem. 274, 15,* 10100-4.

Ferreira-Gomes, J., Neto, F. L. and Castro-Lopes, J. M. (2004). Differential expression of GABA(B(1b)) receptor mRNA in the thalamus of normal and monoarthritic animals. *Biochem. Pharmacol. 68, 8,* 1603-11.

Ferreira-Gomes, J., Neto, F. L. and Castro-Lopes, J. M. (2006). GABA(B2) receptor subunit mRNA decreases in the thalamus of monoarthritic animals. *Brain Res. Bull. 71, 1-3,* 252-8.

Fields, H. L., Bry, J., Hentall, I. and Zorman, G., 1983a. The activity of neurons in the rostral medulla of the rat during withdrawal from noxious heat. *J. Neurosci. 3, 12*, 2545-52.

Fields, H. L., Vanegas, H., Hentall, I.D. and Zorman, G., 1983b. Evidence that disinhibition of brain stem neurones contributes to morphine analgesia. *Nature. 306, 5944*, 684-6.

Finnegan, T. F., Chen, S. R. and Pan, H. L. (2005). Effect of the {mu} opioid on excitatory and inhibitory synaptic inputs to periaqueductal gray-projecting neurons in the amygdala. *J. Pharmacol. Exp. Ther. 312, 2*, 441-8.

Flor, P. J., Van Der Putten, H., Ruegg, D., Lukic, S., Leonhardt, T., Bence, M., Sansig, G., Knopfel, T. and Kuhn, R. (1997). A novel splice variant of a metabotropic glutamate receptor, human mGluR7b. *Neuropharmacology. 36, 2*, 153-9.

Freund, T. F. and Buzsaki, G. (1996). Interneurons of the hippocampus. *Hippocampus. 6, 4*, 347-470.

Fritschy, J. M., Meskenaite, V., Weinmann, O., Honer, M., Benke, D. and Mohler, H. (1999). GABAB-receptor splice variants GB1a and GB1b in rat brain: developmental regulation, cellular distribution and extrasynaptic localization. *Eur. J. Neurosci. 11, 3*, 761-8.

Fundytus, M. E., Osborne, M. G., Henry, J. L., Coderre, T. J. and Dray, A. (2002). Antisense oligonucleotide knockdown of mGluR1 alleviates hyperalgesia and allodynia associated with chronic inflammation. *Pharmacol. Biochem. Behav. 73, 2*, 401-10.

Gage, P. W. (1992). Activation and modulation of neuronal K+ channels by GABA. *Trends Neurosci. 15, 2*, 46-51.

Garraway, S. M., Xu, Q. and Inturrisi, C. E. (2007). Design and evaluation of small interfering RNAs that target expression of the N-methyl-D-aspartate receptor NR1 subunit gene in the spinal cord dorsal horn. *J. Pharmacol. Exp. Ther. 322, 3*, 982-8.

Gassmann, M., Shaban, H., Vigot, R., Sansig, G., Haller, C., Barbieri, S., Humeau, Y., Schuler, V., Muller, M., Kinzel, B., Klebs, K., Schmutz, M., Froestl, W., Heid, J., Kelly, P. H., Gentry, C., Jaton, A. L., Van der Putten, H., Mombereau, C., Lecourtier, L., Mosbacher, J., Cryan, J. F., Fritschy, J. M., Luthi, A., Kaupmann, K. and Bettler, B. (2004). Redistribution of GABAB(1) protein and atypical GABAB responses in GABAB(2)-deficient mice. *J. Neurosci. 24, 27*, 6086-97.

Gear, R. W., Aley, K. O. and Levine, J. D. (1999). Pain-induced analgesia mediated by mesolimbic reward circuits. *J. Neurosci. 19, 16*, 7175-81.

Geiger, J. R., Melcher, T., Koh, D. S., Sakmann, B., Seeburg, P. H., Jonas, P. and Monyer, H. (1995). Relative abundance of subunit mRNAs determines gating and Ca2+ permeability of AMPA receptors in principal neurons and interneurons in rat CNS. *Neuron. 15, 1*, 193-204.

Glitsch, M. and Marty, A. (1999). Presynaptic effects of NMDA in cerebellar Purkinje cells and interneurons. *J. Neurosci. 19, 2*, 511-9.

Gomeza, J., Armsen, W., Betz, H. and Eulenburg, V. (2006). Lessons from the knocked-out glycine transporters. *Handb Exp. Pharmacol. 175*, 457-83.

Grace, C. R., Perrin, M. H., DiGruccio, M. R., Miller, C. L., Rivier, J. E., Vale, W. W. and Riek, R. (2004). NMRstructure and peptide hormone binding site of the first extracellular domain of a type B1 G protein-coupled receptor. *Proc. Natl. Acad. Sci. USA. 101, 35*, 12836-41.

Griebel, G., Perrault, G., Letang, V., Granger, P., Avenet, P., Schoemaker, H. and Sanger, D. J. (1999)a. New evidence that the pharmacological effects of benzodiazepine receptor ligands can be associated with activities at different BZ (omega) receptor subtypes. *Psychopharmacology (Berl). 146, 2,* 205-13.

Griebel, G., Perrault, G. and Sanger, D. J. (1999)b. Study of the modulatory activity of BZ (omega) receptor ligands on defensive behaviors in mice: evaluation of the importance of intrinsic efficacy and receptor subtype selectivity. *Prog. Neuropsychopharmacol. Biol. Psychiatry. 23, 1,* 81-98.

Guan, Y., Guo, W., Zou, S. P., Dubner, R. and Ren, K. (2003). Inflammation-induced upregulation of AMPA receptor subunit expression in brain stem pain modulatory circuitry. *Pain. 104, 1-2,* 401-13.

Guan, Y., Terayama, R., Dubner, R. and Ren, K. (2002). Plasticity in excitatory amino acid receptor-mediated descending pain modulation after inflammation. *J. Pharmacol. Exp. Ther. 300, 2,* 513-20.

Hahm, E. T., Lee, J. J., Min, B. I. and Cho, Y. W. (2004). Opioid inhibition of GABAergic neurotransmission in mechanically isolated rat periaqueductal gray neurons. *Neurosci. Res. 50, 3,* 343-54.

Hammond, D. L., Nelson, V. and Thomas, D. A. (1998). Intrathecal methysergide antagonizes the antinociception, but not the hyperalgesia produced by microinjection of baclofen in the ventromedial medulla of the rat. *Neurosci. Lett. 244, 2,* 93-6.

Han, J. S., Bird, G. C. and Neugebauer, V. (2004). Enhanced group III mGluR-mediated inhibition of pain-related synaptic plasticity in the amygdala. *Neuropharmacology. 46, 7,* 918-26.

Han, J. S., Fu, Y., Bird, G. C. and Neugebauer, V. (2006). Enhanced group II mGluR-mediated inhibition of pain-related synaptic plasticity in the amygdala. *Mol. Pain. 2,* 18.

Han, J. S. and Neugebauer, V. (2005). mGluR1 and mGluR5 antagonists in the amygdala inhibit different components of audible and ultrasonic vocalizations in a model of arthritic pain. *Pain. 113, 1-2,* 211-22.

Hartmann, B., Ahmadi, S., Heppenstall, P. A., Lewin, G. R., Schott, C., Borchardt, T., Seeburg, P. H., Zeilhofer, H. U., Sprengel, R. and Kuner, R. (2004). The AMPA receptor subunits GluR-A and GluR-B reciprocally modulate spinal synaptic plasticity and inflammatory pain. *Neuron. 44, 4,* 637-50.

Hawrot, E., Xiao, Y., Shi, Q. L., Norman, D., Kirkitadze, M. and Barlow, P. N. (1998). Demonstration of a tandem pair of complement protein modules in GABA(B) receptor 1a. *FEBS Lett. 432, 3,* 103-8.

Heinricher, M. M. and Tortorici, V. (1994). Interference with GABA transmission in the rostral ventromedial medulla: disinhibition of off-cells as a central mechanism in nociceptive modulation. *Neuroscience. 63, 2,* 533-46.

Hizue, M., Pang, C. H. and Yokoyama, M. (2005). Involvement of N-methyl-D-aspartate-type glutamate receptor epsilon1 and epsilon4 subunits in tonic inflammatory pain and neuropathic pain. *Neuroreport. 16, 15,* 1667-70.

Hogestatt, E. D., Jonsson, B. A., Ermund, A., Andersson, D. A., Bjork, H., Alexander, J. P., Cravatt, B. F., Basbaum, A. I. and Zygmunt, P. M. (2005). Conversion of acetaminophen to the bioactive N-acylphenolamine AM404 via fatty acid amide hydrolase-dependent arachidonic acid conjugation in the nervous system. *J. Biol. Chem. 280, 36,* 31405-12.

Hough, L. B., Nalwalk, J. W., Leurs, R., Menge, W. M. and Timmerman, H. (2001). Significance of GABAergic systems in the action of improgan, a non-opioid analgesic. *Life Sci. 68, 24,* 2751-7.

Huo, F. Q., Wang, J., Li, Y. Q., Chen, T., Han, F. and Tang, J. S. (2005). GABAergic neurons express mu-opioid receptors in the ventrolateral orbital cortex of the rat. *Neurosci. Lett.* *382, 3,* 265-8.

Hutcheon, B., Morley, P. and Poulter, M. O. (2000). Developmental change in GABAA receptor desensitization kinetics and its role in synapse function in rat cortical neurons. *J. Physiol. 522 Pt 1,* 3-17.

Hutchison, W. D., Harfa, L. and Dostrovsky, J. O. (1996). Ventrolateral orbital cortex and periaqueductal gray stimulation-induced effects on on- and off-cells in the rostral ventromedial medulla in the rat. *Neuroscience. 70, 2,* 391-407.

Inoue, M., Mishina, M. and Ueda, H. (2003). Locus-specific rescue of GluRepsilon1 NMDA receptors in mutant mice identifies the brain regions important for morphine tolerance and dependence. *J. Neurosci. 23, 16,* 6529-36.

Isa, T., Itazawa, S., Iino, M., Tsuzuki, K. and Ozawa, S. (1996). Distribution of neurones expressing inwardly rectifying and Ca(2+)-permeable AMPA receptors in rat hippocampal slices. *J. Physiol. 491 (Pt 3),* 719-33.

Jia, H., Xie, Y. F., Xiao, D. Q. and Tang, J. S. (2004). Involvement of GABAergic modulation of the nucleus submedius (Sm) morphine-induced antinociception. *Pain. 108, 1-2,* 28-35.

Jiang, L., Xu, J., Nedergaard, M. and Kang, J. (2001). A kainate receptor increases the efficacy of GABAergic synapses. *Neuron. 30, 2,* 503-13.

Juttner, R., Meier, J. and Grantyn, R. (2001). Slow IPSC kinetics, low levels of alpha1 subunit expression and paired-pulse depression are distinct properties of neonatal inhibitory GABAergic synaptic connections in the mouse superior colliculus. *Eur. J. Neurosci. 13, 11,* 2088-98.

Kalyuzhny, A. E. and Wessendorf, M. W. (1998). Relationship of mu- and delta-opioid receptors to GABAergic neurons in the central nervous system, including antinociceptive brainstem circuits. *J. Comp. Neurol. 392, 4,* 528-47.

Kaufmann, W. A., Humpel, C., Alheid, G. F. and Marksteiner, J. (2003). Compartmentation of alpha 1 and alpha 2 GABA(A) receptor subunits within rat extended amygdala: implications for benzodiazepine action. *Brain Res. 964, 1,* 91-9.

Kemp, J. A. and McKernan, R. M. (2002). NMDA receptor pathways as drug targets. *Nat. Neurosci. 5 Suppl,* 1039-42.

Kerr, D.I. and Ong, J. (1995). GABAB receptors. *Pharmacol. Ther. 67, 2,* 187-246.

Kim, S. J., Calejesan, A. A. and Zhuo, M. (2002). Activation of brainstem metabotropic glutamate receptors inhibits spinal nociception in adult rats. *Pharmacol. Biochem. Behav. 73, 2,* 429-37.

Kinoshita, A., Shigemoto, R., Ohishi, H., van der Putten, H. and Mizuno, N. (1998). Immunohistochemical localization of metabotropic glutamate receptors, mGluR7a and mGluR7b, in the central nervous system of the adult rat and mouse: a light and electron microscopic study. *J. Comp. Neurol. 393, 3,* 332-52.

Klausberger, T., Roberts, J. D. and Somogyi, P. (2002). Cell type- and input-specific differences in the number and subtypes of synaptic GABA(A) receptors in the hippocampus. *J. Neurosci. 22, 7,* 2513-21.

Knabl, J., Witschi, R., Hosl, K., Reinold, H., Zeilhofer, U. B., Ahmadi, S., Brockhaus, J., Sergejeva, M., Hess, A., Brune, K., Fritschy, J. M., Rudolph, U., Mohler, H. and

Zeilhofer, H. U. (2008). Reversal of pathological pain through specific spinal GABAA receptor subtypes. *Nature. 451, 7176,* 330-4.

Ko, S., Zhao, M. G., Toyoda, H., Qiu, C. S. and Zhuo, M. (2005). Altered behavioral responses to noxious stimuli and fear in glutamate receptor 5 (GluR5)- or GluR6-deficient mice. *J. Neurosci. 25, 4,* 977-84.

LaGraize, S. C. and Fuchs, P. N. (2007). GABAA but not GABAB receptors in the rostral anterior cingulate cortex selectively modulate pain-induced escape/avoidance behavior. *Exp. Neurol. 204, 1,* 182-94.

Lagrange, A. H., Wagner, E. J., Ronnekleiv, O. K. and Kelly, M. J. (1996). Estrogen rapidly attenuates a GABAB response in hypothalamic neurons. *Neuroendocrinology. 64, 2,* 114-23.

Lane, D. A., Patel, P. A. and Morgan, M. M. (2005). Evidence for an intrinsic mechanism of antinociceptive tolerance within the ventrolateral periaqueductal gray of rats. *Neuroscience. 135, 1,* 227-34.

Lauri, S. E., Segerstrale, M., Vesikansa, A., Maingret, F., Mulle, C., Collingridge, G. L., Isaac, J. T. and Taira, T. (2005). Endogenous activation of kainate receptors regulates glutamate release and network activity in the developing hippocampus. *J. Neurosci. 25, 18,* 4473-84.

Laurie, D. J., Boddeke, H. W., Hiltscher, R. and Sommer, B. (1996). Erratum to 'HmGlu1d, a novel splice variant of the human type I metabotropic glutamate receptor' [Eur. J. pharmacol. 296 ((1996)) R1-R3]. *Eur. J. Pharmacol. 302, 1-3,* 229.

Lea, P. M., Custer, S. J., Vicini, S. and Faden, A. I. (2002). Neuronal and glial mGluR5 modulation prevents stretch-induced enhancement of NMDA receptor current. *Pharmacol. Biochem. Behav. 73, 2,* 287-98.

Lehtinen, M. J., Meri, S. and Jokiranta, T. S. (2004). Interdomain contact regions and angles between adjacent short consensus repeat domains. *J. Mol. Biol. 344, 5,* 1385-96.

Lei, L. G., Sun, S., Gao, Y. J., Zhao, Z. Q. and Zhang, Y. Q. (2004). NMDA receptors in the anterior cingulate cortex mediate pain-related aversion. *Exp. Neurol. 189, 2,* 413-21.

Li, W. and Neugebauer, V. (2004). Differential roles of mGluR1 and mGluR5 in brief and prolonged nociceptive processing in central amygdala neurons. *J.Neurophysiol. 91, 1,* 13-24.

Liebman, J. M. and Pastor, G., 1980. Antinociceptive effects of baclofen and muscimol upon intraventricular administration. *Eur. J. Pharmacol. 61, 3,* 225-30.

Liu, S. J. and Lachamp, P. (2006). The activation of excitatory glutamate receptors evokes a long-lasting increase in the release of GABA from cerebellar stellate cells. *J. Neurosci. 26, 36,* 9332-9.

Lourenco Neto, F., Schadrack, J., Platzer, S., Zieglgansberger, W., Tolle, T. R. and Castro-Lopes, J. M. (2000). Expression of metabotropic glutamate receptors mRNA in the thalamus and brainstem of monoarthritic rats. *Brain Res. Mol. Brain Res. 81, 1-2,* 140-54.

Ma, J., Zhang, Y., Kalyuzhny, A. E. and Pan, Z. Z. (2006). Emergence of functional delta-opioid receptors induced by long-term treatment with morphine. *Mol. Pharmacol. 69, 4,* 1137-45.

MacDermott, A. B., Role, L. W. and Siegelbaum, S. A. (1999). Presynaptic ionotropic receptors and the control of transmitter release. *Annu. Rev. Neurosci. 22,* 443-85.

Mahanty, N. K. and Sah, P. (1998). Calcium-permeable AMPA receptors mediate long-term potentiation in interneurons in the amygdala. *Nature. 394, 6694,* 683-7.

Mahmoudi, M. and Zarrindast, M. R. (2002). Effect of intracerebroventricular injection of GABA receptor agents on morphine-induced antinociception in the formalin test. *J Psychopharmacol. 16, 1,* 85-91.

Maione, S., Marabese, I., Leyva, J., Palazzo, E., de Novellis, V. and Rossi, F. (1998). Characterisation of mGluRs which modulate nociception in the PAG of the mouse. *Neuropharmacology. 37, 12,* 1475-83.

Maione, S., Marabese, I., Oliva, P., de Novellis, V., Stella, L., Rossi, F., Filippelli, A. and Rossi, F. (1999). Periaqueductal gray matter glutamate and GABA decrease following subcutaneous formalin injection in rat. *Neuroreport. 10, 7,* 1403-7.

Maione, S., Oliva, P., Marabese, I., Palazzo, E., Rossi, F., Berrino, L. and Filippelli, A. (2000). Periaqueductal gray matter metabotropic glutamate receptors modulate formalin-induced nociception. *Pain. 85, 1-2,* 183-9.

Malherbe, P., Kratzeisen, C., Lundstrom, K., Richards, J. G., Faull, R. L. and Mutel, V. (1999). Cloning and functional expression of alternative spliced variants of the human metabotropic glutamate receptor 8. *Brain Res. Mol. Brain Res. 67, 2,* 201-10.

Manning, B. H. (1998). A lateralized deficit in morphine antinociception after unilateral inactivation of the central amygdala. *J. Neurosci. 18, 22,* 9453-70.

Marabese, I., de Novellis, V., Palazzo, E., Scafuro, M. A., Vita, D., Rossi, F. and Maione, S. (2007). Effects of (S)-3,4-DCPG, an mGlu8 receptor agonist, on inflammatory and neuropathic pain in mice. *Neuropharmacology. 52, 2,* 253-62.

Marowsky, A., Fritschy, J. M. and Vogt, K. E. (2004). Functional mapping of GABA A receptor subtypes in the amygdala. *Eur J Neurosci. 20, 5,* 1281-9.

Marshall, F. H., White, J., Main, M., Green, A. and Wise, A. (1999). GABA(B) receptors function as heterodimers. *Biochem. Soc. Trans. 27, 4,* 530-5.

Matsuda, K., Kamiya, Y., Matsuda, S. and Yuzaki, M. (2002). Cloning and characterization of a novel NMDA receptor subunit NR3B: a dominant subunit that reduces calcium permeability. *Brain Res. Mol. Brain Res. 100, 1-2,* 43-52.

Mayer, M. L. and Miller, R. J. (1990). Excitatory amino acid receptors, second messengers and regulation of intracellular Ca2+ in mammalian neurons. *Trends Pharmacol Sci. 11, 6,* 254-60.

McBain, C. J. and Mayer, M. L. (1994). N-methyl-D-aspartic acid receptor structure and function. *Physiol. Rev. 74, 3,* 723-60.

McCarson, K. E., Ralya, A., Reisman, S. A. and Enna, S. J. (2005). Amitriptyline prevents thermal hyperalgesia and modifications in rat spinal cord GABA(B) receptor expression and function in an animal model of neuropathic pain. *Biochem. Pharmacol. 71, 1-2,* 196-202.

Miguel, T. T. and Nunes-de-Souza, R. L. (2006). Defensive-like behaviors and antinociception induced by NMDA injection into the periaqueductal gray of mice depend on nitric oxide synthesis. *Brain Res. 1076, 1,* 42-8.

Millan, M. J. (1999). The induction of pain: an integrative review. *Prog. Neurobiol. 57, 1,* 1-164.

Millan, M. J. (2002). Descending control of pain. *Prog. Neurobiol. 66, 6,* 355-474.

Miyata, M., Kashiwadani, H., Fukaya, M., Hayashi, T., Wu, D., Suzuki, T., Watanabe, M. and Kawakami, Y. (2003). Role of thalamic phospholipase C[beta]4 mediated by metabotropic glutamate receptor type 1 in inflammatory pain. *J. Neurosci. 23, 22,* 8098-108.

Moechars, D., Weston, M. C., Leo, S., Callaerts-Vegh, Z., Goris, I., Daneels, G., Buist, A., Cik, M., van der Spek, P., Kass, S., Meert, T., D'Hooge, R., Rosenmund, C. and Hampson, R. M. (2006). Vesicular glutamate transporter VGLUT2 expression levels control quantal size and neuropathic pain. *J. Neurosci. 26, 46,* 12055-66.

Monhemius, R., Green, D. L., Roberts, M. H. and Azami, J. (2001). Periaqueductal grey mediated inhibition of responses to noxious stimulation is dynamically activated in a rat model of neuropathic pain. *Neurosci. Lett. 298, 1,* 70-4.

Moreau, J. L. and Fields, H. L., 1986. Evidence for GABA involvement in midbrain control of medullary neurons that modulate nociceptive transmission. *Brain Res. 397, 1,* 37-46.

Morgan, M. M. and Clayton, C. C. (2005). Defensive behaviors evoked from the ventrolateral periaqueductal gray of the rat: comparison of opioid and GABA disinhibition. *Behav. Brain Res. 164, 1,* 61-6.

Morgan, M. M., Clayton, C. C. and Boyer-Quick, J. S. (2005). Differential susceptibility of the PAG and RVM to tolerance to the antinociceptive effect of morphine in the rat. *Pain. 113, 1-2,* 91-8.

Morgan, M. M., Clayton, C. C. and Lane, D. A. (2003). Behavioral evidence linking opioid-sensitive GABAergic neurons in the ventrolateral periaqueductal gray to morphine tolerance. *Neuroscience. 118, 1,* 227-32.

Mott, D.D. and Lewis, D. V. (1994). The pharmacology and function of central GABAB receptors. *Int Rev Neurobiol. 36,* 97-223.

Nakanishi, S. and Masu, M. (1994). Molecular diversity and functions of glutamate receptors. *Annu. Rev. Biophys. Biomol. Struct. 23,* 319-48.

Nakazato, E., Kato, A. and Watanabe, S. (2005). Brain but not spinal NR2B receptor is responsible for the anti-allodynic effect of an NR2B subunit-selective antagonist CP-101,606 in a rat chronic constriction injury model. *Pharmacology. 73, 1,* 8-14.

Nason, M. W., Jr. and Mason, P. (2004). Modulation of sympathetic and somatomotor function by the ventromedial medulla. *J. Neurophysiol. 92, 1,* 510-22.

Neto, F. L., Schadrack, J., Platzer, S., Zieglgansberger, W., Tolle, T. R. and Castro-Lopes, J. M. (2001). Up-regulation of metabotropic glutamate receptor 3 mRNA expression in the cerebral cortex of monoarthritic rats. *J. Neurosci. Res. 63, 4,* 356-67.

Neugebauer, V., Li, W., Bird, G. C., Bhave, G. and Gereau, R. W. t. (2003). Synaptic plasticity in the amygdala in a model of arthritic pain: differential roles of metabotropic glutamate receptors 1 and 5. *J. Neurosci. 23, 1,* 52-63.

Nicol, B., Rowbotham, D. J. and Lambert, D. G. (1996). mu- and kappa-opioids inhibit K+ evoked glutamate release from rat cerebrocortical slices. *Neurosci. Lett. 218, 2,* 79-82.

Nishi, M., Hinds, H., Lu, H. P., Kawata, M. and Hayashi, Y. (2001). Motoneuron-specific expression of NR3B, a novel NMDA-type glutamate receptor subunit that works in a dominant-negative manner. *J. Neurosci. 21, 23,* RC185.

Nyiri, G., Freund, T. F. and Somogyi, P. (2001). Input-dependent synaptic targeting of alpha(2)-subunit-containing GABA(A) receptors in synapses of hippocampal pyramidal cells of the rat. *Eur. J. Neurosci. 13, 3,* 428-42.

Oliva, P., Berrino, L., de Novellis, V., Palazzo, E., Marabese, I., Siniscalco, D., Scafuro, M., Mariani, L., Rossi, F. and Maione, S. (2006). Role of periaqueductal grey prostaglandin receptors in formalin-induced hyperalgesia. *Eur. J. Pharmacol. 530, 1-2,* 40-7.

Palazzo, E., Marabese, I., de Novellis, V., Oliva, P., Rossi, F., Berrino, L., Rossi, F. and Maione, S. (2001). Metabotropic and NMDA glutamate receptors participate in the cannabinoid-induced antinociception. *Neuropharmacology. 40, 3,* 319-26.

Pare, D., Quirk, G. J. and Ledoux, J. E. (2004). New vistas on amygdala networks in conditioned fear. *J. Neurophysiol. 92, 1,* 1-9.

Paronis, C. A., Cox, E. D., Cook, J. M. and Bergman, J. (2001). Different types of GABA(A) receptors may mediate the anticonflict and response rate-decreasing effects of zaleplon, zolpidem, and midazolam in squirrel monkeys. *Psychopharmacology (Berl). 156, 4,* 461-8.

Paulsen, O. and Moser, E. I. (1998). A model of hippocampal memory encoding and retrieval: GABAergic control of synaptic plasticity. *Trends Neurosci. 21, 7,* 273-8.

Pawelzik, H., Hughes, D.I. and Thomson, A. M. (2002). Physiological and morphological diversity of immunocytochemically defined parvalbumin- and cholecystokinin-positive interneurones in CA1 of the adult rat hippocampus. *J. Comp. Neurol. 443, 4,* 346-67.

Pedersen, L. H., Scheel-Kruger, J. and Blackburn-Munro, G. (2007). Amygdala GABA-A receptor involvement in mediating sensory-discriminative and affective-motivational pain responses in a rat model of peripheral nerve injury. *Pain. 127, 1-2,* 17-26.

Pin, J. P. and Duvoisin, R. (1995). The metabotropic glutamate receptors: structure and functions. *Neuropharmacology. 34, 1,* 1-26.

Pinheiro, P. and Mulle, C. (2006). Kainate receptors. *Cell Tissue Res. 326, 2,* 457-82.

Pinheiro, P. S., Perrais, D., Coussen, F., Barhanin, J., Bettler, B., Mann, J. R., Malva, J. O., Heinemann, S. F. and Mulle, C. (2007). GluR7 is an essential subunit of presynaptic kainate autoreceptors at hippocampal mossy fiber synapses. *Proc. Natl. Acad. Sci. USA. 104, 29,* 12181-6.

Poncer, J. C. and Miles, R. (1995). Fast and slow excitation of inhibitory cells in the CA3 region of the hippocampus. *J. Neurobiol. 26, 3,* 386-95.

Potes, C. S., Neto, F. L. and Castro-Lopes, J. M. (2006). Inhibition of pain behavior by GABA(B) receptors in the thalamic ventrobasal complex: effect on normal rats subjected to the formalin test of nociception. *Brain Res. 1115, 1,* 37-47.

Prosser, H. M., Gill, C. H., Hirst, W. D., Grau, E., Robbins, M., Calver, A., Soffin, E. M., Farmer, C. E., Lanneau, C., Gray, J., Schenck, E., Warmerdam, B. S., Clapham, C., Reavill, C., Rogers, D. C., Stean, T., Upton, N., Humphreys, K., Randall, A., Geppert, M., Davies, C. H. and Pangalos, M. N. (2001). Epileptogenesis and enhanced prepulse inhibition in GABA(B1)-deficient mice. *Mol. Cell Neurosci. 17, 6,* 1059-70.

Qu, C. L., Tang, J. S. and Jia, H. (2006). Involvement of GABAergic modulation of antinociception induced by morphine microinjected into the ventrolateral orbital cortex. *Brain Res. 1073-1074,* 281-9.

Quintero, G. C., Erzurumlu, R. S. and Vaccarino, A. L. (2007). Decreased pain response in mice following cortex-specific knockout of the N-methyl-D-aspartate NR1 subunit. *Neurosci. Lett. 425, 2,* 89-93.

Rea, K., Roche, M. and Finn, D. P. (2007). Supraspinal modulation of pain by cannabinoids: the role of GABA and glutamate. *Br. J. Pharmacol. 152, 5,* 633-48.

Renno, W. M. (1998). Microdialysis of excitatory amino acids in the periaqueductal gray of the rat after unilateral peripheral inflammation. *Amino Acids. 14, 4,* 319-31.

Renno, W. M. and Beitz, A. J. (1999). Peripheral inflammation is associated with decreased veratridine-induced release of GABA in the rat ventrocaudal periaqueductal gray: microdialysis study. *J. Neurol. Sci. 163, 2,* 105-10.

Roberts, W. A., Eaton, S. A. and Salt, T. E. (1992). Widely distributed GABA-mediated afferent inhibition processes within the ventrobasal thalamus of rat and their possible relevance to pathological pain states and somatotopic plasticity. *Exp. Brain Res. 89, 2,* 363-72.

Rudolph, U. and Mohler, H. (2006). GABA-based therapeutic approaches: GABAA receptor subtype functions. *Curr. Opin Pharmacol. 6, 1,* 18-23.

Rusakov, D. A., Saitow, F., Lehre, K. P. and Konishi, S. (2005). Modulation of presynaptic Ca2+ entry by AMPA receptors at individual GABAergic synapses in the cerebellum. *J. Neurosci. 25, 20,* 4930-40.

Salt, T. E. and Binns, K. E. (2000). Contributions of mGlu1 and mGlu5 receptors to interactions with N-methyl-D-aspartate receptor-mediated responses and nociceptive sensory responses of rat thalamic neurons. *Neuroscience. 100, 2,* 375-80.

Salt, T. E. and Eaton, S. A. (1994). The function of metabotropic excitatory amino acid receptors in synaptic transmission in the thalamus: studies with novel phenylglycine antagonists. *Neurochem. Int. 24, 5,* 451-8.

Sandkuhler, J., Willmann, E. and Fu, Q. G., 1989. Blockade of GABAA receptors in the midbrain periaqueductal gray abolishes nociceptive spinal dorsal horn neuronal activity. *Eur. J. Pharmacol. 160, 1,* 163-6.

Sartorius, L. J., Nagappan, G., Lipska, B. K., Lu, B., Sei, Y., Ren-Patterson, R., Li, Z., Weinberger, D. R. and Harrison, P. J. (2006). Alternative splicing of human metabotropic glutamate receptor 3. *J. Neurochem. 96, 4,* 1139-48.

Sasaki, Y. F., Rothe, T., Premkumar, L. S., Das, S., Cui, J., Talantova, M. V., Wong, H. K., Gong, X., Chan, S. F., Zhang, D., Nakanishi, N., Sucher, N. J. and Lipton, S. A. (2002). Characterization and comparison of the NR3A subunit of the NMDA receptor in recombinant systems and primary cortical neurons. *J. Neurophysiol. 87, 4,* 2052-63.

Satake, S., Saitow, F., Rusakov, D. and Konishi, S. (2004). AMPA receptor-mediated presynaptic inhibition at cerebellar GABAergic synapses: a characterization of molecular mechanisms. *Eur. J. Neurosci. 19, 9,* 2464-74.

Satake, S., Saitow, F., Yamada, J. and Konishi, S. (2000). Synaptic activation of AMPA receptors inhibits GABA release from cerebellar interneurons. *Nat. Neurosci. 3, 6,* 551-8.

Satake, S., Song, S. Y., Cao, Q., Satoh, H., Rusakov, D. A., Yanagawa, Y., Ling, E. A., Imoto, K. and Konishi, S. (2006). Characterization of AMPA receptors targeted by the climbing fiber transmitter mediating presynaptic inhibition of GABAergic transmission at cerebellar interneuron-Purkinje cell synapses. *J. Neurosci. 26, 8,* 2278-89.

Sbrenna, S., Marti, M., Morari, M., Calo, G., Guerrini, R., Beani, L. and Bianchi, C. (1999). L-glutamate and gamma-aminobutyric acid efflux from rat cerebrocortical synaptosomes: modulation by kappa- and mu- but not delta- and opioid receptor like-1 receptors. *J Pharmacol. Exp. Ther. 291, 3,* 1365-71.

Schuler, V., Luscher, C., Blanchet, C., Klix, N., Sansig, G., Klebs, K., Schmutz, M., Heid, J., Gentry, C., Urban, L., Fox, A., Spooren, W., Jaton, A. L., Vigouret, J., Pozza, M., Kelly, P. H., Mosbacher, J., Froestl, W., Kaslin, E., Korn, R., Bischoff, S., Kaupmann, K., van der Putten, H. and Bettler, B. (2001). Epilepsy, hyperalgesia, impaired memory, and loss of pre- and postsynaptic GABA(B) responses in mice lacking GABA(B(1)). *Neuron. 31, 1,* 47-58.

Schulz, H. L., Stohr, H. and Weber, B. H. (2002). Characterization of three novel isoforms of the metabotrobic glutamate receptor 7 (GRM7). *Neurosci. Lett. 326, 1,* 37-40.

Schwarz, D. A., Barry, G., Eliasof, S. D., Petroski, R. E., Conlon, P. J. and Maki, R. A. (2000). Characterization of gamma-aminobutyric acid receptor GABAB(1e), a GABAB(1) splice variant encoding a truncated receptor. *J. Biol. Chem. 275, 41,* 32174-81.

Shafizadeh, M., Semnanian, S., Zarrindast, M. R. and Hashemi, B. (1997). Involvement of GABAB receptors in the antinociception induced by baclofen in the formalin test. *Gen. Pharmacol. 28, 4,* 611-5.

Shin, M. C., Jang, M. H., Chang, H. K., Kim, Y. J., Kim, E. H. and Kim, C. J. (2003). Modulation of cyclooxygenase-2 on glycine- and glutamate-induced ion currents in rat periaqueductal gray neurons. *Brain Res. Bull. 59, 4,* 251-6.

Sieghart, W., Fuchs, K., Tretter, V., Ebert, V., Jechlinger, M., Hoger, H. and Adamiker, D. (1999). Structure and subunit composition of GABA(A) receptors. *Neurochem. Int.* 34, 5, 379-85.

Silva, E. and Hernandez, L. (2007). [Extracellular aminoacids in the amygdala and nucleus accumbens in the rat during acute pain]. *Invest. Clin. 48, 2,* 213-24.

Silva, E., Hernandez, L., Contreras, Q., Guerrero, F. and Alba, G. (2000). Noxious stimulation increases glutamate and arginine in the periaqueductal gray matter in rats: a microdialysis study. *Pain. 87, 2,* 131-5.

Silva, E., Hernandez, L., Quinonez, B., Gonzalez, L. E. and Colasante, C. (2004). Selective amino acids changes in the medial and lateral preoptic area in the formalin test in rats. *Neuroscience. 124, 2,* 395-404.

Silva, E., Quinones, B., Freund, N., Gonzalez, L. E. and Hernandez, L. (2001). Extracellular glutamate, aspartate and arginine increase in the ventral posterolateral thalamic nucleus during nociceptive stimulation. *Brain Res. 923, 1-2,* 45-9.

Spinella, M., Cooper, M. L. and Bodnar, R. J. (1996). Excitatory amino acid antagonists in the rostral ventromedial medulla inhibit mesencephalic morphine analgesia in rats. *Pain. 64, 3,* 545-52.

Squires, R. F., Klepner, C. A. and Benson, D.I., 1980. Multiple benzodiazepine receptor complexes: some benzodiazepine recognition sites are coupled to GABA receptors and ionophores. *Adv. Biochem. Psychopharmacol. 21,* 285-93.

Stefani, A., Pisani, A., Mercuri, N. B., Bernardi, G. and Calabresi, P. (1994). Activation of metabotropic glutamate receptors inhibits calcium currents and GABA-mediated synaptic potentials in striatal neurons. *J. Neurosci. 14, 11 Pt 1,* 6734-43.

Steiger, J. L., Bandyopadhyay, S., Farb, D. H. and Russek, S. J. (2004). cAMP response element-binding protein, activating transcription factor-4, and upstream stimulatory factor differentially control hippocampal GABABR1a and GABABR1b subunit gene expression through alternative promoters. *J. Neurosci. 24, 27,* 6115-26.

Stiller, C. O., Linderoth, B., O'Connor, W. T., Franck, J., Falkenberg, T., Ungerstedt, U. and Brodin, E. (1995). Repeated spinal cord stimulation decreases the extracellular level of gamma-aminobutyric acid in the periaqueductal gray matter of freely moving rats. *Brain Res. 699, 2,* 231-41.

Sugita, S. and North, R. A. (1993). Opioid actions on neurons of rat lateral amygdala in vitro. *Brain Res. 612, 1-2,* 151-5.

Suh, H. W., Song, D. K., Choi, Y. S. and Kim, Y. H. (1995). Multiplicative interaction between intrathecally and intracerebroventricularly administered morphine for antinociception in the mouse: involvement of supraspinal NMDA but not non-NMDA receptors. *Life Sci. 56, 8,* PL181-5.

Suzuki, T., Aoki, T., Ohnishi, O., Nagase, H. and Narita, M. (2000). Different effects of NMDA/group I metabotropic glutamate receptor agents in delta- and mu-opioid receptor agonist-induced supraspinal antinociception. *Eur. J. Pharmacol. 396, 1,* 23-8.

Tai, Y. H., Wang, Y. H., Tsai, R. Y., Wang, J. J., Tao, P. L., Liu, T. M., Wang, Y. C. and Wong, C. S. (2007). Amitriptyline preserves morphine's antinociceptive effect by regulating the glutamate transporter GLAST and GLT-1 trafficking and excitatory amino acids concentration in morphine-tolerant rats. *Pain. 129, 3,* 343-54.

Tanaka, E. and North, R. A. (1994). Opioid actions on rat anterior cingulate cortex neurons in vitro. *J. Neurosci. 14, 3 Pt 1,* 1106-13.

Thomas, D. A., McGowan, M. K. and Hammond, D. L. (1995). Microinjection of baclofen in the ventromedial medulla of rats: antinociception at low doses and hyperalgesia at high doses. *J. Pharmacol. Exp. Ther. 275, 1,* 274-84.

Thomas, D. A., Navarrete, I. M., Graham, B. A., McGowan, M. K. and Hammond, D. L. (1996). Antinociception produced by systemic R(+)-baclofen hydrochloride is attenuated by CGP 35348 administered to the spinal cord or ventromedial medulla of rats. *Brain Res. 718, 1-2,* 129-37.

Tiao, J. Y., Bradaia, A., Biermann, B., Kaupmann, K., Metz, M., Haller, C., Rolink, A. G., Pless, E., Barlow, P. N., Gassmann, M. and Bettler, B. (2008). The sushi domains of secreted GABA(B1) isoforms selectively impair GABA(B) heteroreceptor function. *J. Biol. Chem. 283, 45,* 31005-11.

Tortorici, V., Vasquez, E. and Vanegas, H. (1996). Naloxone partial reversal of the antinociception produced by dipyrone microinjected into the periaqueductal gray of rats. Possible involvement of medullary off- and on-cells. *Brain Res. 725, 1,* 106-10.

Traynelis, S. F., Burgess, M. F., Zheng, F., Lyuboslavsky, P. and Powers, J. L. (1998). Control of voltage-independent zinc inhibition of NMDA receptors by the NR1 subunit. *J. Neurosci. 18, 16,* 6163-75.

Urca, G., Nahin, R. L. and Liebeskind, J. C., 1980. Glutamate-induced analgesia: blockade and potentiation by naloxone. *Brain Res. 192, 2,* 523-30.

Valerio, A., Ferraboli, S., Paterlini, M., Spano, P. and Barlati, S. (2001). Identification of novel alternatively-spliced mRNA isoforms of metabotropic glutamate receptor 6 gene in rat and human retina. *Gene. 262, 1-2,* 99-106.

van Praag, H. and Frenk, H. (1990). The role of glutamate in opiate descending inhibition of nociceptive spinal reflexes. *Brain Res. 524, 1,* 101-5.

Vasquez, E. and Vanegas, H. (2000). The antinociceptive effect of PAG-microinjected dipyrone in rats is mediated by endogenous opioids of the rostral ventromedical medulla. *Brain Res. 854, 1-2,* 249-52.

Vaughan, C. W. (1998). Enhancement of opioid inhibition of GABAergic synaptic transmission by cyclo-oxygenase inhibitors in rat periaqueductal grey neurones. *Br. J. Pharmacol. 123, 8,* 1479-81.

Vaughan, C. W. and Christie, M. J. (1997). Presynaptic inhibitory action of opioids on synaptic transmission in the rat periaqueductal grey in vitro. *J. Physiol. 498 (Pt 2),* 463-72.

Vaughan, C. W., Ingram, S. L., Connor, M. A. and Christie, M. J. (1997). How opioids inhibit GABA-mediated neurotransmission. *Nature. 390, 6660,* 611-4.

Vazquez, E., Escobar, W., Ramirez, K. and Vanegas, H. (2007). A nonopioid analgesic acts upon the PAG-RVM axis to reverse inflammatory hyperalgesia. *Eur. J. Neurosci. 25, 2,* 471-9.

Wagner, E. J., Bosch, M. A., Kelly, M. J. and Ronnekleiv, K. (1998). A powerful GABA(B) receptor-mediated inhibition of GABAergic neurons in the arcuate nucleus. *Neuroreport. 9, 18,* 4171-7.

Waldmeier, P. C., Kaupmann, K. and Urwyler, S. (2008). Roles of GABAB receptor subtypes in presynaptic auto- and heteroreceptor function regulating GABA and glutamate release. *J. Neural. Transm. 115, 10,* 1401-11.

Wei, F., Wang, G. D., Kerchner, G. A., Kim, S. J., Xu, H. M., Chen, Z. F. and Zhuo, M. (2001)a. Genetic enhancement of inflammatory pain by forebrain NR2B overexpression. *Nat. Neurosci. 4, 2,* 164-9.

Wei, K., Eubanks, J. H., Francis, J., Jia, Z. and Snead, O. C., 3rd, (2001)b. Cloning and tissue distribution of a novel isoform of the rat GABA(B)R1 receptor subunit. *Neuroreport. 12, 4,* 833-7.

Whiting, P. J., Bonnert, T. P., McKernan, R. M., Farrar, S., Le Bourdelles, B., Heavens, R. P., Smith, D. W., Hewson, L., Rigby, M. R., Sirinathsinghji, D. J., Thompson, S. A. and Wafford, K. A. (1999). Molecular and functional diversity of the expanding GABA-A receptor gene family. *Ann.NY Acad Sci.868,* 645-53.

Wilding, T. J. and Huettner, J. E. (2001). Functional diversity and developmental changes in rat neuronal kainate receptors. *J. Physiol. 532, Pt 2,* 411-21.

Williams, J. T., Christie, M. J. and Manzoni, O. (2001). Cellular and synaptic adaptations mediating opioid dependence. *Physiol. Rev. 81, 1,* 299-343.

Williams, K. (1993). Ifenprodil discriminates subtypes of the N-methyl-D-aspartate receptor: selectivity and mechanisms at recombinant heteromeric receptors. *Mol. Pharmacol. 44, 4,* 851-9.

Winkler, C. W., Hermes, S. M., Chavkin, C. I., Drake, C. T., Morrison, S. F. and Aicher, S. A. (2006). Kappa opioid receptor (KOR) and GAD67 immunoreactivity are found in OFF and NEUTRAL cells in the rostral ventromedial medulla. *J. Neurophysiol. 96, 6,* 3465-73.

Wollmuth, L. P. and Sobolevski, A. I. (2004). Structure and gating of the glutamate receptor ion channel. *Trends Neurosci. 27(6),* 321-328.

Xiao, D. Q., Zhu, J. X., Tang, J. S. and Jia, H. (2005). GABAergic modulation mediates antinociception produced by serotonin applied into thalamic nucleus submedius of the rat. *Brain Res. 1057, 1-2,* 161-7.

Yokoro, C. M., Pesquero, S. M., Turchetti-Maia, R. M., Francischi, J. N. and Tatsuo, M. A. (2001). Acute phenobarbital administration induces hyperalgesia: pharmacological evidence for the involvement of supraspinal GABA-A receptors. *Braz. J. Med Biol. Res. 34, 3,* 397-405.

Youn, D. H., Royle, G., Kolaj, M., Vissel, B. and Randic, M. (2008). Enhanced LTP of primary afferent neurotransmission in AMPA receptor GluR2-deficient mice. *Pain. 136, 1-2,* 158-67.

Zambotti, F., Zonta, N., Tammiso, R., Conci, F., Hafner, B., Zecca, L., Ferrario, P. and Mantegazza, P. (1991). Effects of diazepam on nociception in rats. *Naunyn Schmiedebergs Arch. Pharmacol. 344, 1,* 84-9.

Zarrindast, M., Valizadeh, S. and Sahebgharani, M. (2000). GABA(B) receptor mechanism and imipramine-induced antinociception in ligated and non-ligated mice. *Eur. J. Pharmacol. 407, 1-2,* 65-72.

Zhang, S., Tang, J. S., Yuan, B. and Jia, H. (1997). Involvement of the frontal ventrolateral orbital cortex in descending inhibition of nociception mediated by the periaqueductal gray in rats. *Neurosci. Lett. 224, 2,* 142-6.

Zhu, H., Rockhold, R. W. and Ho, I. K. (1998). The role of glutamate in physical dependence on opioids. *Jpn. J. Pharmacol. 76, 1,* 1-14.

Zhu, H., Ryan, K. and Chen, S. (1999). Cloning of novel splice variants of mouse mGluR1. *Brain Res. Mol. Brain Res. 73, 1-2,* 93-103.

Zhuo, M. (2002). Glutamate receptors and persistent pain: targeting forebrain NR2B subunits. *Drug Discov. Today. 7, 4,* 259-67.

Zhuo, M. and Gebhart, G. F. (1997). Biphasic modulation of spinal nociceptive transmission from the medullary raphe nuclei in the rat. *J. Neurophysiol. 78, 2,* 746-58.

In: Biological Aspects of Human Health and Well-Being   ISBN: 978-1-61209-134-1
Editor: Tsisana Shartava   © 2011 Nova Science Publishers, Inc.

Chapter XII

MULTIPHOTON MICROSCOPY OF INTRAVITAL DEEP OCULAR TISSUES

Bao-Gui Wang[*] and Karl-Jürgen Halbhuber[†]

Lasermicroscopy Research Unit, Institute of Microscopic Anatomy (Anatomy II),
Medical Faculty, Friedrich-SchillerUniversityJena, Germany

ABSTRACT

Currently, femtosecond lasers (femtolasers) are being extensively employed in diverse research and application fields. Femtolasers-mediated multiphoton excitation laser scanning microscopy is one of the most exciting recent developments in biomedical imaging and becomes more and more an inspiring imaging technique in the intact bulk tissue examinations. In this review, this non-linear excitation imaging technique including two-photon autofluorescence (2PF) and second harmonic generated signal imaging (SHG) was employed to investigate the microstructures of whole-mount corneal, retinal, and scleral tissues in their native environment. Image acquisition was based on intense ultrafast femtosecond near-infrared (NIR) laser pulses, which were emitted from a mode-locked solid-state Ti: sapphire system. By integrating high-numerical aperture diffraction-limited objectives, multiphoton microscopy/tomography of ocular tissues was performed at a high light irradiance order of MW-GW/cm^2, where two or more photons were simultaneously absorbed by endogenous molecules located in the thick tissues. As a result, the cellular and fibrous components of intact scleral and corneal tissues were selectively displayed by in-tandem detection with 2PF and SHG without the assistance of any exogenous dye. High-resolution optical images of keratocytes in cornea, fibroblasts, mature elastic fibers and blood capillaries in sclerae as well as of the retina radial Müller glial cells, ganglion cells, bipolar cells, photoreceptors, and retina pigment epithelial (RPE) cells were acquired. Furthermore, this promising technique has been proved to be an indispensable tool in assisting femtolasers intratissue surgery, especially for *in situ* assessing the obtained microsurgical effects. Most remarkably, the activated keratocytes, also named myofibroblasts during wound repair, were *in vivo* detected using the

---

[*] Address: Institute of Anatomy, Teichgraben 7, 07743 Jena, Germany. Phone:++49 3641 938550; fax: ++49 3641 938552. Email: Baogui.Wang@mti.uni-jena.de

[†] Email: Karl.Halbhuber@mti.uni-jena.de

multiphoton excitation imaging in the treated animals twenty-four hours after the intrastromal surgery. Data show that the in-tandem combination of 2PF and SHG allows for *in situ* co-localization imaging of various microstructural components in the whole-mount ocular tissues. Qualitative and quantitative assessment of microstructures was obtained. The selective displaying merits of tissue components only with the excitation of different wavelengths is the most exciting development for bulk tissue imaging, which allows to selectively studying of three-dimensional (3-D) architecture of cellular microstructures and extracellular matrix arrangement at a substantial depth. Using the laser power within the threshold value, the bulk tissues can be imaged numerous times without visible photodisruption. Intrinsic emission multiphoton microscopy/tomography is consequently confirmed to be an efficient and sensitive non-invasive imaging approach, featured with high contrast and subcellular spatial resolution. The non-linear optical imaging yields vivid insights into biological specimens that may ultimately find its clinical application in optical pathological diagnostics. We believe that this promising technique will also find more applications in the biological and medical basic research in the near future.

Keywords: 2PF, SHG, autofluorescence, tomography, optical, non-linear, near-infrared (NIR), cornea, sclera, retina, epithelium, Bowman's layer, keratocytes, endothelium, fibroblasts, capillary, elastic fibers, collagen, Müller cells, bipolar cells, RPE, laser surgery, myofibroblasts

1. INTRODUCTION

The advent of femtolaser sources led to the experimental implementation of multiphoton excitation microscopy, providing a non-invasive approach to examine the structures and functions of living cells and tissues. This optical imaging technique, based on simultaneous absorption of two or more photons under high light irradiance, is one of the most exciting recent technical developments in biomedical imaging [Denk *et al.* 1990, So *et al.* 1998, Vogel *et al.* 1999, König 2000a, Campagnola *et al.* 2002, Zoumi *et al.* 2002, Halbhuber *et al.* 2003, Zipfel *et al.* 2003a, Williams *et al.* 2005, Han *et al.* 2006]. This image acquisition is based on non-linear optical response to endogenous and exogenous fluorophores arising from their asymmetric distribution of the electronic charges in the molecular ground and excited state.

In this review, we briefly describe the physical principles and history of the multiphoton microscopy as well as its imaging advantages in comparison to single-photon confocal laser scanning microscopy (CLSM). The multiphoton excitation imaging of vital tissues is here focused. The microstructures of whole-mount corneal, retinal, and scleral tissues are explicitly elucidated in their native environment. Considerable attentions to delineating the cellular and fibrous components with subcellular spatial resolution, especially to quantitative and qualitative acquisition of microstructures in different layers of the intact bulk tissues were paid. Further on, the merits of this promising imaging technique in assisting femtolaser intratissue surgery and in detecting the activated keratocytes during wound repair are reviewed.

The feasibility of simultaneous absorption of two photons was predicted by Goeppert–Mayer in her dissertation in 1931. Multiphoton microscopy, including two-photon fluorescence (2PF) and second harmonic generation (SHG), only occurs when the irradiance reaches an order of MW-GW/cm^2 [Göppert, 1931]. This technique met with the interests of biomedical scientists after the introduction of convenient lasers. The first two-photon excitation of electronic states was found for blue fluorescence of CaF_2:Eu^{2+} crystals based on the availability of lasers in 1961 [Kaiser et al. 1961]. To date, tunable mode-locked Ti:sapphire lasers operating at high repetition rate (typically 80–90 MHz) and a ultrashort pulse duration (less than 200 fs) are typically used as NIR sources in multiphoton microscopy [Zipfel et al. 2003b, König et al. 2003, Rubart 2004, Wang et al. 2006a]. The required transient high irradiance can be obtained by tightly focusing ultrashort laser pulses within a subfemtoliter focal volume provided by a special objective with high numerical aperture.

Endogenous fluorophores such as NAD(P)H, flavin, lipofuscin, and porphyrins in biological tissues were revealed. In single-photon confocal laser scanning microscopy (CLSM), most of the fluorophores can be stimulated by UV radiation to emit longer wavelengths with majority of the spectral intensity between 400 and 600 nm. Multiphoton microscopy is well suited for the high-resolution imaging of intrinsic molecular signals in living specimens. The NAD(P)H in mitochondria is the predominant source of endogenous fluorophore in cells. This intrinsic fluorophore can be evoked at wavelengths of 720–780 nm in a two-photon absorption process [Zipfel et al. 2003b]. In addition, the lipofuscin found in retina pigment epithelial (RPE) cells [Han et al. 2006] and the elastin located in bulk tissues [König et al. 2005a] can also be evoked under two-photon excitation condition.

However, multiphoton microscopy requires enormously high light intensities that, if continuous, would almost instantly vaporize the specimen. This problem has been circumvented by using such lasers that produce amazingly brief pulses (fs, $\sim 10^{-15}$ s) at a high repetition rate, thus generating a very high instantaneous energy but low average energy. In a two-photon excitation process, the rate of excitation is proportional to the average squared photon density; the fluorophore must simultaneously absorb two photons per excitation process. Only in the focused volume is there sufficient intensity to generate appreciable excitation [König et al. 2000b, Williams et al. 2001, Halbhuber et al. 2003, Vogel et al. 2003, Helmchen et al. 2005]. As a result, the multiphoton imaging is spatially confined to the minute sub-femtoliter focal volume resulting from the focusing of special microscope objectives; consequently is thereby no essential out-of- focus photobleaching.

This multiphoton imaging technique, in conjunction with the pulsed infrared light, provides other distinct advantages over single-photon CLSM: (i) No pinhole aperture is required for multiphoton microscopy. Under CLSM microscope, not only the in-focus and also the out-of-focus molecules can be excited [Halbhuber et al. 1998]. The emitted fluorescent signals of the both areas are collected. The pinhole aperture, an indispensable element of CLSM, functions to rejecting the out-of-focus fluorescence signals. The development of multiphoton excitation greatly minimized this problem because only the in-focus molecules can be stimulated under two-photon excitation condition. Meanwhile, the focused volume can be further adjusted through the depth of the sample and thus to achieve non-linear optical sectioning in three-dimensions; (ii) Samples can be examined with subcellular resolution in ex vivo even in vivo states [König et al. 1996, Halbhuber et al. 2003, Wang et al. 2007a]. Conventional fluorescence microscopy yields poor images, because in-focus structures are washed out by the fluorescence signals outside plane of focus. In general, most of the biological one-photon based microscopic techniques require labelling with conventional fluorophores or fluorescent proteins. Thus, the *in situ* microstructural basis for intact tissue properties without any histological assistance is difficult to be revealed under single-photon CLSM. By contrast, the endogenous molecules in deep bulk tissues can be probed under multiphoton microscope. Therefore, this new technology provides an encouraging examination approach for living cells and tissues; (iii) In addition, owing to the use of short (high-energy) excitation wavelengths, photodamage and photobleaching unavoidably takes place in CLSM; the use of UV light also limited imaging depth penetrating into biological tissues due to its high absorption and scattering coefficients of the incident radiation. By contrast, multiphoton microscopy in conjunction with NIR light allows for reduced photodamage and deeper sensing depth.

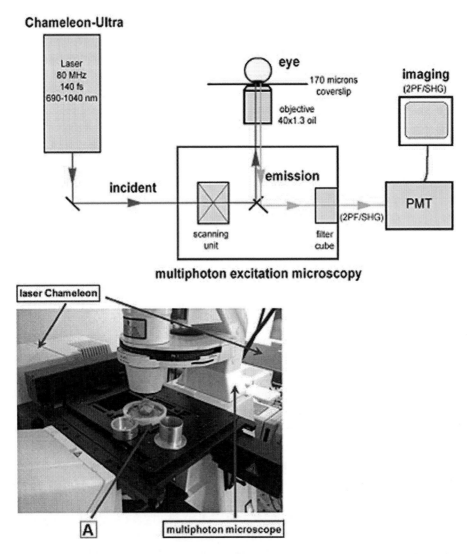

Figure 1. *Ex vivo* 2PF/SHG signal imaging tomography in diverse ocular tissues. The schematic (upper figure) shows that the multiphoton imaging system included an inverted confocal microscope and the laser Chameleon. The optical apparatus deflecting laser beams is not shown in this drawing. The photogram (lower figure) taken from upper-side focused on the sample alignment stage depicting on-site laser microscopy of a porcine eye globe. (A). A special experimental sample holder including a tissue chamber, a 170-micron-thick glass, and two parts of metal covers which fulfilled an essential function to keeping the eyeball moist during image acquisitions (Reprint Lasers Surg Med 2008).

These advantages of multiphoton microscopy, along with its 3D demonstration, has been employed as an encouraging imaging method in studying of living cells [König et al. 1996, Piston 1999], mapping ligand-gated ion channel distributions [Denk 1994], organelles and DNA imaging [Malak et al. 1997, König et al. 2000c, Dedov et al. 2001], imaging mammalian embryos [Squirrell et al. 1999], skin probing [Masters et al. 1997, So et al. 1998, König et al. 2003], arterial imaging [Zoumi et al. 2004], detecting of microtubules in neuronal cell [Dombeck et al. 2003] and deep brain tissue [Levene et al. 2004, Brecht et al. 2004, Nitsch et al. 2004]. Cellular interactions were as well revealed using this non-linear image acquisition technique [Stoll et al. 2002]. Membrane imaging and visualizing of biomolecular array were reported [Moreaux et al. 2000, Campagnola et al. 2003]. Fluorescence spectroscopy of NAD(P)H and flavoprotein under two-photon absorption were furthermore addressed [Huang et al. 2002]. The benefits of this emerging optical technique indicate that it may provide significant progress in the assessment of cellular and extracellular components in ocular tissues.

2. MULTIPHOTON IMAGING SYSTEM

Multiphoton microscopy of ocular tissues was performed with an inverse laser scanning microscope (Carl Zeiss, Jena, Germany), to whicha tunable compact mode-locked Ti:sapphire laser *Chameleon*(Coherent Inc., Santa Clara, CA, USA) was coupled (Figure 1).The laser pulses were featured by 80 MHz to 90 MHz repetition rate, 690 nm to 1040 nm wavelength spectrum, and 140 fs pulse duration (Figures 1 and 3).

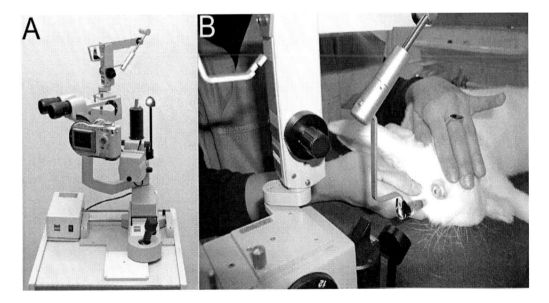

Figure 2. Examination of eye globes before multiphoton excitation imaging using a Zeiss slit lamp. (A). Overview of the all-rounder Zeiss slit lamp 120 equipped with a digital camera; (B). On-site photogram of corneal examination pre-treatment. Only the clear corneas were forwarded to take further examinations.

A diffraction-limited 40x objective (1.3 N.A., oil, Zeiss Plan Neofluar) with a work distance penetrating into the samples up to 140 µm and a 20x objective (0.95 N.A., water) with a work distance of more than 1000 µm in bulk tissue were used. A 650 nm beam splitter (Chroma Technology, BrattleboroVT, USA) in the backward direction was indispensable for separating the emitted light from the reflected incident light. Another band-pass filter 435-485 nm for 2PF imaging was additionally built up before a photomultiplier tube (PMT). For SHG signal imaging, additional narrower band-pass 420 ± 10 nm was used. PMT was an essential component in the image acquisitions. The pixel-wise detecting photosensitive device, rapidly responding with high sensitivity and efficiency, was externally installed under the baseport of the multiphoton microscope to modulate and magnify the emission fluorescence and SHG signals.

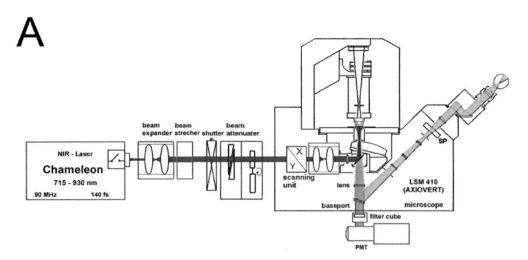

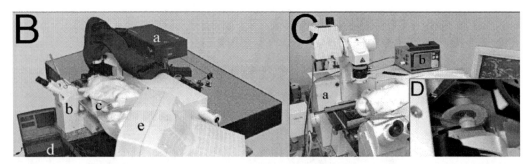

Figure 3. Corneal multiphoton microscopy with living animals. (A). As illustrated on the schematic drawing, the ultrashort laser pulses were deflected into the multiphoton microscope and then conveyed into samples through a scanning unit built up in the microscopic system. The emitted lights including autofluorescence and second harmonic generated signals were separated from the incident lights by a filter cube and then guided into PMT in a backward direction; (B). On-site photogram for overview of multiphoton imaging system. The animal was being imaged under the inverse multiphoton microscope. The optical mirrors deflecting the laser beams are illustrated in a simplified way on this photogram. a) compact mode-locked tunable 90 MHz Ti: sapphire laser Chameleon; b) multiphoton microscope based on the inverse Zeiss LSM; c) animal during imaging; d) wavelength tuner; e) imaging monitor; (C). Frontal photogram showing a rabbit being imaged under the multiphoton microscope. The animal was narcotized by intravenous narcotics ketamine and xylacine. The blood oxygen concentration and pulse frequency of the being treated animal was totally monitored from begin to end during the non-linear optical imaging by a Draeger Dialogue 2000. Insert (D). Image taken from underside of the sample stage showing the animal eye being treated.

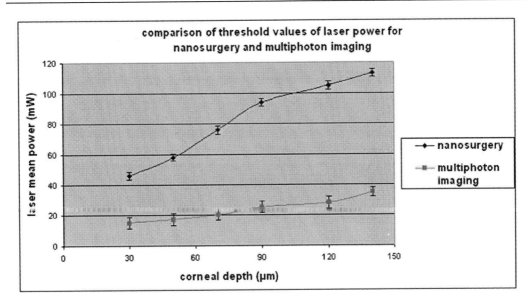

Figure 4. Diagram illustrating the discrepancy of optimized laser power for multiphoton excitation imaging and of laser threshold power for intrastromal nanosurgery in different depths of cornea with living animals. The laser powers set for multiphoton imaging were between 20 mW and 35 mW in dependence on the imaging depth. If the laser intensity goes beyond threshold value, photodisruptive effects could occur. Photodisruptive effects mean nanosurgical effects in this paper. For instance, at a depth of 90 micrometers in cornea, the optimized laser power for multiphoton imaging was 26 mW and the threshold value for photodisruption was 98 mW (Reprint Cell Tissue Res 2007).

The emission radiances of autofluorescence and SHG were measured with the assistance of another highly sensitive polychromatic META-detector and of a special lambda-scanning program installed in this microscopic system (Carl Zeiss, Jena, Germany). The emission radiances were separated into various wavelengths and then the entire spectrum was projected onto 32 channels of the META detector. LSM software can display the images acquired from the META detector channels as an image series from shortest to longest wavelengths. The lambda-scanning fulfilled a function to determining the emission wavelength optimum of a fluorophore.

Image acquisition of cellular autofluorescence and collagen SHG was performed at different excitation wavelengths respectively. The wavelengths are mentioned in the following sessions with the corresponding images. Owing to the high NIR penetration and the minute focus volume (0.1 fl), serial optical transversal sections with low phototoxicity to the surrounding region were obtained in certain fine z-steps. The high numerical aperture objectives used allowed the emission lights to be efficiently collected.

The laser powers used for multiphoton imaging were set between 20 mW and 35 mW with respect to different depths (Figure 4) [Wang et al. 2007a].

For image preparations, Zeiss LSM 510 Image Brower (Carl Zeiss, Jena, Germany) and Adobe Photoshop 5.0 (Adobe Systems Incorporated, San Jose, CA, USA)were used for image transfer and image improvement in an acceptable manner.

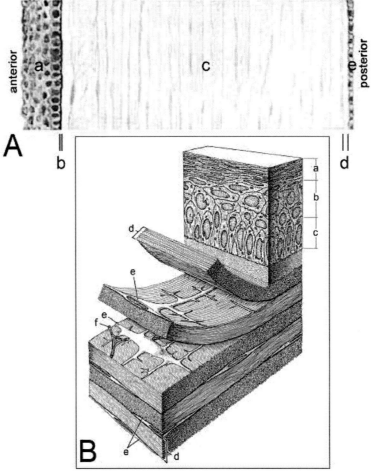

Figure 5. Histological cross-section (A) and 3-D illustration of microstructural components (B) in cornea. The figure (A) is an imitated drawing according to the histological cross-section stained with HE showing the microstructure five layers: a, epithelium; b, Bowman's layer; c, stroma; d, Descemet's layer; e, endothelium. Bowman's layer is located underneath the basal cells and connects to the upperside of stroma. The thin Descemet's layer functions as the normal basement membrane of endothelial cells and forms the scaffolding on which the endothelial cells spread themselves. The figure (B): a, epithelial squamous cells; b, epithelial wing cells; c, epithelial basal cells; d, collagenous lamellas; e, stromal keratocytes; f, nerve fibres. Epithelium involves three sublayers: (i) 2-3 layers of superficial, flat cells named epithelial squamous cells; (ii) 3-4 wing-shaped polygonal cells named wing cells; (iii) single layer of columnar basal cells named epithelial basal cells. Keratocyte is the predominant cell and collagen is the main protein structure in extracellular matrix of corneal stroma. The cells of endothelium are arranged in a continuous monolayer. The subbasal nerve fibres stem from the peripheral limbus and run through stroma and Bowman's layer and finally formthe nerve terminal endings between epithelial cells (Reprint, Dissertation, Wang, 2006).

3. SUBJECTS AND EXAMINATIONS

The *ex vivo* multiphoton imaging was performed with porcine eyes since the porcine eyeballs are of a size and structure similar to the human ones. The fresh samples were removed from ocular orbit 20 minutes after slaughtering and transported in a nutrition solution (Sigma-Aldrich, UK) at a constant temperature (+ 4°C) to the experimental laboratory.

In general, the whole-mount scleral, corneal, and retinal tissues were not stained or fixed before multiphoton excitation imaging. In order to minimize the autolysis post mortem, examination time with each sample was kept not longer than 10 minutes. Glucose was especially added to PBS in order to keep the ocular tissues better in vitality.

3.1. Corneal and Sclera Examinations

The examinations of cornea and sclera were performed with whole eye globes. The entire globe was laid on the round 170-micron-thick glass set in a 65 mm-diameter dish. In order to hold eye globes in a physiologically similar state, two drops of the nutrition solution (Sigma-Aldrich, UK) were rendered to the entire globes during the examinations. In addition, the entire eye globe was covered with a special chamber, which was made of rustless steel and comprised two parts: one was a cylinder with 2.6 cm-diameter and 2.5 cm-height, the other was a cover set to the cylinder. During the imagings, the laser beams were directed into sample tissues from their outer surfaces. The emission radiance was subsequently collected and then immediately projected into a photomultiplier tube (PMT, Hamamatsu, Bridgewater, New Jersey, USA) [Wang *et al*, 2005a and 2008a].

3.2. Retinal Examinations

To make a preparation for retinal specimens, a posterior portion of eye globe was explanted transsclerally with the assistance of a 8 mm-diameter biopsy punch (Stiefel Inc., Offenbach, Germany), then mounted onto the central point of a 65 mm-diameter plastic dish bottomed with a 170-micron-thick glass coverslip (Karl Hecht KG, Germany); meanwhile, the retinal side of the explant was laid downwards to contact the glass. One drop of the nutrition solution (Sigma-Aldrich, UK) was rendered to the whole sample in order to keep its moisture. One drop of Aqua bidest was added between the glass and eye globe in order to minimize the light dispersion. The laser beams penetrated the internal limiting membrane into retina. The retinal tissues examined were located *in situ* 2 mm from the optic disk on the temporal side of the eyeballs [Wang *et al*. 2006b and 2008a].

The *in vivo* corneal multiphoton imaging was performed with rabbits (Figures 2 and 3). New Zealand Albino rabbits (Charles River Wiga, Sulzfeld, Germany) with an average age of 6 ± 0.8 months and a mean weight of 4.1 ± 0.6 kg were used.Animal care and procedures were in accordance with the guidelines of the "German Law on the Protection of Animals" and the "Principles of Laboratory Animal Care". Animal permission was granted by the Thuringian Ministry for Research and Culture (Nr. 02-17/04 Thuringia).

Corneas were examined preoperatively by a Zeiss SL 120 Slit lamp (Carl Zeiss, Jena, Germany) and a Zeiss Stemi DV 4 stereomicroscope (Carl Zeiss, Jena, Germany). Only the clear corneas were forwarded to take optical imaging. The conjunctival sac of the treated eye was rinsed with phosphate-buffered saline (PBS) pre- and post-treatment. After narcotization by the intravenous narcotics ketamine (Atarost, Twistringen, Germany) and xylacine (Bayer Vital, Leverkusen, Germany), the animals were positioned on an animal bed linked to the multiphoton microscope. The eye globe was laid on a 170 µm–thick glass coverslip lying in a specimen chamber specialized for *in vivo* animal experiments. The life parameters of the being treated animals such as the blood oxygen concentration and pulse frequency were monitored on a back extremity with a Draeger dialogue 2000 (Draegerwerke, Lübeck, Germany) [Wang *et al.* 2006c and 2007c].

4. CORNEAL MULTIPHOTON IMAGING

In vivo microscopic evaluation of ocular structures has always been a challenge for ophthalmic clinicians and microscopical researchers. Although the inherent transparency of cornea was exploited initially by clinic instruments such as the slit lamp biomicroscope and ophthalmoscope, high magnification ocular observations still presented many difficulties. Clinical confocal microscopy is thought to improve the *in vivo* ocular imaging. While confocal microscopy can achieve cellular imaging in the cornea, it cannot detect the collagenous component of stroma. In addition, this technique uses a visible light source, which causes more scattering through the media. Generally, its image quality including spatial resolution and signal to noise is unsatisfied. So far, the micro- and nanostructures of corneal tissue have been mostly limited to investigations with conventional histological and electron microscopical methods [Wilson 1970, Boyde 1974, Maser *et al.* 1977, Sekundo *et al.* 1994, Bentley *et al.* 2001]. In these methods, fixation, sectioning and staining have to be prepared before imaging examination. The needs to the exhaustive procedures are prone to introduce artefacts; in the case of pathological examination, the methods do not permit *in situ* sequential images of disease evolving. The *in vivo* studies of biological tissues using the conventional imaging approaches are not meet the needs for basic research and clinical diagnostics.

The introduction of multiphoton microscopy promises to lead to significant progress *in situ* even *in vivo* investigation of tissue component with high precision and contrast [Brakenhoff *et al.* 1996, König *et al.* 1996, Diaspro *et al.* 2000, Zipfel *et al.* 2003a, Levene *et al.* 2004, Wang *et al.* 2005b]. Based on this technique, visualization of the intracellular mitochondria [Barzda *et al.* 2005] and quantification of intracellular serotonin [Maiti *et al.* 1997] and NAD(P)H [Patterson *et al.* 2000] have been reported. It makes high-resolution imaging of endogenous fluorophores possible and is well suited to intrinsic molecular imaging in biological specimens.

4.1. Optical Overview of Corneal Architecture with Subcellular Resolution Based on Multiphoton Excitation Microscopy/Tomography

In order to make a better understanding of corneal optical imaging, histology of cornea is briefly described in the following words. Generally, cornea comprises five layers: epithelium, Bowman's layer, stroma, Descemet's layer and endothelium (Figure 5). Epithelium involves three sublayers: (i) 2-3 layers of superficial, flat squamous cells named epithelial squamous cells; (ii) 3-4 wing-shaped polygonal cells named epithelial wing cells; (iii) single layer of columnar basal cells named epithelial basal cells. Oxygen can diffuse across the three sublayers and penetrate into stroma. Bowman's layer is located directly under the basal cells and connects to the upperside of stroma. Stroma is comprised of stromal cells and extracellular matrix involving collagens and amorphous substance. Keratocyte, a flat cell sandwiched between the collagen lamellas, is the predominant cell responsible for maintaining the integrity and regeneration of corneal fibrous components. Each cell always extends five to seven cell processes connecting to adjacent keratocytes. Descemet's layer is the normal basement membrane of endothelial cells and forms the scaffolding on which the endothelial cells spread themselves. The cells of endothelium are arranged in a continuous monolayer, functioning predominantly to maintaining the hydration and dehydration balance in stroma. Hence, endothelium is an essential part for keeping the corneal transparency [Scharenberg 1955, Maurice 1984]. The in situ non-linear optical detecting of the deepest-located endothelial cells was reviewed.

The 2PF of intrinsic fluorophores contained in corneal cells and the SHG of stromal collagens can be realized when the photon intensity reaches to MW-GW/cm^2 [Fine et al. 1971, So et al. 1998, Campagnola et al. 2002]. The high photon intensity was induced in the following studies by the ultrashort laser pulses in a sub-femtoliter intrastromal focus volume obtained with diffraction-limited objectives

4.1.1. Multiphoton Autofluorescence Tomography of Corneal Epithelial Layers (At Depths of 10, 25, and 41 Microns, Respectively)

Intact cornea imaging was obtained under multiphoton microscope. The three layers can be optically differentiated based on their different cells morphology. The epithelial cells, replenished by cell moving centrally from the limbus and anteriorly from the basal layer of epithelium, represent a barrier against fluid loss and exogenous pathogens. As illustrated in epithelial cells (Figure 8), the cells are closely packed and tightly adhered to each another. Based on this feature, the cellular outline can be defined clearly. A smooth, intact, and healthy corneal epithelium is a prerequisite for maintaining corneal transparency.

The cells can be studied with intracellular observations under the multiphoton microscopy and show following characteristics: the fluorescent cytoplasm and the non-fluorescent nuclei. The fluorescent cytoplasmatic granules of mitochondria encircle non-fluorescent nuclei. Owing to the absence of suitable fluorophore at the excitation wavelength of 760 nm, the nuclei are darkly displayed. The intracellular autofluorescence was based on two-photon excitation of NAD(P)H, and with an emission wavelength range between 460 nm and 480 nm [Wang et al. 2008a].

The 3-D tomography of epithelial cells was depicted (Figures 6, 7, 9). The squamous cells are determined by flat cell bodies and oval to round non-fluorescent nuclei surrounded by the cytoplasmic fluorescent granules stemmed from NAD(P)H located in mitochondria. The wing cells have 3-4 polygonal cellular contours and are located in the middle epithelial layer. The single columnar basal cells contact the submicron-scale basal membrane, which connects the upperside of the Bowman's layer. The black portion on this optical section was the basement membrane of epithelial basal cells [Wang et al. 2007a]. The keratocytes in stroma can be detected in a further depth under the Bowman's layer (Figure 10).

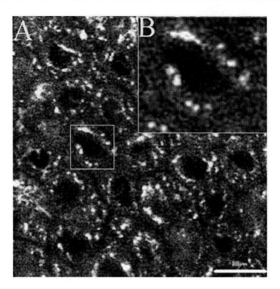

Figure 6. *In vivo* intrinsic 2PF image of epithelial squamous cells in cornea. (A). Epithelium at a corneal depth of 10 micrometers with high subcellular spatial resolution and high signal-to-noise ratio at an excitation wavelength of 760 nm.The squamous cells, possessing flat cell bodies and oval to round nuclei, are characterized by the fluorescent cytoplasmatic granules stemming from mitochondria and non-fluorescent nuclei displayed darkly on the optical section. The fluorescent granules encircle non-fluorescent nuclei; (B). Observation of one cell with high magnification. The endogenous fluorophore was stimulated substantially under two-photon excitation. *Scale bar: 10 microns* (ReprintLasers Surg Med 2007).

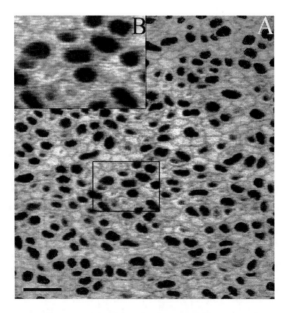

Figure 7. *In vivo* intrinsic 2PF imaging of epithelial wing cells in cornea. (A). Epithelial cells at a corneal depth of 25 microns excited at a wavelength of 760 nm. The fluorescence stemming from mitochondria encircle non-fluorescent nuclei. Based on the multiphoton tomography, the cells were determined to have 3-4 wing-shaped polygonalcellular contours and were located in the middleepithelial layer. The fluorescence based on NAD(P)H in mitochondria was verified by emission spectrum. (B). Observation of one cell with high magnification. *Scale bar: 25 microns* (Reprint Lasers Surg Med 2007).

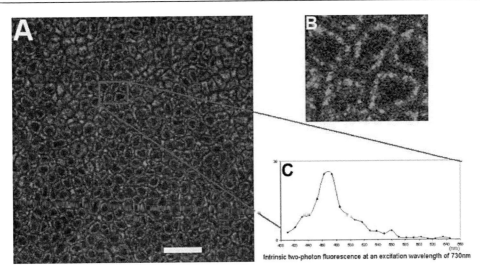

Figure 8. Emission spectrum of intrinsic 2PF images of epithelial cells at a depth of 28.4 microns in a cornea with high signal-to-background ratio. (A).The fluorescent granules of mitochondria and the non-fluorescent dark nuclei can be identified easily without any staining assistance; (B).The high magnification of cells contour and fluorescent granules, revealing the bioenergetics of mitochondria in a certain living period; Histogram (C) reveals that the predominant spectral profile was located between 460 nm and 480 nm (y-axes: amplitude), which corresponds to the eliciting intracellularNAD(P)H fluorescence. Knowledge of the spectrum provide a novel signature for further analysis of molecular entities emitting these fluorescent signatures and would have a use in differentiating the NAD(P)H autofluorescence from other intrinsic fluorophores such as flavin, lipofuscin, and porphyrins in tissues and cells. *Scale bar: 30 microns* (Reprint Lasers Surg Med 2008).

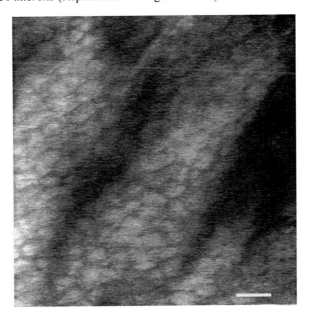

Figure 9. *In vivo*intrinsic 2PF image of epithelial basal cells at a corneal depth of 41 micrometers excited at a wavelength of 760 nm. The single columnar basal cellscontact the basement membrane, which connects theupperside of the stroma.The non-fluorescent portions of this image are corresponding to the collagen in basement membrane, which forms the scaffolding on which the basal cells spread themselves. *Scale bar: 20 microns* (ReprintLasers Surg Med 2007).

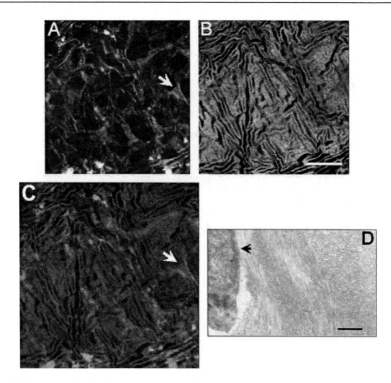

Figure 10. Multiphoton excitation images of corneal stroma performed with different excitation wavelengths at a depth of 118.7 micrometers (A, B, C). (A).Keratocytes image based on the cytoplasmatic NAD(P)H in mitochondria at an excitation wavelength of 730 nm; Owing to the absence of collagen lamellas, the stromal cells can be selectively studied. One keratocyte with three cellular processes is marked with an arrow head. Using this type of high-resolution non-linear optical imaging, the cellular details were *in vivo* studied. The mean size of keratocytes based on a measurement of more than 100 cells: 18.48 ± 3.46 x 13.53 ± 2.87 x 9.7 ± 1.8 microns (measurement of cell body without processes, up to 140 μm depth in stroma, rabbits); (B).Collagen lamellas based on SHG detection at an excitation wavelength of 840 nm. SHG enables the non-centrosymmetric biological structures to be directly imaged. The intrastromal collagen lamellas and filaments can be selectively visualized by SHG signal imaging due to the deficiency of keratocytes autofluorescence at this excitation wavelength;(C).Overlaying of 730 nm and 840 nm excitation images discloses the microstructural co-localization and topography of keratocytes and their hosting collagen lamellas *in situ* (with pseudocolours). Green: autofluorescence; Red: SHG. The keratocytes, playing an important role in the metabolic and physiologic homeostasis, are folded into the labyrinths of collagen fibrils. Only based on two different wavelengths, keratocytes and collagen lamellas yield two distinct features for a same region;(D). A transmission electron microgram (TEM) of corneal stroma was enclosed to complementarily reveal the ultrastructural topography between cellular and extracellular components. Keratocytes, incorporating into corneal stroma, are displayed that they are sandwiched loosely in part among the collagen lamellas. *Scale bars: 40 microns (upper) for images A,B, C; 500 nm (lower) for EM* (ReprintLasers Surg Med 2007).

The thickness of epithelium of New Zealand rabbit was determined to be approximately 38.6 ± 5.8 microns with the assistance of the multiphoton microscope.

The emission spectra of intracellular autofluorescence were measured. Knowledge of the spectrum provides an interesting signature for further analysis of the molecular entities emitting these fluorescent signatures and would have a use in differentiating the NAD(P)H autofluorescence from other intrinsic fluorophores such as flavin, lipofuscin, and porphyrins.

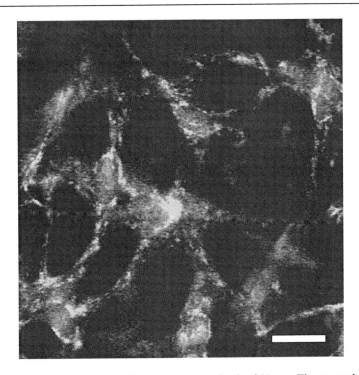

Figure 11. High spatially-resolved stromal keratocytes at a depth of 80 μm. The stromal cell can be easily identified through the fluorescent cytoplasm and dark non-fluorescent nuclei. Based on the deficiency of collagen signal at this excitation wavelength, cellular boundaries can be clearly determined. The cellular density can be studied. **Eight cells in 100 x 120 μm² area are counted.** Thanks to the high spatial resolution, each cell can be morphologically studied in details. The cells possess large, flattened and eccentric nuclei; the cell bodies had a well-defined satellite shape with long, fine and generally non-branched cell processes projecting from their apices. The processes of neighbouring fibroblasts came into contact with each other to form a continuous network structure. Remarkably, no cellular photobleaching or photodisruptive phenomena took place during the high-zoomed imaging under multiphoton microscope. *Scale bar: 20 microns* (Reprint Ann Anat 2006).

4.1.2. Revealing of Stromal Intact Cellular and Collagenous Networks Based on Multiphoton Excitation Imaging

(i) Selective Displaying of Intratissue Keratocyte Networks in Stroma with Intracellular Spatial Resolution

Keratocytes are the principal corneal stromal cell. By far, the morphology of the cells was examined by electron microscopy [Nishida *et al.* 1988]. Keratocyte networks were visualized using vital dyes [Poole *et al.* 1993]. However, none of these methods allows direct visualization of keratocytes in the intact bulk corneal tissue.

With the assistance of the excited endogenous cytoplasmatic fluorophore, the *in situ* visualization of intratissue cells becomes realizable, which overcomes the artefacts always occurring during the histological preparation. The artefacts would involve disturb of the topography and morphology of microstructures. More interesting, the sample tissues can be scanned numerous times without observable morphological photo-impairment under multiphoton microscope.

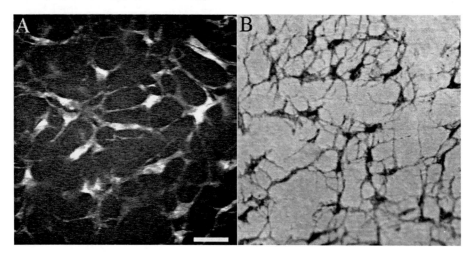

Figure 12. Network structure of keratocytes formed by the cellular processes elucidated by two approaches: non-linear optical imaging (A) and conventional histology of gold impregnation (B). The cell bodies had a well-defined satellite shape with long, fine and generally non-branched cell processes projecting from their apices. The processes of neighbouring fibroblasts contacted with each other to form an extensive and continuous network structure. The *in situ* selectivity of non-linear optical imaging nicely coincides with the *in vitro* conventional histology. The cellular network appears to constitute an integrated system which may be involved in the synchronized regulation of the metabolic and physiologic homeostasis for the maintenance of corneal transparency. *Scale bar: 30 microns* (Reprint Histochem Cell Biol 2006).

Owing to the absence of collagen lamellas at the excitation wavelength of 730 nm, the stroma cells can be studied selectively (Figures 10_A, 11, 12, 13).The keratocytes can be identified through the fluorescent cytoplasm and dark non-fluorescent nuclei. Owing to the high spatial resolution, each cell can be studied with morphological details. The cells are connected with each other through the fluorescent processes. Keratocytes, the modified tissue fibroblasts, possess large, flattened and eccentric nuclei (Figure 11). The cell bodies had a well-defined satellite shape with long, fine and generally non-branched cell processes projecting from their apices [Wang *et al.* 2006d]. The processes of neighbouring fibroblasts contacted with each other to form an extensive and continuous cellular network structure. The cellular processes functioning as communication of signal and exchange of nutritive substances displayed in Figure 12_A are comparable to the conventional histology in Figure 12_B. The cellular network appears to constitute an integrated system, which may be involved in the synchronized regulation of the metabolic and physiologic homeostasis for the maintenance of corneal transparency. Based on the ultrahigh spatial resolution and 3-D visualization, the size of keratocytes was determined. Mean value of keratocytes based on measurement of more than 1000 cells in 50 *ex vivo* porcine eye globes: 23.48 ± 3.66 x 16.58 ± 2.68 x 11.8 ± 2.8 microns (only the cell bodies were measured, up to 140 µm depth of stroma with z-adjusting; eye globes of adult porcine; measurement within one hour after slaughtering). The cell morphology was also *in vivo* studied with the high-resolution optical images. The mean volume value of keratocytes was determined based on measurements of more than 100 cells: 18.48 ± 3.46 x 13.53 ± 2.87 x 9.7 ± 1.8 microns (only cell bodies were measured, at depths of up to 140 microns, rabbits) [Wang *et al.* 2007a and 2008a].

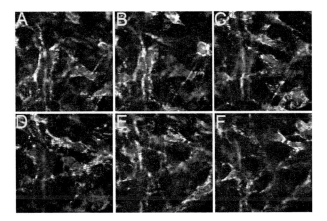

Figure 13. Three-dimensional non-linear optical tomography of stromal keratocytes with vertical z-interval of 5 micrometers based on the two-photon autofluorescence imaging at a wavelength of 760 nm. Mean value of keratocytes based on 3-D measurement of more than 1000 cells in *ex vivo* porcine eye globes: 23.48 ± 3.66 x 16.58 ± 2.68 x 11.8 ± 2.8 microns (only the cell bodies were measured, up to 140 micrometers depth of stroma with z-adjusting; eye globes of adult porcine; measurement within one hour after slaughtering). *Scale bar: 15 microns* (Reprint Ann Anat 2006).

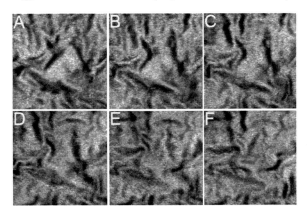

Figure 14. *In vivo* corneal collagen lamellas 3-D tomography in conjunction with a z-interval of 5 micrometers based on the SHG signal imaging in a backward direction. SHG takes place in a minute focal volume, permitting high vertical resolution of collagenous component in thick intact tissues.The distribution and orientation of the collagen lamellas is clearly revealed. In correlation with the z-positioning, the 3-D architecture of intratissue collagen can be studied. The lamellas are densely packed with characteristic fluctuations and run parallel to the corneal surface. However, they show different orientations in the x-y dimension, especially on the surface of keratocytes. In comparison with SHG image in sclera (Figure 18), corneal collagen is arranged in an organized pattern of lamellas. As a rather simple and straightforward imaging method, the SHG tomography can be used clinically for evaluating the cornea fibrosis and stromal edema.*Scale bar: 10 microns* (Reprint Ann Anat 2006).

(ii) SHG Signal Imaging of Collagen Lamellas in Corneal Stroma

Collagen is the main protein structure in corneal stroma. The collagen filaments can be excited under multiphoton microscope at wavelengths between 800 and 860 nm, with emission lights equal to half of the incident lights [Williams *et al.* 2005]. The intrastromal collagen lamellas and filaments can be selectively visualized by second harmonic imaging [Wang *et al.* 2005c]. SHG enables the anisotropic non-centrosymmetric biological structures, possessing large hyperpolarizabilities under high laser irradiance, to be directly imaged [Roth *et al.* 1979, Freund *et al.* 1986]. Owing to its own non-centrosymmetric structure of collagen fibrils, a triple helix composed of three protein chains, collagen is thereby a well-documented source of tissue SHG imaging. Corneal stroma contains various types of collagen, each one being distinguished from the others by the biochemical composition of three protein chains. The predominant collagen in normal cornea is type I and VI; type IV is characteristic of basement membranes; type VII constitutes the anchoring fibrils of epithelium; type III becomes much more prominent during wound healing. Some other types, such as collagens XI and XVIII, which do not exist in the normal physiological state of cornea, appear only during the wound reparative phase [Light 1979, Tseng *et al.* 1982, Nakayasu *et al.* 1986, Zimmermann *et al.* 1986, Kato *et al.* 2003]. Collagen has the intrinsic capability of frequency doubling under high light irradiance based on its special biochemical composition. Membrane imaging and visualizing of biomolecular array with the assistance of SHG signal imaging were reported [Moreaux *et al.* 2000, Campagnola *et al.* 2002]. Similar to the occurring conditions of two-photon excited fluorescence, SHG takes place in a minute focal volume at high laser irradiance, permitting high resolution 3-D optical sectioning of collagenous component in thick intact tissues [Campagnola *et al.* 2003].

As illustrated in Figure 10_B, distribution and orientation of the collagen lamellas are clearly revealed. The lamellas are densely packed with characteristic fluctuations. The *in vivo*optical imaging provides reliable information about the characteristics of the cornea, the details of which coincide with the conventional histological architecture [Freemann *et al.*1978]. With the assistance of SHG, the *in vivo* 3-D tomography of collagen lamella in cornea with high resolution becomes feasible for the first time (Figure 14). The 3-D architecture of collagen lamellas can be visualized in conjunction with the z-positioning. The collagen can be selectively visualized due to the deficiency of keratocyte autofluorescence at the wavelength 830 nm. The lamellas run parallel to the corneal surface. However, in the x-y dimension, they show different orientations, especially on the surface of keratocytes [Wang *et al.* 2007a].

As a rather simple and straightforward imaging method, SHG signal imaging is invaluable for studying the assembly of collagen fibrils in bulk tissues. In addition to the fundamental study of collagen fibrils in different tissues, SHG imaging may have a clinical use for early diagnostics of cornea-related diseases such as stromal edema, which is characterized by disturbing the regularity of collagen fibrils and the groove-like order of collagen lamellas. Based on its selectively displaying function, SHG could also be of great merit to *in vivo* monitor collagen regeneration in the wound healing process after laser refr Spectral imaging verified that both cornea and sclera have the SHG emission wavelength at 420 nm (Figure 18). In comparison with sclera, collagens in cornea are arranged in a rather organized fashion of lamellas packing with a depth-dependent variation (Figure 14 and Figure 18). The spaces of inter- and intra-collagenous lamellas are always filled with keratocytes and their processes. The imaging specificity of SHG allowed to differentiating the both ocular tissues. More interesting, it offered a high degree of contrast enhancement in the topography between cellular and collagenous components.

The *in situ* knowledge of collagen is of crucial importance in the anatomy and pathology of the connective tissues. SHG microscopy has promise for imaging collagen organization and provides high resolution and more detailed images of collagen distribution. Multiphoton microscopy has thereby the potential to become one of the most valuable tools *in situ* even *in*

vivo microscopy of tissue collagen and could be even employed to inspect collagen-related alterations such as corneal fibrosis and cellular infiltration in melanoma.
active surgery.

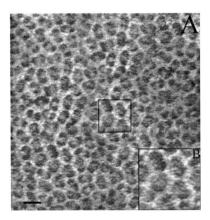

Figure 15. *In vivo* intrinsic 2PF imaging of endothelial cells. (A). Autofluorescence image of endothelial cells taken at an excitation wavelength of 760 nm. The objective (20x 0.95 N.A., water) with a work distance of more than 1000 microns in bulk tissue used in this paper allowed the deepest endothelial layer of cornea to be clearly visualized. The cells are featured by the filmy cytoplasm and large round to oval nuclei. The paving-stone mosaic pattern of the hexagonal endothelial cells shows only slight variation in size and shape based on the *in vivo*non-linear optical tomography. The thickness of the continuous monolayer endothelial cells was determined to be 8.4 ± 2.1 microns. The homogenous cells have an average value of 7.38 ± 1.25 microns in diameter. The optical imaging of endothelial cells makes it possible to on-line monitor endothelial cells during the intrastromal surgery;(B). Observation with high magnification. *Scale bar: 10 microns* (Reprint Lasers Surg Med 2007).

(iii) Microstructural Co-localization and Topography of Keratocytes and their Hosting Collagen Lamellas in Corneal Stroma Based on 2PF and SHG

The microstructural co-localization of corneal stroma was displayed with different excitation wavelengths in a depth of 118.7 microns (Figure 10_C). Keratocytes and collagen lamellas can be selectively displayed only with the in-tandem assistance of two different laser wavelengths based on the non-linear optical effects. The keratocytes, folded into the labyrinths of collagen fibrils and playing an important role in the metabolic and physiologic homeostasis, are major cellular components of the corneal stroma. They can be identified through the fluorescent cytoplasm and non-fluorescent nuclei in the multiphoton imaging. The fluorescent cytoplasm was based on intrinsic NAD(P)H in mitochondria excited at 730 nm. Thus, the high-resolution optical images acquisition allowed the cell morphology to be *in situ* studied precisely. In details, their long cytoplasmic processes came into contact with those of neighboring keratocytes. The cytoplasmatic autofluorescence revealed the bioenergetics of mitochondria in their different living periods [Wang *et al*. 2008a].

As seen in Figure 10_D, a transmission electron microgram (TEM) of corneal stroma was enclosed to complementarily elucidate the ultrastructural topography between cellular and extracellular components in this relationship. Keratocytes, incorporating into corneal stroma, are displayed in this figure that they are sandwiched loosely in part among the collagen lamellas.

4.1.3. Multiphoton Excitation Detection of Endothelial Cells at a Corneal Depth of 369 Microns with Living Animals

Endothelium is located in the deepest layer of cornea. The thickness of cornea of the New Zealand rabbit was determined to be 380 ± 18 microns measured with a pachometer (Corneo-Gagetm plus 2, Sonogage, U.S.). The objective (20x 0.95 N.A., water) with a work distance of more than 1000 micrometers in bulk tissue allowed the deep endothelial layer of cornea to be visualized (Figure 15). The paving-stone mosaic pattern of the hexagonal endothelial cells under multiphoton microscope*in vivo* shows only slight variation in size and shape. The volume of endothelial cells was measured with the non-linear optical tomography. The thickness of the continuous monolayer endothelial cells was determined to be 8.4 ± 2.1 microns. The homogenous cells have an average value of 7.38 ± 1.25 microns in diameter. However, the values of *in situ* optical measurement here are not in agreement with the conventional histological results. The endothelial cells with thickness of 4–6 micrometers and diameter of 8–11 micrometers were reported [Maurice, 1984]. The reasons for this difference might involve (i) different values between *in vivo* (non-linear optical tomography) and *in vitro* (histology); (ii) different species between human (histology) and rabbit (in this review); (iii) the refractive index mismatch of optical measurement. As to the phenomenon of refractive index mismatch, geometric distortions between the sample and the medium play a role in the image deterioration. In the fact, this optical mismatch interferes with the real size of cells. The vertical mismatch is yet not corrected automatically by the microscopic operation system. We measured the thickness of cells located in tissues between utmost-upper point and utmost-lower point of the cells. The size of cells presented here provides the examination outcomes based on laser microscopy[Wang *et al.* 2007a].

The optical imaging of endothelial cells makes it possible to on-line monitor endothelial cells during the intrastromal surgery and to conduct follow-up observations on these cells postoperatively. To date, deep lamellar endothelial keratoplasty (DLEK) has been used as an innovative approach to treat Fuchs' endothelial dystrophy (Terry *et al.* 2005). The endothelial cells from a donor are implanted to a recipient, whose endothelium is functionally insufficient to keep the corneal hydration and dehydration balance. The grafting endothelial cells should be pre- and post-operatively monitored. Assisted by multiphoton microscopy, the *in vivo* evaluation and monitoring of the implanted endothelium with subcellular spatial resolution become easier.

4.2. Optical Determination of Bowman's Layer with In-tandem Combination Assistance of 2PF and SHG Signal Imaging

Bowman's layer was described in humans by William Bowman in 1947. The author defined it as a homogeneous layer between epithelium and substantia propria in light microscopy, a lusterless but perfectly transparent, continuous, homogenous lamina that did not appear to have any internal structure. Camber *et al.* (1987) found that in the case of the pig, Bowman's layer was unequal and corrugated in form, thus making it difficult to distinguish it from the surrounding fine collagenous fibres. Stained with hematoxylin-eosin (HE-staining), Merindano *et al.* (2002) reported that the pig appears to possess no Bowman's layer. On contrary, Hayashi *et al.* (2002) reported that pig possess Bowman's layer in cornea.

The border between Bowman's layer and the substantia propria is clearly defined in guinea pig. The diameters of the collagen fibrils in Bowman's layer and the substantia propria are almost the same. However, the existing morphological studies of cornea offer little information about the morphometry of Bowman's layer. The nonlinear optical imaging was used here to detect this layer and conduct the morphometric study.

Bowman's layer is comprised of collagen[Maurice 1984]. The epithelial basal cells connecting to the upper side of Bowman's layer are fluorescent at the excitation wavelength of 730 nm (Figure 16_A). When the laser focus was adjusted further deeper to the stromal layer, the keratocytes would be displayed based on its autofluorescence. On Figure 16_B, the collagen components in Bowman's layer between basal cells and stroma are demonstrated based on SHG signal imaging of collagen at the incident wavelength of 840 nm. With the assistance of tissue optical tomography based on in-tandem detection with 2PF and SHG, the Bowman's layer can be optically defined, where SHG is there but 2PF absent. By contrast, basal cells show 2PF but no SHG; Stroma has SHG and also 2PF. As a result, the collagen depth without cellular structure should be the Bowman's layer located between basal cells and keratocytes. The thickness of this collagen component between epithelial basal cells and stroma was determined to be approximately 18.6 ± 3.6 microns (porcine eye ex-vivo, cornea*in situ*).

4.3. Optical Observations of Nerve Fibres in Cornea

The nerve fibers in cornea were examined with three approaches: reflexion, 2PF and SHG. Through comparison of the three figures (Figures 16 _A, B, C), it has been postulated that neither at the incident wavelength of 730 nm nor at 840 nm can the nerve fibres of subbasal plexus be displayed under the condition of multiphoton absorption. The beading-like subbasal nerve fibres are displayed only in the reflexion image of Figure 16_C. The nerve fibres can be visualized three-dimensionally based on this imaging approach. They stem from the peripheral limbus and run through Bowman's layer and formthe nerve terminal endings between the epithelial cells. Most of these visible fibres distribute in the anterior one third corneal thickness. In the posterior half of cornea, these fibres were sparsely seen. The subbasal nerve fibres comprise beading area and interbeading area (Figure 16_D). The fibre diameter in the areas of beading and in the areas between beadings was measured in this work. The size of fibres in beading area is proximately 1.5-2 folds larger than that of the interbeading parts. The mean diameter of subbasal nerve fibres was measured with mean value of 1.88 ± 0.39 microns (subbasal nerve fibres, 50 porcine eyes *ex vivo*, reflexion imaging). To be concluded, the nerve fibers in cornea can not be displayed under non-linear optical imaging condition.

Non-linear optical images of the full-thickness of cornea were obtained with multiphoton microscopy based on 2PF and SHG signal imaging. Epithelial cells, stroma, and endothelial cells were identified and displayed with the assistance of the non-linear optical imaging technique. Qualitative and quantitative studies of the corneal cells were conducted. In particular, corneal Bowman's layer was optically determined *in situ*. Our results of non-destructive selective visualization of various microstructural components in corneal tissue show that multiphoton microscopy is becoming a powerful tool in advancing our understanding of corneal optical tomography and biomechanics. By means of the *in vivo* feasible microscopy, the cell reaction during wound repair in laser intrastromal nanosurgery is investigated and reported in another following session. In further studies, the fluctuation distribution of the strain collagen in tissue will be computerized and tried to find the correlation with the morphometry of individual filaments and collagenous lamellas in cornea

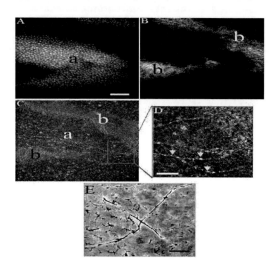

Figure 16. Optical determination of Bowman's layer and nerve fibres in cornea. The determination of Bowman's layer based on the in-tandem detection with 2PF (A) and SHG signal imaging (B). (C). The nerve fibres of subbasal plexus can be displayed only based on the reflection optical imaging. The three figures (A-C) were taken at a same intracorneal plane at a depth of 48.88 μm. Portion a in (C) corresponds to the portion a in (A); portion b in (C) corresponds to the portion b in (B). (D). High magnification of nerve fibres showing the beadings (arrow heads on D) of subbasal nerve fibres. With the assistance of the three figures (A-C) based on the different imaging approaches,the collagen layer between the epithelial basal cells and stroma can be definitely determined. The predominant component of Bowman's layer is collagen. The epithelial basal cells are fluorescent at the excitation wavelength of 730 nm (Figure A). When the laser focus was adjusted further deeper to the stromal layer, the keratocytes would be imaged based on its autofluorescence. In Figure B, the collagen component in Bowman's layer between basal cells and stroma is demonstrated based on SHG signal imaging of collagen at the incident wavelength of 840 nm. With the assistance of tissue optical tomography based on in-tandem detection with the both approaches, the Bowman's layer can be optically defined, where SHG is there but 2PF absent. By contrast, basal cells have 2PG but no SHG; Stroma has SHG and also 2PF. As a result, the collagen layer without cellular structure should be the Bowman's layer located between basal cells and keratocytes. The thickness of this collagenous component between epithelial basal cells and stroma was determined to be approximately 18.68 ± 3.66 microns (porcine eye ex-vivo, cornea in-situ). The nerve fibres in cornea were studied with 2PF, SHG, and reflection (C, D) imaging approaches as well as gold chloride-staining (E). Arrow heads in figure (E) show the nerve fibres in stroma. It was determined that the nerve fibers in cornea can not be excited under non-linear optical imaging (A and B). The beading-like subbasal nerve fibres are displayed on the reflexion image (Figure C). The nerve fibres can be followed three-dimensionally based on optical tomography. They stem from the peripheral limbus and run through stroma and Bowman's layer and finally formthe nerve terminal endings between epithelial cells which can no more be observed under this light technique. Most of these visible fibres distribute in the anterior one third corneal thickness. In the posterior half of cornea, these fibres were seen sparsely. The subbasal nerve fibres comprise beading area and interbeading area (Figure D). The size of the fibres in beading area is proximately 1.5-2 folds larger than that of the interbeading area. The mean diameter of subbasal nerve fibres was measured with a mean value of 1.88 $\pm$ 0.39 microns (subbasal nerve fibres, 50 porcine eyes ex-vivo, reflexion imaging). *Scale bars: 30 microns for A-C; 50 microns for D; 20 microns for E.*

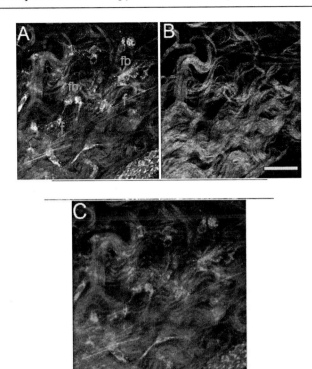

Figure 17. Multiphoton excitation imaging at a depth of 93.9 micrometers in scleral tissue performed with different excitation wavelengths to elucidate fibrocytes, capillaries, elastic fibres, and collagen filaments. The images give snapshots of cell morphology and extracellular matrix organization. The imaging plane was parallel to the sclera surface. The emitted lights, including 2PF and SHG, were deflected backward in the direction towards PMT so that the non-linear optical images were recorded. (A). Fibroblasts and mature elastic fibers based on its blue/green autofluorescence at an excitation wavelength of 730 nm were detected. The non-fluorescent nuclei of fibroblasts displayed darkly on the optical sections are surrounded by the fluorescent granules. The cellular boundary can be determined based on the fluorescent cytoplasm and the non-stimulated extracellular collagens. Using z-positioning, the vertical dimension of the cells was determined. The size of the cells on this site was 17.68 ± 2.36 x 14.33 ± 3.89 x 8.7 ± 1.5 microns (z dimension). Besides the fibrocytes, the intratissue elastic fibers and capillaries were also *in situ* displayed based on 2PF signal imaging; (B). Collagenous fibers image based on SHG detection at an excitation wavelength of 840 nm were obtained by ultraviolet/violet radiation in a backward direction. Most of collagen bundles run parallel to the sclera's surface. They can be selectively demonstrated due to the autofluorescence deficiency of fibroblasts and elastic fibres at the excitation wavelength of 840 nm. The imaging specificity of SHG allowed collagen orientation to be well identified; (C). Co-localization of cellular and fibrous microstructures. Overlaying of 730 nm and 840 nm excitation images reveals the *in situ* state topography of elastic fibers, fibroblasts and collagen filaments without the assistance of any staining or slicing (with pseudocolours). Green: autofluorescence at the wavelength of 730 nm; Red: SHG at 840 nm. The rare elastic fibers were embedded in the hosting framework of collagen filaments. The two figures (A and B) were imaged at a same optical plane with a time-shifting of 8 seconds. *Scale bars: 40 microns.* The symbols on Figure A: fb:fibroblasts; f:elastic fibers. The novel non-invasive imaging opens the general possibility of high-resolution *in situ* 3-D imaging of elastic fibers and collagen structures in connective tissues. This paper demonstrated that the contrast for both intratissue cells and collagen fibers was sufficiently generated only based on the in-tandem detection with 2PF and SHG. The high-degree contrast offers a snapshot of clear topography between cellular and collagenous components. This proof-of-principle examination has already started to reveal intricate features that are otherwise lost in conventional histology studies, suggesting the potential of using non-linear optical imaging as a routine tool for more advanced studies (Reprint Lasers Surg Med 2008).

5. Scleral Multiphoton Imaging

5.1. Selective Displaying of Fibrocytes Based on 2PF at an Excitation Wavelength of 730 nm

Two-photon autofluorescence image of fibroblasts, the main cellular composition in scleral tissue, was acquired at an excitation wavelength of 730 nm (Figure 17_A). The imaging plane was parallel to the sclera surface. The intracellular autofluorescence was based on the two-photon process of NAD(P)H, which was located in the mitochondria of fibroblasts. With intracellular observations, the non-fluorescent nuclei displayed darkly on the optical sections surrounded by the fluorescent mitochondria can be clearly seen. Thus, the cellular boundary can be clearly determined based on the fluorescent cytoplasmatic granules and the non-stimulated extracellular collagens. Furthermore, the cytometric measurement of fibroblasts based on these image features was done. Using z-positioning, the vertical dimension of cells could be determined. The cells on this site have a average size of 17.68 ± 2.36 x 14.33 ± 3.89 x 8.7 ± 1.5 microns (z) [Wang *et al*. 2008a].

In addition to the usual cellular component of fibroblasts, a few melanocytes are generally present in the deep layer of sclerae. These rare pigment-containing cells, interspersed between amounts of fibroblasts, have similar morphology to fibroblasts such as dendritic or tripolar cell bodies and prominent nuclei with little cytoplasm. These cells could be identified with the assistance of the special spectral imaging based on the contained pigment granules. The cells displayed in Figure 17 were determined to be the fibroblasts since our optical plane was located at a one-third depth of sclera.

5.2. Selective Displaying of Collagen Based on SHG at an Excitation Wavelength of 840 nm

SHG signal image of the collagen, in which the fibroblasts host, was obtained at an excitation wavelength of 840 nm (Figure 17_B). The sclera SHG image acquisition at a depth of 93.9 microns in a backward direction is displaying the *in situ* collagen filaments and bundles. Most of collagen bundles run parallel to the sclera's surface. They can be selectively demonstrated due to the autofluorescence deficiency of fibroblasts at the excitation wavelength of 840 nm. With the assistance of SHG signal imaging, the *in vivo* tomography of collagen fibres in sclerae with high resolution can also be performed (Figure 18). Thus, the 3-D of collagen architectures can be studied in correlation with z-positioning [Wang *et al.* 2008a].

3-D tomography of collagenous component in sclerae with a vertical distance of 5 microns was acquired (Figure 18). SHG images of collagen fibrils were taken in a backward direction at depths of between 76.9 and 91.9 microns in the whole-mount scleral tissue. Distribution and orientation of the collagen bundles were revealed clearly. The cross-linking collagen fibres, lacking of lamellar arrangement like in cornea, are packed randomly and densely.

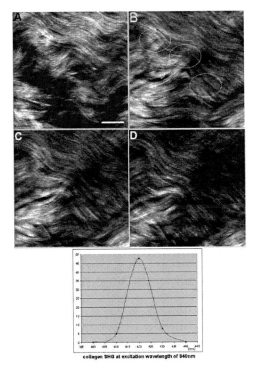

Figure 18. SHG signal imaging tomography and emission spectrum in sclera. The tomogram was performed with a vertical interval of 5 microns in a backward direction showing collagenous filaments at an excitation wavelength of 840 nm. A:at a depth of 76.9 micrometers. Histogram showing the SHG emission wavelength of 420 nm (y-axes: SH amplitude). *Scale bars: 20 microns.* SHG 3-D tomography of collagenous filaments was taken at depths of between 76.9 μm and 91.9 μm in the whole-mount scleral tissue. The optical sectioning was performed parallel to the coverslip surface. Distribution and orientation of the collagen bundles were revealed clearly. The cross-linking collagen fibres (marking with rounds in Figure B) with different extending directions (arrows in Figure B) are packed randomly and densely. Comparing to the SHG images in cornea (Figure 14), the morphological difference is significant. The non-linear optical tomography shows that no collagen lamellas are imaged in sclerae. The non-homogeneous tubelike structures of collagen filaments and irregular arrangement of collagen fibers in sclerae would most likely contribute to this striking SHG signal image discrepancy between sclera and cornea (Reprint Lasers Surg Med 2008).

5.3. Non-linear Optical Comparison of Collagen Organisation in Scleral and Corneal Tissues

Spectral imaging verified that both cornea and sclera have the SHG emission wavelength at 420 nm (Figure 18). However, significant SHG image discrepancies between the collagenous structure in sclera and cornea were observed (comparing Figure 18 in sclera with Figure 14 in cornea). In contrast to collagen lamellas in cornea, the backward SHG signals in sclerae were more significant. The cornea collagen is arranged in an organized fashion packing with a depth-dependent variation. The spaces of inter- and intra-collagenous lamellas are always filled with keratocytes and their processes. The collagen fibres in sclera showing different direction connect to each other.

The SHG radiation pattern is mainly determined by the phase-matching condition. The part of the backward SHG signals may be attributed to the backscattering of the forward SHG signals from the highly scattered scleral tissue [Han *et al.* 2005]. The non-homogeneous collagen fibrils and the irregular arrangement of collagen fibers in sclerae would most likely contribute to this striking non-linear optical image discrepancy between sclera and cornea.

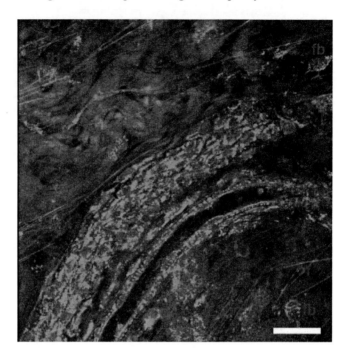

Figure 19. Intrinsic 2PF imaging of intratissue blood capillaries. The image was performed at a depth of 93.9 microns with an excitation wavelength of 730 nm. The elastic fibers and fibrocytes are also displayed under the multiphoton microscopy.The sparse elastic fibers distributed **perivascularly**with a density of less than 20 pieces in a 200 x 200 μm² area. Collagen filaments were absent in this figure. The capillaries imaged were embedded in scleral tissue located in a physiologic-similar circumstance. The two-photon autofluorescence imaging provides us with qualitativeand quantitative assessment of the intratissue capillaries. The intact capillary displayed has a diameter of 40.63 ± 4.65 microns and the other optically-sliced one, in which some of the blood cells were located and fluorescent, has a diameter of 25.53 ± 2.98 microns. The wall of each capillary comprises monolayer of endothelial cells. The flat and pancake-shaped endothelial cells, a layer of fairly thin specialized epithelium, are featured by the filmy cytoplasm and the large round nuclei. The size of the endothelial cells was measured with size of 10.46 ± 1.34 x 8.35 ± 1.67 x 5.7 ± 0.7 microns (z dimension). The symbols: f:elastic fibers; e:endothelial cells of blood capillaries; fb:fibroblasts.*Scale bar: 30 microns* (Reprint Lasers Surg Med 2008).

5.4. *In situ* Studies of Intratissue Capillary Based on Multiphoton Excitation Imaging

The capillaries imaged were embedded in scleral tissue located in a physiologic-similar circumstance.

Two-photon autofluorescence imaging provides us with qualitativeand quantitative assessment of the capillaries (Figure 19). As regards to a peripheral circulation, blood flowing from arteries narrows into arterioles, and then narrows further into capillaries. The emitted 2PF lights were deflected backward in the direction towards the PMT so that the non-linear optical images could be recorded. The intact capillary displayed has a diameter of 40.63 ± 4.65 microns and the other optically-sliced one, in which some of the blood cells are located and fluorescent, has a diameter of 25.53 ± 2.98 microns. The flat and pancake-shaped monolayer endothelial cells, a layer of fairly thin specialized epithelium, are featured by the filmy cytoplasm and the large round nuclei. The size of the endothelial cells was measured with values of 10.46 ± 1.34 x 8.35 ± 1.67 x 5.7 ± 0.7 microns (z) [Wang *et al.* 2007a].

The *in situ* imaging of intratissue capillary in intact bulk tissue would be interesting for the morphological study of microangiopathy such as in the pathophysiology of diabetes mellitus.

5.5. *In situ* Optical Differentiation of Intrascleral Elastic Fibers from Collagens Based on Multiphoton Excitation Imaging

The sparse mature elastic fibres embedded in collagenous networks were also discernible based on its emission blue/green autofluorescence under multiphoton microscope. By contrast, collagen in the tough layer had an emission of violet radiation (420 nm) by SHG imaging (Figures 17 and 20).

Elastic fibers are one of the main fibrous components of connective tissues. They, providing elastic recoil to tissues, are an abundant and integral part of many extracellular matrices, in which they also provide resilience and deformability to tissues. The major component of elastic fibers is the protein elastin, an insoluble rubberlike fibrous polymerresponsible for the characteristic elastic properties. It plays an important role in providing mechanical strength for connective tissues. Collagens are the framework of connective tissues and provide the main determinants of its mechanical behaviour in tension. The elastic function complements collagen fibrils, which impart the connective tissues tensile strength. Ultrastructurally, elastic fibers possess complex structures comprising an amorphous core of extensively cross-linked elastin, surrounded by fibrillin-rich microfibrils. Cross-linking between collagen and elastin provides necessity for the maintenance of proper function of different tissues [Brown-Augsburger *et al.* 1994].

By far, most of our knowledge about elastic fibers is limited to *in vitro* studies, which require the enucleated biopsies to be further processed: histological slicing, embedding, fixation and finally staining with different procedures such as Silverman-Movat pentachrome [Elbadawi 1976], Weigert's resorcin–fuchsin, Hart's elastica-staining or immunogold antibody labelling [Farguharson *et al.* 1991]. With the advent of fluorescence microscope and laser scanning microscopic technique, the autofluorescence of elastic fibers can be recorded using ultraviolet (UV) excitation radiation based on single photon absorption. However, this approach is unable to differentiate the elastic fibers in bulk connective tissues from the framework of collagen filaments based on their intrinsic fluorescence. The novel non-invasive multiphoton imaging technology opens the general possibility of high-resolution*in situ* imaging of elastic fibers in connective tissues without any assistance of staining. Recently, it is reported that the intratissue elastic fibers can be detected in heart valves based on multiphoton autofluorescence by near infrared femtosecond laser pulses [Schenke-Layland *et al.* 2004, König *et al.* 2005a]. We attempted to circumvent this limitation of single photon laser scanning microscopy by using this high technology to differentiate the elastic fibers from the collagen fibrils distributing in bulk scleral tissue.

The co-localization of elastic fibers and collagenfilaments has been displayed under multiphoton microscope (Figure 17_C). High quality images of intraocular tissues were produced, enabling both qualitative and quantitative analysis of the fibrous microstructures within eye globes. As shown in Figure 17_A, the elastic fibers in scleral tissue are detected at an excitation wavelength of 730 nm based on its blue/green autofluorescence emission, meanwhile the collagen fibrils are displayed at an excitation wavelength of 840 nm based on the SHG signal imaging [Wang *et al.* 2008a]. The elastic fibers can be clearly distinguished from the collagenous structures because SHG emits ultraviolet/violet radiation. In a sample sentence, the image selectivity was based that the collagen SHG can be obtained at 840 nm excitation instead of at 730 nm excitation and the 2PF of elastic fibers can be acquired at 730 nm excitation instead of at 840 nm excitation. The rare elastic fibers were embedded in the framework of collagen fibrils and distributed perivascularly with a density of less than 20 pieces in a 200 x 200 μm^2 area (Figure 19). Based on the same settings, such fibrous component in cornea was not observed at the excitation wavelength of 730 nm.

Investigations by multiphoton microscopy have the advantages of enabling examination of the elastic fibers in its physiological state, avoiding the artefacts induced by histological processing, and allowing multiple examinations with the same sample tissue. After the optical imaging, the same scleral samples were then prepared for cryosection or paraffin-embedding and then stained with both Weigert's resorcin–fuchsin and Hart's elastica methods. Unexpectedly, the elastic fibers were not found under light microscopic observations. Presumably, the reason would involve that the *in situ* non-linear optical approach is more sensitive for detection of sparse microstructures in bulk tissues than the histological staining.

Our research suggests that the multiphoton microscopy is a highly sensitive imaging technique for studying fibrous microstructures in bulk tissues. Based on the 2PF imaging of fibrocytes, endothelial cells of capillary, elastic fibres and the SHG imaging of collagen in scleral tissue, the optical co-localization and microstructral topography of different components in dense connective tissue became feasible. The 3-D images give snapshots of cell morphology and cell-extracellular matrix organization. The novel multiphoton imaging approach opens the general possibility of highly-resolved *in situ* co-localization of elastic fibers and collagen structures in connective tissues without staining [Wang *et al.* 2007a].

In a conclusion, the fibrocytes and their hosting collagen can be selectively displayed only with the assistance of two different laser wavelengths. The intratissue elastic fibers can be *in situ* differentiated from the collagen filaments with the assistance of multiphoton excitation imaging.

6. RETINAL MULTIPHOTON IMAGING

A clear view of the retinal cellular microstructures is essential for the prophylaxis and early diagnosis of retinal diseases. In terms of the retinal image acquisition, an optical coherence tomography (OCT) system is being currently presented in clinic. OCT is based on the detection of backscattering photons from the ocular structures.

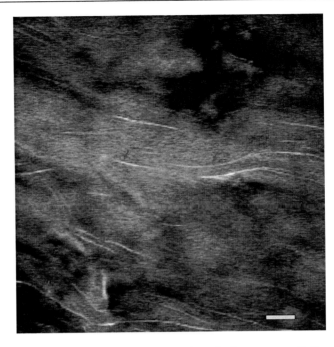

Figure 20. Autofluorescence of elastic fibers in cryosection of sclera at an excitation wavelength of 730 nm. Elastic fibers stemming from sclera run centrically and penetrate finally into corneal limbus. Arrow heads just show the sparse elastic fibers. *Scale bar; 20 microns.*

It provides retina imaging in axial sectioning. The actual resolution of the commercially available OCT systems is typically limited to 10-15 micrometers, which is insufficient to resolve the nerve fibers and the intracellular microstructures of retinal cells [Huang *et al.* 1991; Li *et al.* 2001; Fujimoto *et al.* 2003]. Advances in non-linear optical imaging are now allowing direct observations of retinal structure including cells and nerve fibres. 2PF, selectively conducting the transversal images of retinal cells with subcellular resolution, represents a more interesting tool for the investigation of quantitative and spectral characteristics of retinal cells. Based on the experience with the corneal and scleral tissues using multiphoton microscopy, further investigations on retinal tissue with the same imaging settings were performed. The coronal non-linear optical sectioning of retinal layers with subcellular spatial resolution was acquired.

Taking advantage of the layered structure of retina, most of the cells in retina can be identified in different depths with the multiphoton imaging technique.A paraffin cross-section of retina (Figure 21_E) reveals that the ganglion cells (the output neurons of the retina) lie innermost in the retina closest to the lens, and the photoreceptors (rods and cones) lie outermost against the RPE cells. Owing to this cellular arrangement, light must first pass through the whole retina before striking and activating the cone photoreceptors. And thereafter, the photons are absorbed by the photoreceptors and translated further into a biochemical message. Subsequently, an electrical signal that can stimulate all the succeeding neurons of the retina is transformed.

The nerve fibers, the axon of ganglion cells functioning as the third ganglion of optic nerves, guide signals produced in the photoreceptor further to the visual processing center located in the occipital lobe of cerebral cortex. These nerve fibers are presented (Figure 21_A). Müller cells are the radial glial cells of retina. The internal limiting membrane of the retina is composed of the contacting Müller cells' end feet, which are located between the nerve fibers and the associated basement membrane. The mitochondria-enriched Müller glial cells can be clearly seen based on its intrinsic fluorescence (Figures 21_A and B). Despite the high resolution and contrast, the interplexiform cells such as dopaminergic starburst amacrine cells and horizontal cells can not be differentiated from bipolar cells in this figure. Rods and cones are displayed in Figure 21_C. As showing in Figure 21_D, most of the RPE cells were found to have regular hexagon-shaped cell bodies and to possess six immediate neighboring cells. The mosaic pattern of these cells showed only slight variation in size and shape. The distribution of cytoplasmic mitochondrial granules allows the non-fluorescent cell membranes of RPE cells to be accurately delineated. The fluorescent cytoplasm surrounding the dark nuclei was based on the mitochondrial NAD(P)H autofluorescence. With the assistance of optical tomography and based on their morphological features, the thickness of the continuous monolayer pigment cells was determined to be 6.6 ± 1.3 microns (z dimension). The homogenous cells had an average value of 9.88 ± 1.65 microns in diameter (x-y dimension) [Wang et al. 2008a].

The multiphoton excitation imaging of retinal lipofuscin has been recently reported [Han et al. 2006]. Lipofuscin, a ubiquitous material presenting in the RPE cell, appears to be the product of the peroxidation of unsaturated fatty acids, and may be symptomatic of membrane damage, or damage to mitochondria and lysosomes [Eldred et al. 1993]. Lipofuscin accumulation is a major risk factor implicated in macular degeneration. Once formed, the RPE cell apparently has no means either to degrade or release the lipofuscin granular into the extracellular space. Excessive accumulation of lipofuscin occurs as a result of lifelong phagocytosis of photoreceptor outer segments. The accumulation represents a common pathogenetic pathway in various monogenetic and complex retinal diseases including the age-related macular degeneration (AMD). Under one-photon CLSM, RPE lipofuscin can be excited at 364, 488, 568, and 633 nm [Marmorstein et al. 2002]. The distribution of human RPE lipofuscin granules was also recorded under multiphoton microscope excited by wavelength of 800 nm with emission spectrum of 500 nm to 550 nm wavelengths. The typical diameter of lipofuscin granules was found to be below one micrometer [Han et al. 2006, Bindewald-Wittich et al. 2006]. The two-photon excitation of melanin was examined typically with the wavelength of 800 nm (Teuchner et al. 2000, Hoffmann et al. 2001). In this paper, the RPE cells were excited and imaged with the same settings for imaging of corneal and scleral cells (excitation wavelength of 760 nm using 435-485 nm filter cube). The autofluorescence was mainly based on the cytoplasmatic NAD(P)H.

To the best of our knowledge, ours is the first report on the non-linear imaging of Müller cells in different layers of intact retinal tissue based on 2PF with intracellular observation [Wang et al. 2008a]. The 3-D co-localization images give snapshots of cell-extracellular matrix organisation, providing highly-resolved visualization of the Müller cells in their native environment. Multiphoton microscopy proved to be a novel tool suitable for visualizing retinal autofluorescence with subcellular spatial resolution and would contribute to assessing the pathophysiology of age- and disease-associated organelles alterations in retina.

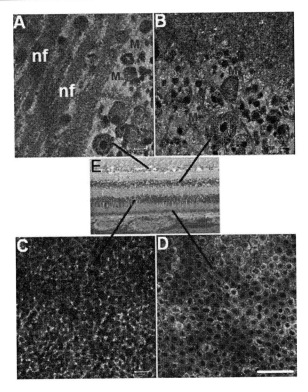

Figure 21. Intrinsic 2PF tomography of retinal layers at an excitation wavelength of 760 nm. Each panel (A, B, C, D) shows a non-linear optical transversal section of retina at different depths. Most of the cells in different layers of retina can be identified with this two-photon imaging technique. The imaging plane was parallel to the outer layer of tunica fibrosa bulbi. (A).Starting at a depth of 15.8 micrometers showing the nerve fiber layer (nf), ganglion cells and Müller glial cells(M); (B). Bipolar cells and Müller glial cells at 42.9 micrometers in depth; Müller cells are the radial glial cells of retina. The internal limiting membrane of the retina is composed of the contacting Müller cells' end feet, which are located between the nerve fibers and the associated basement membrane. To the best of our knowledge, ours is the first report on the Müller cells in different layers of intact retinal tissue based on 2PF with intracellular resolution (Wang *et al.* 2008, Lasers Surg Med); (C).Photoreceptor cells at 96.3 micrometers in depth; (D).Mosaic pattern of the hexagonal RPE cells at 108.8 micrometers in depth. This figure was acquired with the laser beams penetrating through the whole layers of retina (ganglion cells, bipolar cells, photoreceptors) and then reaching the RPE cells using the high-numerical aperture diffraction-limited objective 40 x 1.3 oil, which had a work distance up to 140 micrometers in tissue. The sample examined was an intact whole-mounted eye globe wall involving retina, choroid and sclera. A same precise filter set as used in cornea was used to acquire the intrinsic fluorescence of unstained RPE cells. Most of the RPE cells were found to have regular hexagon-shaped cell bodies and possess six immediate neighboring cells. These cells showed only slight variation in size and shape. In particular, the distribution of mitochondrial granules allows the non-fluorescent cellular membranes of RPE cells to be accurately delineated. The fluorescent cytoplasm surrounding the dark nuclei was based on mitochondrial NAD(P)H autofluorescence. With the assistance of 3-D optical tomography, the thickness of the continuous monolayer pigment cells was determined to be 6.6 ± 1.3 microns (z dimension). The homogenous cells had an average value of 9.88 ± 1.65 microns in diameter (x-y dimension). The middle figure (E) is a paraffin cross-section of a porcine retina located at periphery of fovea, stained by conventional histological technique of Hematoxylin-Eosin (HE). *Scale bars: 50 microns for the four optical images*(Reprint Lasers Surg Med 2008).

7. USE OF MULTIPHOTON EXCITATION IMAGING IN ASSISTING AND ASSESSING INTRASTROMAL SURGERY

The NIR femtosecond pulsed lasers were used as a combined tool for multiphoton excitation imaging and microsurgery [Wang *et al.* 2007d]. Femtolasers provide transient high peak power that can be used in conjunction with focusing optics possessing high numerical aperture to produce light intensities at high terawatt values per square centimetre enough for intratissue photoablation. In the low-repetition-rate regime, material processing is induced by a single pulse based on the formation of high-density plasma. By contrast, the focused laser pulses in a high-repetition-rate femtosecond laser system result in an increase in temperature through heat accumulation, which induces the formation of low-density plasma. As the number of pulses increases, the threshold energy for photodisruption decreases due to a cumulative effect [Stern *et al.* 1989, Stuart *et al.* 1996, Juhasz *et al.* 1996, Vogel *et al.* 1999, König *et al.* 2002].

The use of a femtosecond oscillator with high repetition-rate enables laser surgery at low pulse energy on an order of sub-nanojoule and nanojoule [König *et al.* 2005b]. A femtosecond laser pulse can produce localized energy absorption because the pulse width is shorter than the time required for heat to diffuse out of the focal volume. Thus, the photodisruption occurs only in the focal volume and no photon damage takes place outside the focal femtoliter volume. Another reason for the no-photon-damage outside focus is that biological tissues virtually possess no efficient cellular absorbers in the 700–840 nm spectral region. The non-amplified femtolasers show its advantages in nanoprocessing and multiphoton absorption without compromising viability in the surrounding region. NIR femtolasers have recently attracted a large amount of attention in cellular nanosurgery and nanoprocessing in biological materials. Several researchers have adopted this non-amplified NIR femtosecond laser pulses for precise nanosurgery of cells and sub-cellular organelles such as knock-out single organelles in living cells [König *et al.* 1999, Watanabe *et al.* 2004], nanodissection of human chromosomes [König *et al.* 2001], membrane optoporation [Kohli *et al.* 2005]. Photoablation in human chromosomes with a full width at half maximum (FWHM) cut size of less than 70 nm have been demonstrated with femtolaser pulses of less than 3 nJ pulse energy [König *et al.* 2001]. NJ femtolasers possess the capability to perform precise intratissue and intracellular photodissection on a submicrometer precision.

Intrastromal surgery was performed with the same multiphoton imaging system but with a higher laser power in comparison with the laser power for imaging. In general, the laser power used for multiphoton imaging was set between 20 mW (mean power, corresponding to 0.25 nJ pulse energy and 1.60 kW peak power) and 35 mW (mean power, corresponding to 0.44 nJ pulse energy and 2.80 kW peak power) with respect to different depths in the cornea with a frame scan (2PF and SHG). For corneal microsurgery, laser power increased to between 130 mW (mean power, corresponding to 1.64 nJ pulse energy and 10.40 kW peak power) and 165 mW (mean power, corresponding to 2.08 nJ pulse energy and 13.20 kW peak power) with a line scan at a wavelength of 800 nm.

Intrastromal microsurgery at a depth of 100 micrometers in cornea was visualized using the multiphoton excitation imaging (Figure 22). The *in situ* differentiation of corneal layers makes it possible to determine the target region before the intrastromal microsurgery. Non-linear optical imaging helped making it possible to determine the targeted region for intrastromal surgery. Guided by the optical overview of cornea (Figures 6, 7, 9, 10, 15), the laser was directed on a region of interest for further surgical procedures. After intrastromal treatment was performed with a laser power of 130mW at a wavelength of 800 nm by a line scan (zoom=1.3, t=1s, at a depth of 100 microns), the optical images (SHG at an excitation wavelength of 830 nm in Figure 22_A and autofluorescence images at 760 nm in Figure 22_B) were immediately taken (within 10s). As seen in the optical sections, four laser cuts were acquired with submicron size scale of 0.4-0.6 microns. Some transient subtle bubbles were detected (mean diameters of 2-3 microns, lifetimes less than 2 seconds) along the cuts. With the multiphoton excitation imaging, the surgical performance can be clearly visualized and precisely evaluated within seconds. Multiphoton microscopy/tomography consequently proved capable of optically evaluating the intrastromal laser surgical effects [König *et al.* 2004, Wang *et al.* 2007a].

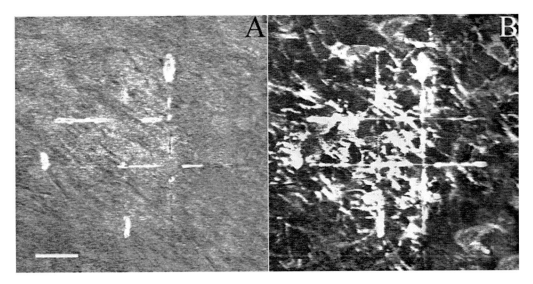

Figure 22. Postoperative multiphoton optical evaluating of intrastromal femtolasers surgery performed at a depth of 100 micrometers. A. SHG image at an excitation wavelength of 830 nm in a backward direction; B. Autofluorescence image at an excitation wavelength of 760 nm. The non-linear optical imaging/tomography of cornea was used as a precise means for definition of interest of region preoperation as well as for inspection and confirmation of the surgical results in real time immediately after the intrastromal nanosurgery.*Scale bar: 45 microns.*After visualizing of cornea with assistance of the 3-D non-linear optical tomography, the targeted region for intrastromal surgery was determined. On accomplishment of the intrastromal nanosurgery performed with a laser power of 130 mW, at a wavelength of 800 nm, with line scan (zoom=1.3, t =1 s, at a depth of 100 μm), the optical images were immediately taken (within 10 s).As seen in the optical sections, four laser cuts were acquired at sub-micron size of 0.4–0.6 microns. The overlying and underlying layers remain intact. During the treatment, some transient subtle bubbles were detected (mean diameters of 2-3 microns, lifetimes less than 2 seconds) along the incision. The surrounding keratocytes and collagen lamellas remain intact and are even comparable to those microstructures in physiological stats (Figure 10). The surgical performance can be simultaneously visualized and evaluated in real time, which makes the multiphoton excitation imaging an indispensable assessing component of the laser intratissue surgery (Reprint Lasers Surg Med 2007).

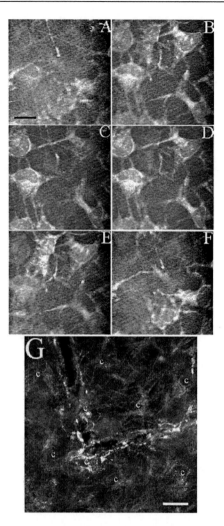

Figure 23. *In vivo* multiphoton excitation imaging tomography of myofibroblasts based on their intrinsic 2PF at an excitation wavelength of 760 nm with a vertical z-interval of 2 micrometers twenty-four hours after the laser operation (A-F). (G). *In vivo* reflection imaging of the treated stromal region twenty-four hours after the intrastromal surgery.The activated cells are marked with the symbol c in Figure G. *Scale bars: 20 microns for A-F and 35 microns for G.* Optical imaging (reflexion) of the treated cornea 24 h postoperation showing the ablated area and the myofibroblasts at a distance of 50 micrometers from the intrastromal laser lesions. Surgery was performed within the cornea at depths of between 90 μm and 100 μm with vertical z-interval of 2 μm using high-powered laser beams between 150 mW (at the depth of 90 μm) and 165 mW (at the depth of 100 μm) with line scanning. Observed 24 h after laser treatment, the luminescence along the cut edges became noticeably weaker compared to that immediately after the laser operation. 3-D 2PF tomography of the activated cells was performed to study the cellular response to the lesions in the laser-treated corneas during the postoperative wound healing process. Remarkable, cell aggregates were detected in the surrounding area of treated region. Due to the deficiency of collagen SHG signals at the excitation wavelength of 760 nm, the morphological details of the activated cells can be selectively studied with three dimensions and intracellular observations. Based on their altered morphology observed -- (i) hypertrophic cell bodies with diameters of 25-30 microns (the normal cells have a size of less than 20 microns, Figure 10); (ii) the appearance of augmented cell processes; and (iii) nuclei round or oval in shape -- these cells were determined to be the activated intrastromal keratocytes, called myofibroblasts. Control corneas exhibited no such cell aggregates.The rapid transformation of keratocytes seems to be the first cellular reaction after intrastromal laser surgery (Reprint Lasers Surg Med 2007).

The data obtained in this paper suggest distinct advantages of nJ femtosecond lasers compared with microjoule (μJ) femtosecond lasers. First, the absence of large intratissue bubbles by nJ femtolasers makes it feasible to ablate the tissue more precisely. The sharp boundary between the treated and untreated areas indicates the high precision of the surgical treatment (Figure 22 and Figure 23_G). Collateral effects, such as tissue streaks induced by the μJ femtolasers in intracorneal surgery, were reported [Lubatschowski et al. 2000]. Streaks, created inside the corneal tissue irradiated with 160 fs laser pulses at a pulse energy of 2 μJ, were visualized on histological sections. In TEM check, the streaks can be noticed as a dark staining, crossing the picture in vertical direction. The diameter of the streaks was in the range of 200–500 nm. The intrastromal streaks, still seen for some days postoperatively, would have an effect on the wound healing process [Arnold et al. 2005]. Such collateral phenomenon was not observed with nJ femtolasers based on histology, optical imaging or EM examinations.

The nJ ultrashort lasers were employed not only as probing beams but as also a surgical tool. The outcomes of intratissue microsurgery in animal studies have been on-line displayed with the most convenient imaging combination being available. Conventional histological methods have difficulty in showing the intratissue minute cuts because the two cut edges merge immediately after the absorption of the tiny bubbles. Although electron microscopy can show the microsurgical results, the procedure is complicated and time-consuming. The non-linear optical tomography of the cornea is perfectly suited for determining the targeted region preoperatively. The corneal layers and depth confirmed by this optical tomography ensure the precision of femtosecond laser surgery. This imaging technique is indispensable for immediately evaluating intratissue microsurgery. Therefore, integrating the two techniques in one system, as was done in this paper, is currently the most efficient and convenient way to carry out intratissue surgery with non-amplified femtolasers.

This paper reviews the *in vivo* application of optical corneal imaging based on multiphoton microscopy/tomography in the intrastromal microsurgery performed with NIR nJ femtolasers. The surgical performance can be simultaneously visualized and evaluated in real time, which makes the multiphoton excitation imaging an indispensable component for laser intratissue surgery. The use of multiphoton imaging was found also in the generation of intratissue ablation [Wang et al. 2007b] and intracorneal lenticules [Wang et al. 2008b]. With the assistance of the non-linear optical imaging, multiphoton-mediated intratissue ablation was generated while excluding severe damage and mechanical effects in the surrounding tissue. In conclusion, multiphoton microscopyis an outstanding tool in assisting in performing intratissue laser surgery and ensures predictable and reproductive surgical outcomes. This non-linear optical imaging technique continues to find an increasing number of applications in biological and medical laser surgery. A most interesting potential application of the multiphoton excitation imaging technique would be in assisting the intracellular organelle nanosurgery.

8. Multiphoton Microscopy of Activated Cells in Stroma after Surgical Treatment

Keratocyte is the predominant cell responsible for maintaining the integrity and regeneration of corneal components. The cells are connected to each other by gap junctions through their own cell processes. This cell in physiological state is inactive in phagocytosis. Under stress such as laser surgery, the cells can transform to myofibroblasts in the area around the corneal wound, and secret new ECM that ultimately reconstitutes the components of cornea [Jester *et al.* 1999, Wilson 2002, Mohan *et al.* 2003]. The factors, inducing cascades of wound repair, are secreted firstly by the activated cell [Funderburgh *et al.* 2001, Berryhill *et al.* 2002].

Multiphoton excitation imaging was used to detect the stromal cell activation. Surgery was performed within the cornea at depths of between 90 μm and 100 μm with vertical z-step of 2 micrometers using high-powered laser beams between 150 mW (at a depth of 90 μm) and 165 mW (at a depth of 100 μm) assisted by the line scanning (Figure 23_G). Observed 24 hours after the laser treatment, part of the cut spaces were filled with migrating cells. Remarkable, cell aggregates were detected in the surrounding region within 50 μm of the intrastromal lesion sites. Based on their special morphology observed -- (i) hypertrophic cell bodies with diameters of 25-30 microns (the normal cells have a size of less than 20 microns, Figure 10); (ii) the appearance of augmented cell processes; and (iii) the round or oval nuclei were surrounded by the intensified cytoplasmic autofluorescence granules -- these cells most likely seemed to be activated intrastromal keratocytes (myofibroblasts) (Figure 23). Control corneas exhibited no such cell aggregates [Wang *et al.* 2007a].

Interestingly, we have been able to *in vivo* detect stromal myofibroblasts after intratissue ultrashort laser surgery. The myofibroblasts are defined in this research by their special morphology and intensifier intracellular endogenous fluorescence. There are a large number of studies related to the transformation of fibroblasts. Masur *et al.* [1996] have studied the transformation of keratocytes with a smooth muscleα-actin stain. Whereas transforming growth factor-β1 stimulates myofibroblasts differentiation, fibroblast growth factor-2 and platelet-derived growth factor inhibit such transformation [Maltseva *et al.* 2001, Baldwin *et al.* 2002]. Cells transformed from keratocytes resemble satellite cells, which are highly reflective and characterized by intense cytoplasmic fluorescence. The rapid transformation of keratocytes during the course of laser-induced wound repair seems to be the first cellular reaction after the intrastromal laser surgery. Thereafter, myofibroblasts synthesize chemokines and chemotaxis, leading to characteristic inflammatory cascade reactions during wound repair. The contributions of myofibroblasts involving wound contraction and phagocytosis in the healing process have been discussed elsewhere [Funderburgh *et al.* 2001, Chakravarti *et al.* 2004].

2PF signal imaging was performed to study the cellular response to the lesions in the laser-treated corneas during the postoperative wound healing. *In vivo* imaging allowed the intracorneal myofibroblasts to be detected based on its intrinsic NAD(P)H autofluorescence at the wavelength of 760 nm.

CONCLUSION

Our aim was to use spectrally sensitive multiphoton excitation microscopy to elucidate microstructures of the whole-mount ocular tissues. Here, data show that 2PF can be used to identify cellular components, whereas SHG is valuable in displaying of collagenous microstructure in bulk tissues. Thick, dense, and non-fixed bulk tissues can be easily probed based on their endogenous fluorophore and collagen. Contrast specifically for both intratissue cells and their hosting ECM was sufficiently generated only based on the in-tandem detection with 2PF and SHG signal imaging. The 3-D co-localization images give snapshots of cell morphology and cell-extracellular matrix organization. The quantitative analysis of tissue microstructures including cellular and fibrous compositions was obtained. Multiphoton microscopy allows for the *in vivo*non-invasive study of biological specimens in three dimensions with submicron resolution at depths of up to some hundreds of micrometers.Because the examinations in this paper do not involve any assistance of exogenous dyes or the need for tissue preparation or fixation, it avoids artefacts induced by additional sample staining and slicing procedures. This proof-of-principle examination has already started to reveal the intricate features in bulk tissues that are otherwise lost in conventional histology studies, suggesting the potential of using non-linear optical imaging as a routine tool for more advanced optical histology.

Several other advantages of multiphoton microscopy can be immediately confirmed and inferred from our experimental data: (i) Improved depth penetration in thick specimens due to greatly reduced scattering of longer wavelengths. In the NIR illumination wavelengths of 700–840 nm, there are virtually no other efficient tissue absorbers except for water. With a suitable objective 20x 0.9 N.A. used in the review paper, the deepest layer of corneal tissue can be even optically visualized *in situ* [Wang *et al.* 2007a]); (ii) A promising imaging approach for *in vivo*examinations with biomedical samples since no cellular and tissue staining or slicing is required; (iii) Tremendous improvements in optical localization and z-resolution because the majority of all multiphoton excitations was confined to a minute volume of 0.1 fl focal point; the smallest z-step reaches to the high precision of 25 nm; (iv) Dramatically promoted signal-to-noise ratio since the large separation between the incident NIR excitation and the subsequent visible emission wavelengths, which makes it straightforward to achieve a high signal to background ratio.

More recently, a novel organ image acquisition technique called two-photon tissue cytometry has been posed. The criteria for the development of 3-D tissue cytometry include spatial resolution, depth penetration and molecular specificity. The imaging features of multiphoton microscopic technique are perfectly in coincidence with these requests. Two-photon tissue cytometry, imaging the microstructural components in organs and tissues with sufficient details to reveal 3-D sample morphology, offers all the benefits of fluorescence-based approaches including high specificity and sensitivity as well as appropriateness for molecular imaging. Coupled with automated histological sectioning, an entire heart has been imaged with subcellular resolution based on two-photon microscopy [Ragan *et al.* 2007]. This new sensitive imaging technique was demonstrated to be capable of quantifying 3-D morphology of cardiac myocytes and of evaluating microvasculature remodelling. In an analogous manner, the development of imaging entire eye globes based on multiphoton excitation microscopy would also be encouraging. The data achieved in this review paper provide the primary anatomical knowledge of eye globes for the further development of two-photon ocular tissue cytometry.

Multiphoton excitation images obtained from various tissues show providing excellent structural and biomechanical characterization. Intrinsic 2PF has been used to distinguish dysplastic and neoplastic tissues from normal tissues, to identify tumorous tissue [Guo *et al.* 1999, Brown *et al.* 2003] and the decrepit brain tissues associated with Alzheimer's disease [Hensley *et al.* 1998, Christie *et al.* 2001]. The nonlinear optical tomography also provides a new approach to study of dysplasia in squamous epithelium and abnormal collagen in keratoconus [Lyubovitsky *et al.* 2006, Morishige *et al.* 2007]. Owing to emerging technologies such as fluorescence resonance energy transfer (FRET) and fluorescence correlation spectroscopy, multiphoton technique is now being further widely applied to probe single protein molecules and to investigate the interactions of their functions [Xu *et al.*1996, Combs *et al.* 2001, Biskup *et al.* 2007]. Imaging of SHG signals was employed as a sensitive tool for detection of corneal fibrosis and then further to be used to assess the effects of antifibrotic therapy on corneal wound healing [Farid *et al.* 2008].

Data show that the combination of 2PF and SHG allows for the selective visualization of various microstructural components in the whole-mount scleral, corneal and retinal tissues. *In situ* quantitative and qualitative assessment of microstructures in different layers of the intact bulk tissues was obtained. It enables high-resolution deep tissue imaging of endogenous fluorophores and is well suitable for molecular imaging in living biological specimens. Intrinsic emission multiphoton microscopy/tomography is consequently confirmed to be an efficient and sensitive non-invasive imaging approach. The non-linear optical imaging can be used as a versatile biomedical tool for optical diagnostic imaging, yielding vivid insights into biological specimens, which will ultimately find its clinical application in optical pathological diagnostics. We also believe that this promising technique will find more applications in the biological and medical basic research in the near future.

ACKNOWLEDGMENTS

The cooperation with Drs. Karsten König and Iris Riemann from the Fraunhofer Institute of Biomedical Technology, Harald Schubert from the Institute of Laboratory Animals Science, Shuping Song from the Institute of Molecular Cell Biology, Dietrich Schweitzer and Martin Hammer from the Eye Clinic, Annett Eitner from the Institute of Anatomy II, as well as with the companies of Carl Zeiss Jena and JenLab in Germany is gratefully acknowledged.

REFERENCES

Arnold, CL; Heisterkamp, A; Ertmer, W; Lubatschowski, H. Streak formation as side effect of optical breakdown during processing the bulk of transparent Kerr media with ultra-short laser pulses. *Appl Phys B*, 2005; 80(2):247-253.

Baldwin, HC; Marshall, J. Growth factors in corneal wound healing following refractive surgery: a review. *Acta OphthalmolScand*, 2002; 80 (3):238-247.

Barzda, V; Greenhalgh, C; Aus der Au, J; Elmore, S; Van Beek, J; Squier, J. Visualization of mitochondria in cardiomyocytes by simultaneous harmonic generation and fluorescence microscopy. *Opt Express*, 2005; 13:8263-8276.

Bentley, E; Abrams, GA; Covitz, D; Cook, CS; Fischer, CA; Hacker, D; Stuhr, CM; Reid, TW; Murphy, CJ. Morphology and immunohistochemistry of spontaneous chronic corneal epithelial defects in dogs. *Invest Ophthal Vis Sci*, 2001; 42(10):2262-2269.

Berryhill, BL; Kader, R; Kane, B; Birk, DE; Feng, J; Hassell, JR. Partial restoration of the keratocyte phenotype to bovine keratocytes made fibroblastic by serum. *Invest Ophthalmol Vis Sci*, 2002; 43 (11):3416-3421.

Bindewald-Wittich, A; Han, M; Schmitz-Valckenberg, S; Snyder, SR; Giese, G; Bille, JF; Holz, FG. Two-photon-excited fluorescence imaging of human RPE cells with a femtosecond Ti:Sapphire laser.*Invest Ophthalmol Vis Sci,*2006; 47:4553-4557.

Biskup, C; Zimmer, T; Kelbauskas, L; Hoffmann, B; Klöcker, N; Becker, W; Bergmann, A; Benndorf, K. Multi-dimensional fluorescence lifetime and FRET measurements. *Microsc Res Tech*, 2007; 70(5):442-451.

Boyde, A. (1974). *Scanning Electron Microscopy: Three-dimensional aspects of SEM images*(Well, O,C,) (ed,). New York, McGraw-Hill.

Brakenhoff, GJ; Squier, J; Norris, T; Bliton, AC; Wade, WH; Athey, B. Real-time two-photon confocal microscopy using a femtosecond, amplified Ti:sapphire system. *J Microsc*, 1996; 181:253-259.

Brecht, M; Fee, MS; Garaschuk, O; Helmchen, F; Margrie, TW; Svoboda, K; Osten, P. Novel approaches to monitor and manipulate single neurons*in vivo*. *J Neurosci*, 2004; 24(42):9223-9227.

Brown-Augsburger, P; Broekelmann, T; Mecham, L; Mercer, R; Gibson, MA; Cleary, EG; Abrams, WR; Rosenbloom, J; Mecham, RP. Microfibril-associated glycoprotein binds to the carboxyl-terminal domain of tropoelastin and is a substrate for transglutaminase. *J Biol Chem*, 1994; 269(45):28443-28449.

Brown, E; McKee, T; Ditomaso, E; Pluen, A; Seed, B; Boucher, Y; Jain, RK. Dynamic imaging of collagen and its modulation in tumors*in vivo* using second-harmonic generation. *Nat Med*, 2003; 9:796-800.

Camber, O; Rehbinder, C; Nikkila, T; Edman, P. Morphology of the pig cornea in normal conditions and after incubation in a perfusion apparatus. *Acta Vet Scand*, 1987; 28(2):127-134.

Campagnola, PJ; Millard, AC; Terasaki, M; Hoppe, PE; Malone, CJ; Mohler, WA. Three-dimensional high-resolution second harmonic generation imaging of endogenous structural proteins in biological tissues. *Biophys J*, 2002; 82(1):493–508.

Campagnola, PJ; Loew, LM. Second-harmonic imaging microscopy for visualizing biomolecular arrays in cells, tissues and organisms.*Nat Biotechnol,*2003; 21(11):1356-1360.

Chakravarti, S; Wu, F; Vij, N; Roberts, L; Joyce, S. Microarray studies reveal macrophage-like function of stromal keratocytes in the cornea. *Invest Ophthalmol Vis Sci*, 2004; 45 (10):3475-3484.

Christie, RH; Bacskai, BJ; Zipfel, WR; Williams, RM; Kajdasz, ST; Webb, WW; Hyman, BT. Growth arrest of individual senile plaques in a model of Alzheimer's disease observed by *in vivo* multiphoton microscopy.*J Neurosci*, 2001;21(3):858-864.

Combs, CA; Balaban, RS. Direct Imaging of Dehydrogenase Activity within Living Cells Using Enzyme-Dependent Fluorescence Recovery after Photobleaching (ED-FRAP). *Biophys J*, 2001; 80(4):2018-2028.

Dedov, VN; Cox, GC; Roufogalis, BD. Visualisation of mitochondria in living neurons with single- and two-photon fluorescence laser microscopy. *Micron*, 2001; 32(7):653-660.

Denk, W; Strickler, JH; Webb, WW. Two-photon laser scanning fluorescence microscopy. *Science*, 1990; 248 (4951):73-76.

Denk, W. Two-photon scanning photochemical microscopy: mapping ligand-gated ion channel distributions.*PNAS,* 1994; 91(14):6629-6633.

Diaspro, A; Robello, M. Two-photon excitation of fluorescence for three-dimensional optical imaging of biological structures. *J Photochem Photobiol B,* 2000; 55(1):1-8.

Dombeck, DA; Kasischke, KA; Vishwasrao, HD; Ingelsson, M; Hyman, BT; Webb, WW. Uniform polarity microtubule assemblies imaged in native brain tissue by second-harmonic generation microscopy. *PNAS,* 2003; 100(12):7081-7086.

Elbadawi, A. Hexachrome modification of Movat's stain. *Biotechnic and Histochemistry,* 1976; 51(5):249-253.

Eldred, GE; Lasky, MR. Retinal age pigments generated by self-assembling lysosomotropic detergents. *Nature,* 1993; 361(6414):724-726.

Farguharson, C; Robins, SP. Immunolocalization of collagen type I, III, and IV, elastin and fibronectin within the heart of normal and copper deficient rats. *J Comp Pathol,* 1991;104(3):245-255.

Farid, M; Morishige, N; Lam, L; Wahlert, A; Steinert, RF; Jester, JV. Detection of Corneal Fibrosis by Imaging Second Harmonic Generated Signals (SHG) in Rabbit Corneas Treated with Mitomycin C Following Excimer Laser Surface Ablation. *Invest Ophthalmol Vis Sci,* 2008; DOI: 10.1167/iovs.08-1983.

Fine, S; Hansen, WP. Optical second harmonic generation in biological systems. *Appl Opt,* 1971; 10(10):2350–2353.

Freeman, IL. Collagen polymorphism in mature rabbit cornea. *Invest Ophthalmol Vis Sci,* 1978; 17(2):171-177.

Freund, I; Deutsch, M; Sprecher, A.Connective tissue polarity. Optical second-harmonic microscopy, crossed-beam summation, and small-angle scattering in Rat-tail tendon. *Biophys J,* 1986; 50:693-712.

Funderburgh, JL; Funderburgh, ML; Mann, MM; Corpuz, L; Roth, MR. Proteoglycan expression during transforming growth factor beta-induced keratocyte-myofibroblast transdifferentiation. *J Biol Chem,* 2001; 276(47):44173–44178.

Fujimoto, JG. Optical coherence tomography for ultrahigh resolution*in vivo* imaging. *Nat Biotechnol,* 2003; 21(11):1361-1367.

Göppert-Mayer, M. Über Elementarakte mit zwei Quantensprüngen. *Ann Phys,* 1931; 9:273-294.

Guo, Y; Savage, HE; Liu, F; Schantz, SP; Ho, PP; Alfano, RR. Subsurface tumor progression investigated by non-invasive optical second harmonic tomography. *PNAS,* 1999; 96(19):10854-56.

Halbhuber, KJ; Krieg, W; König, K. Laser scanning microscopy in enzyme histochemistry. Visualization of cerium-based and DAB-based primary reaction products of phosphatases, oxidases and peroxidases by reflectance and transmission laser scanning microscopy. *Cell Mol Bio (Noisy-le-grand),* 1998; 44: 807-826.

Halbhuber, KJ; König, K. Modern laser scanning microscopy in biology, biotechnology and medicine. *Ann Anat,* 2003; 185(1):1-20.

Han, M; Giese, G; Bille, JF. Second harmonic generation imaging of collagen fibrils in cornea and sclera. *Opt Express,* 2005; 13:5791-5797.

Han, M; Bindewald-Wittich, A; Holz, FG; Giese, G; Niemz, MH; Snyder, S; Sun, H; Yu, J; Agopov, M; Schiazza, OL; Bille, JF. Two-photon excited autofluorescence imaging of human retinal pigment epithelial cells. *J Biomed Opt,* 2006; 11(1):010501.

Hayashi, S; Osawa, T; Tohyama, K. Comparative observations on corneas, with special reference to Bowman's layer and Descemet's membrane in mammals and amphibians. *J Morphol,* 2002; 254(3):247-58.

Helmchen, F; Denk, W. Deep tissue two-photon microscopy. *Nat Methods,* 2005; 2(12):932-940.

Hensley, K; Maidt, ML; Yu, Z; Sang, H; Markesbery, WR; Floyd, RA. Electrochemical analysis of protein nitrotyrosine and dityrosine in the Alzheimer brain indicates region-specific accumulation. *J Neurosci,* 1998; 18(20):8126-8132.

Hoffmann, K; Stücker, M; Altmeyer, P; Teuchner, K; Leupold, D. Selective Femtosecond Pulse-Excitation of Melanin Fluorescence in Tissue. *J Invest Dermatol,* 2001; 116:629-630.

Huang, D; Swanson, EA; Lin, CP; Schuman, JS; Stinson, WG; Chang, W; Hee, MR; Flotte, T; Gregory, K; Puliafito, CA; Fujimoto, JG. Optical coherence tomography. *Science,* 1991; 254(5035):1178-1181.

Huang, S; Heikal, AA; Webb, WW. Two-photon fluorescence spectroscopy and microscopy of NAD(P)H and flavoprotein. *Biophys J,* 2002; 82(5):2811-2825.

Jester, JV; Petroll, WM; Cavanagh, HD. Corneal stromal wound healing in refractive surgery: the role of myofibroblasts. *Prog Retin Eye Res,* 1999; 18(3):311-356.

Juhasz, T; Kastis, GA; Suárez, C; Bor, Z; Bron, WE. Time-resolved observations of shock waves and cavitation bubbles generated by femtosecond laser pulses in corneal tissue and water. *Lasers Surg Med,* 1996; 19:23-31.

Kaiser, W; Garrett, CGB. Two-Photon Excitation in CaF2: Eu2+. *Phys Rev Lett,* 1961; 7:229-231.

Kato, T; Chang, JH; Azar, DT. Expression of type XVIII collagen during healing of corneal incisions and keratectomy wounds. *Invest Ophthalmol Vis Sci,* 2003; 44:78-85.

Kohli, V; Elezzabi, AY; Acker, JP. Cell nanosurgery using ultrashort laser pulses: Applications to membrane surgery and cell isolation. *Lasers Surg Med,* 2005; 37:227-230.

König, K; Simon, U; Halbhuber, KJ. 3D-resolved two-photon fluorescence microscopy of living cells using a modified confocal laser scanning microscope. *Cell Mol Biol (Noisy-le-grand),* 1996; 42(8):1181-1194.

König, K; Riemann, I; Fischer, P; Halbhuber, KJ. Intracellular nanosurgery with near infrared femtosecond laser pulses. *Cell Mol Biol (Noisy-le-grand),* 1999; 45(2):195-201.

König, K. Multiphoton Microscopy in Life Sciences. *J Microscopy,* 2000(a); 200(Pt 2):83-104.

König, K. Laser tweezers and multiphoton microscopes in life sciences. *Histochem Cell Biol,* 2000(b); 114(2):79-92.

König, K; Riemann, I; Fischer, P; Halbhuber, KJ. Multiplex FISH and three dimensional DNA imaging with near infrared femtosecond laser pulses. *Histochem Cell Biol,* 2000(c); 114(4): 337-345.

König, K; Riemann, I; Fritzsche, W. Nanodissection of human chromosomes with near-infrared femtosecond laser pulses. *Opt Lett,* 2001; 26(11):819-821.

König, K; Krauss, O; Riemann, I. Intratissue surgery with 80 MHz nanojoule femtosecond laser pulses in the near infrared. *Opt Express,* 2002; 10:171-176.

König, K; Wang, BG; Krauss, O; Riemann, I; Schubert, H; Kirste, S; Fischer, P. First *in vivo* animal studies on intraocular nanosurgery and multiphoton tomography with low-energy

80 MHz near infrared femtosecond laser pulses. In: Manns F, Soderberg PG, Ho A (editors). *Title: Ophthalmic Technologies XIV.*San Jose: SPIE; 2004; 5314:262-269.

König, K; Schenke-Layland, K; Riemann, I; Stock, UA. Multiphoton autofluorescence imaging of intratissue elastic fibers. *Biomaterials,* 2005(a); 26(5):495–500.

König, K; Wang, BG; Riemann, I; Kobow, J. Cornea surgery with nanojoule femtosecond laser pulses. In: Manns F, Soederberg PG, Ho A, Stuck BE, Belkin M (editors). *Title: Ophthalmic Technologies XV.* San Jose: SPIE; 2005(b); 5688:288-293.

König, K; Riemann, I. High-resolution multiphoton tomography of human skin with subcellular spatial resolution and picosecond time resolution. *J Biomed Opt,* 2003; 8(3):432-439.

Levene, MJ; Dombeck, DA; Kasischke, KA; Molloy, RP; Webb, WW. *In vivo* multiphoton microscopy of deep brain tissue.*J Neurophysiol,* 2004; 91(4):1908-1912.

Li, Q; Timmers, AM; Hunter, K; Gonzalez-Pola, C; Lewin, AS; Reitze, DH; Hauswirth, WW.Noninvasive imaging by optical coherence tomography to monitor retinal degeneration in the mouse. *Invest Ophthalmol Vis Sci,* 2001; 42:2981-2989.

Light, ND. Bovine type I collagen: A study of cross-linking in various mature tissues. *Biochim Biophys Acta,* 1979; 581(1):96-105.

Lubatschowski, H; Maatz, G; Heisterkamp, A; Hetzel, U; Drommer, W; Welling, H; Ertmer, W. Application of ultrashort laser pulses for intrastromal refractive surgery. *Graefe's Arch Clin Exp Ophthalmol,* 2000; 238(1):33-39.

Lyubovitsky, JG; Spencer, JA; Krasieva, TB; Andersen, B; Tromberg, BJ. Imaging corneal pathology in a transgenic mouse model using nonlinear microscopy. *J Biomed Opt,* 2006; 11(1):014013.

Maltseva, O; Folger, P; Zekaria, D; Petridou, S; Masur, SK. Fibroblast growth factor reversal of the corneal myofibroblast phenotype. *Invest Ophthalmol Vis Sci,* 2001; 42(11):2490-2495.

Malak, H; Castellano, FN; Gryczynski, I; Lakowicz, JR. Two-photon excitation of ethidium bromide labeled DNA. *Biophys Chem,* 1997; 67:35-41.

Marmorstein, AD; Marmorstein, LY; Sakaguchi, H; Hollyfield, JG. Spectral profiling of autofluorescence associated with lipofuscin, Bruch's Membrane, and sub-RPE deposits in normal and AMD eyes.*Invest Ophthalmol Vis Sci,*2002; 43:2435–2441.

Masur, SK; Dewal, HS; Dinh, TT; Erenburg, I; Petridou, S. Myofibroblasts differentiate from fibroblasts when plated at low density. *PNAS,* 1996; 93:4219-4223.

Maiti, S; Shear, JB; Williams, RM; Zipfel, WR; Webb, WW. Measuring serotonin distribution in live cells with three-photon excitation. *Science,* 1997; 275 (5299):530-532.

Maurice, D.M. (1984). *The eye: The cornea and sclera* (Davson, H.) (ed.). New York, London: Academic Press.

Maser, MD; Trimble, JJ. Rapid chemical dehydration of biological samples for scanning electron microscopy using 2,2-dimethoxypropane. *J Histochem Cytochem,* 1977; 25(4):247-251.

Masters, BR; So, PT; Gratton, E. Multiphoton excitation fluorescence microscopy and spectroscopy of in vivo human skin. *Biophys J,* 1997; 72(6):2405-2412.

Merindano, MD; Costa, J; Canals, M; Potau, JM, Ruano, D. A comparative study of Bowman's layer in some mammals: Relationships with other constituent corneal structures. *European Journal of Anatomy,* 2002; 6(3):133-139.

Mohan, RR; Hutcheon, AE; Choi, R; Hong, J; Lee, J; Mohan, RR; Ambrósio, R Jr; Zieske, JD; Wilson, SE. Apoptosis, necrosis, proliferation, and myofibroblast generation in the stroma following LASIK and PRK. *Exp Eye Res,* 2003; 76(1):71-87.

Morishige, N; Wahlert, AJ; Kenney, MC; Brown, DJ; Kawamoto, K; Chikama, T; Nishida, T; Jester, JV. Second-harmonic imaging microscopy of normal human and keratoconus cornea. *Invest Ophthalmol Vis Sci,* 2007; 48(3):1087-1094.

Moreaux, L; Sandre, O; Mertz, J. Membrane imaging by second-harmonic generation microscopy. *J Opt Soc Am B,* 2000; 17:1685-1694.

Nakayasu, K; Tanaka, M; Konomi, H; Hayashi, T. Distribution of types I, II, III, IV and V collagen in normal and keratoconus corneas. *Ophthalmic Res,* 1986; 18(1):1-10.

Nishida, T; Yasumoto, K; Otori, T; Desaki, J. The network structure of corneal fibroblasts in the rat as revealed by scanning electron microscopy.*Invest Ophthalmol Vis Sci,* 1988; 29(12):1887-90.

Nitsch, R; Phol, EE; Smorodchenko, A; Infante-Durarte, C; Aktas, O; Zipp, F. Direct impact of T cells on neurons revealed by two-photon microscopy in living brain tissue. *J Neurosci,* 2004; 24(10):2458-2464.

Patterson, GH; Knobel, SM; Arkhammar, P; Thastrup, O; Piston, DW. Separation of the glucose stimulated cytoplasmic and mitochondrial NAD(P)H responses in pancreatic islet cells. *PNAS,* 2000; 97(10):5203-5207.

Piston, DW. Imaging living cells and tissues by two-photon excitation microscopy. *Trends Cell Biol,* 1999; 9(2):66-69.

Poole, CA; Brookes, NH; Clover, GM. Keratocyte networks visualised in the living cornea using vital dyes.*J Cell Sci,* 1993; 106(Pt 2):685-691.

Ragan, T; Sylvan, JD; Kim, KH; Huang, H; Bahlmann, K; Lee, RT; So, PT. High-resolution whole organ imaging using two-photon tissue cytometry. *J Biomed Opt,* 2007; 12(1):014015.

Roth, S; Freund, I. Second harmonic generation in collagen. *J Chem Phys,*1979; 70:1637-1643.

Rubart, M. Two-photon microscopy of cells and tissue. *Circ Res,* 2004; 95(12):1154-1166.

So, PTC; Kim, H; Kochevar, I.Two-photon deep tissue ex vivo imaging of mouse dermal and subcutaneous structures. *Opt Express,* 1998; 3(9):339-350.

Scharenberg, K. The cells and nerves of the human cornea. *Am J Ophthal,* 1955; 40: 368-379.

Schenke-Layland, K; Riemann, I; Opitz, F; König, K; Halbhuber, KJ; Stock, UA. Comparative study of cellular and extracellular matrix composition of native and tissue engineered heart valves. *Matrix Biol,* 2004; 23(2):113-25.

Sekundo, W; Marshall, GE; Lee, WR; Kirkness, CM. Immuno-electron labelling of matrix components in congenital hereditary endothelial dystrophy. *Graefes Arch Clin Exp Ophthalmol,* 1994; 232(6):337-346.

Squirrell, JM; Wokosin, DL; White, JG; Bavister, BD. Long-term two-photon fluorescence imaging of mammalian embryos without compromising viability. *Nat Biotechnol,*1999; 17:763–767.

Stern, D; Schoenlein, RW; Puliafito, CA; Dobi, ET; Birngruber, R; Fujimoto, JG. Ablation by nanosecond, picosecond, and femtosecond lasers at 532 nm and 625 nm. *Arch Ophthalmol,* 1989; 107(4):587-592.

Stoll, S; Delon, J; Brotz, TM; Germain, RN. Dynamic imaging of T cell-dendritic cell interactions in lymph nodes. *Science,* 2002; 296(5574):1873-1876.

Stuart, BC; Feit, MD; Herman, S; Rubenchik, AM; Shore, BM; Perry, MD. Optical ablation by high-power short-pulse lasers. *J Opt Soc Am (B),* 1996; 13:459-468.

Terry, MA; Ousley, PJ; Will, B. A practical femtosecond laser procedure for DLEK endothelial transplantation: cadaver eye histology and topography. *Cornea,* 2005; 24(4):453-459.

Teuchner, K; Ehlert, J; Freyer, W; Leupold, D; Altmeyer, P; Stücker, M; Hoffmann, K. Fluorescence studies of melanin by stepwise two-photon femtosecond laser excitation. *J Fluorescence,* 2000; 10(3):275-281.

Tseng, SC; Smuckler, D; Stern, R. Comparison of collagen types in adult and fetal bovine corneas. *J Biol Chem,* 1982; 257(5):2627-2633.

Vogel, A; Noack, J; Nahen, K; Theisen, D; Busch, S; Parlitz, U; Hammer, D; Noojin, G; Rockwell, B; Birngruber, R. Energy balance of optical breakdown in water at nanosecond to femtosecond time scales. *Appl Phys B,* 1999; 68(2):271-280.

Vogel, A; Venugopalan, V. Mechanisms of pulsed laser ablation of biological tissues. *Chem Rev,* 2003; 103(2):577-644.

Williams, RM; Zipfel, WR; Webb, WW. Multiphoton microscopy in biological research. *Curr Opin Chem Biol,* 2001; 5(5): 603-608.

Williams, RM; Zipfel, WR; Webb, WW. Interpreting second-harmonic generation images of collagen I fibrils. *Biophy J,* 2005; 88:1377-1386.

Wang, BG; Riemann, I; Halbhuber, KJ; Schubert, H; Kirste, S; König, K. In-vivo animal follow-up studies on intrastromal surgery with near-infrared nanojoule femtosecond laser pulses. In: Jacques SL; Roach WP (Editors). *Title: Optical Interactions with Tissue and Cells XVI.* San Jose: SPIE; 2005(a); 5695(47):292-302.

Wang, BG; Halbhuber, KJ; Riemann, I; König, K. In-vivo Corneal Nonlinear Optical Tomography based on Second Harmonic and Multiphoton Autofluorescence Imaging induced by Near-Infrared Femtosecond Lasers with Rabbits. In: Chatard JP, Dennis PNJ (Editors). *Title: Detectors and Associated Signal Processing II.* Jena: SPIE; 2005(b); 5964(41):1-11.

Wang, BG; Halbhuber, KJ; Riemann, I; Schubert, H; Kirste, S; König, K. Application of multiphoton autofluorescence imaging (MAI) and second harmonic generation (SHG) to intrastromal surgery with rabbits. abstract, *FOM,* 2005(c). www.focusonmicroscopy. org/2005/PDF/073_Wang.pdf.

Wang, BG; König, K; Riemann, I; Krieg, R; Halbhuber, KJ. Intraocular Multiphoton Microscopy with Subcellular Spatial Resolution by Infrared Femtosecond Lasers. *Histochem Cell Biol,* 2006(a); 126(4):507-515.

Wang, BG; König, K; Riemann, I; Schubert, H; Halbhuber, KJ. Multiphoton Imaging of Corneal Tissue with Near-Infrared Femtosecond Laser Pulses: Corneal Optical Tomography and Its Use in Refractive Surgery. In: Periasamy A, So PTC (Editors). *Title: Multiphoton Microscopy in the Biomedical Sciences VI.* San Jose: SPIE; 2006(b); 6089(59):1-12.

Wang, BG; Halbhuber, KJ. Corneal Multiphoton Microscopy and Intratissue Optical Nanosurgery by Nanojoule Femtosecond Near-Infrared Pulsed Lasers. *Ann Anat,* 2006(c); 188(5):395-409.

Wang, BG. *Kornea Laserchirurgie und Multiphoton Mikroskopie mittels Naher-Infrarot Nanojoule Femtosekunden Laser mit Kaninchen.* Dissertation. Friedrich-Schiller-

Universität Jena, Medizinische Fakultät. 2006(d); urn:nbn:de:gbv:27-20061009-144159-3.

Wang, BG; Riemann, I; Schubert, H; Schweitzer, D; König, K; Halbhuber, KJ. Multiphoton microscopy for monitoring intratissue femtosecond laser surgery effects. *Lasers Surg Med,* 2007(a); 39(6):527-533.

Wang, BG; Riemann, I; Schubert, H; Halbhuber, KJ; König, K. In-vivo intratissue ablation by nanojoule femtosecond near-infrared lasers. *Cell Tissue Res,* 2007(b); 28(3):515-520.

Wang, BG; Eitner, A; Lindenau, J; Halbhuber, KJ. Cellular two-photon microscopy in ocular tissues. abstract, *Cytometry A,* 2007(c); 71A (7):511-531.

Wang, BG; König, K; Halbhuber, KJ. Intraocular nonlinear optical tomography and corneal flap generation using nanojoule femtosecond near-infrared lasers. abstract, *Clinical & Surgical Ophthalmology,* 2007(d); 25(7):246.

Wang, BG; Eitner, A; Lindernau, J; Halbhuber, KJ. High-Resolution Two-Photon Excitation Microscopy of Ocular Tissues in Porcine Eye. *Lasers Surg Med,* 2008(a); 40(4):247-256.

Wang, BG; Lohmann, CP; Riemann, I; Schubert, H; Halbhuber, KJ; König, K. Multiphoton-mediated Corneal Flap Generation Using the 80 MHz Nanojoule Femtosecond Near-Infrared Laser. *J Refract Surg,* 2008(b); accepted.

Watanabe, W; Arakawa, N; Matsunaga, S; Higashi, T; Fukui, K; Isobe, K; Itoh, K. Femtosecond laser disruption of subcellular organelles in a living cell. *Opt Express,* 2004; 12:4203-4213.

Wilson, MJ. Structure of the corneal stroma. *Vision Res,* 1970; 10(6):519-520.

Wilson, SE. Analysis of the keratocyte apoptosis, keratocyte proliferation, and myofibroblast transformation responses after photorefractive keratectomy and laser in situ keratomileusis. *Trans Am Ophthalmol Soc,* 2002; 100:411-433.

Xu, C; Williams, RM; Zipfel, WR; Webb, WW. Multiphoton Excitation Cross-Sections of Molecular Fluorophores. *Bioimaging,* 1996; 4(3):198-207.

Zimmermann, DR; Trueb, B; Winterhalter, KH; Witmer, R; Fischer, RW. Type VI collagen is a major component of the human cornea.*FEBS Lett,* 1986; 197(1-2):55-58.

Zipfel, WR; Williams, RM; Webb, WW. Nonlinear magic: multiphoton microscopy in the biosciences. *Nat Biotechnol,*2003(a); 21(11):1369-1377.

Zipfel, WR; Williams, RM; Christie, R; Nikitin, AY; Hyman, BT; Webb, WW. Live tissue intrinsic emission microscopy using multiphoton-excited native fluorescence and second harmonic generation. *PNAS,* 2003(b); 100(12):7075-7080.

Zoumi, A; Yeh, A; Tromberg, BJ. Imaging cells and extracellular matrix*in vivo* by using second-harmonic generation and two-photon excited fluorescence. *PNAS,* 2002; 99(17):11014-11019.

Zoumi, A; Lu, X; Kassab, GS; Tromberg, BJ. Imaging coronary artery microstructure using second-harmonic and two-photon fluorescence microscopy. *Biophys J,* 2004; 87(4):2778-2786.

INDEX

B

C

D

E

G

H

I

N

O

P

Q

R

S

U

V

W

X

Y

Z